P9-DGB-570

# The Wilson Chronology of Science and Technology

**Other titles in the Wilson Chronology Series:**

ROLLING MEADOWS LIBRARY
3110 MARTIN LANE
ROLLING MEADOWS, IL 60008

# The Wilson Chronology of Science and Technology

### George Ochoa and Melinda Corey

509
OCH

The H. W. Wilson Company

New York • Dublin

1997

ROLLING MEADOWS LIBRARY
3110 MARTIN LANE
ROLLING MEADOWS, IL 60008

Library of Congress Cataloging-in-Publication Data

Ochoa, George.
    The Wilson chronology of science and technology / by George
    Ochoa and Melinda Corey.
        p.    cm.
    Includes bibliographical references and index.
    ISBN 0-8242-0933-8
    1. Science—History—Chronology. I. Corey, Melinda. II. Title.
    Q125.026   1997
    509'.02—dc21                                          97-22060
                                                               CIP

Copyright © 1997 by The H. W. Wilson Company.

Original edition copyright ©1995 by The Stonesong Press, Inc.

All rights reserved. No part of this work may be reproduced or
copied in any form or by any means, including but not restricted to
graphic, electronic, and mechanical—for example, photocopying,
recording, taping, or information and retrieval systems—without
the express written permission of the publisher, except that a review-
er may quote and a magazine, newspaper, or electronic information
service may print brief passages as part of a review written specifically for inclusion in that magazine, newspaper, or electronic service.

Printed in the United States of America

06 05 04 03 02 01 00 99 98 97  10 9 8 7 6 5 4 3 2 1

The H. W. Wilson Company
950 University Avenue
Bronx, NY  10452

http://www.hwwilson.com

*In memory of*
*José Verdi Cevallos, physician and poet,*
*and*
*Harriet Griggs Guild, physician and pathfinder*

# Contents

# Acknowledgments

We are indebted to Kerry Benson and Risa Schneider, diligent researchers and contributors. We thank Tom Brown for keyboarding much of the manuscript and Mudit Tyagi for his editorial assistance. We also thank Paul Fargis and Sheree Bykofsky of The Stonesong Press, Ginny Faber of Ballantine, and Michael Schulze, Hilary Claggett, Lynn Amos, Joseph Sora, Frank McGuckin, and John O'Sullivan of H. W. Wilson. Finally, we thank Mary F. Tomaselli for indexing the book.

# Introduction

A chronology is a record of change; *The Wilson Chronology of Science and Technology* is a record of stunning change. In no field of human endeavor have things been altered more quickly, dramatically, and irreversibly than in our knowledge of the physical universe and our ability to manipulate it. The most rapid and visible changes have taken place in just the last two centuries, but many of the deepest changes took place much earlier. It is obvious that microwave ovens, television, computers, the theory of relativity, and the model of the atom are products of scientific knowledge; it may be less obvious that so are bread, the dog, the pipeline, the concept of angles, and the year.

*The Wilson Chronology of Science and Technology* charts, from prehistory to the present, how we came to know what we know about nature and how we built upon that knowledge. It is a story with many twists and turns, frequent shifts of locale, and a decidedly uneven pace. For millions of years the tale hardly moves at all; after the development of agriculture it moves gradually but slowly; then in the last few hundred years it moves at a blinding pace. A physician from ancient Rome would have had much in common, technologically, with a physician in 17th-century Italy, but both would have great difficulty figuring out what to do with a CAT scanner in a late-20th-century hospital.

It is no accident, then, that most of this book deals with the few hundred years since the 16th and 17th centuries, when people like Copernicus, Galileo, and Francis Bacon brought about a fundamental shift in humanity's approach to studying nature, a change known as the scientific revolution. Yet *The Wilson Chronology of Science and Technology* also outlines other important stories of scientific progress: how early humans spread out of Africa across the world and developed such technologies as fish hooks, sewing needles, and the use of fire; how farmers and herders domesticated wild plants and animals; how Chinese experimenters invented silk, paper, and gunpowder while Chinese astronomers first recorded what became known as Halley's comet; how the Hindus developed a system of numerals that was passed to the Arabs and from them to Europeans; and how the Maya in Central America independently developed their own system of numerals.

*The Wilson Chronology of Science and Technology* ranges over as many different topics as it does periods and locales. It charts the growth of the disciplines usually taught in school—biology, chemistry, physics, and the earth sciences. But it also maps the rise of technology, engineering, archaeology, paleontology, mathematics, medicine, psychology, computer science, and the exploration of earth and space. It concentrates on the physical sciences but nevertheless includes entries on linguistics and the social sciences—anthropology, sociology, economics, and political science.

Throughout, the timeline reports action. It tells what was discovered, invented, suggested, argued, and disproved—by whom, when, where, and why. It notes wrong steps as well as right ones, established ideas as well as controversial ones. There are such exploded theories as the four humors and phlogiston (material supposedly lost in combustion); such outmoded procedures as medical bleeding; and such obsolete inventions as the Stanley Steamer. There are the sometimes meandering, sometimes steady processes by which modern theories were developed (chemical bonding, molecular genetics) and modern technologies created (electronics, satellites). There are controversies mostly restricted to earlier times—the scientific debate over Darwin's theory of evolution—and the controversies of the present day—who first settled the Americas; where is the mass that astronomers call "missing"?

Lest the accumulation of events become overwhelming, *The Wilson Chronology of Science and Technology* includes sidebars that spotlight moments in the story. Some of these offer more detail on how a discovery was made and why it was significant, while others bring out the human side of scientists and inventors (including Charles Goodyear, who never made a penny from his invention of vulcanized rubber).

In Stanley Kubrick's film *2001: A Space Odyssey* (1968) there is a sequence in which a hairy ancestor of humans, having learned how to use a bone as a weapon, hurls it into the air; the next thing we know, a spacecraft is orbiting Earth. If the devil is in the details, then this account of our development, however evocative, leaves much to be explained. How do we know what we know about the universe and ourselves? What are we still trying to discover? How did we come to own the technologies we possess; why is it that they sometimes seem to own us? *The Wilson Chronology of Science and Technology* is written to help answer these questions.

George Ochoa and Melinda Corey
Dobbs Ferry, New York
September 1997

# A Note to the Reader

*The Wilson Chronology of Science and Technology* is arranged by year and within each year by category. The categories are as follows:

ARCH   Archaeology

ASTRO  Astronomy, space science, space exploration

BIO      Biology, biochemistry, agriculture, ecology

CHEM   Chemistry

EARTH  Earth sciences (geology, oceanography, meteorology), earth exploration

MATH   Mathematics

MED    Medicine

MISC    Miscellaneous

PALEO  Paleontology

PHYS   Physics

PSYCH Psychology, neuroscience, artificial intelligence

SOC     Social sciences (anthropology, sociology, economics, political science), linguistics

TECH   Technology, engineering

In the timeline, B.C. dates are indicated by negative numbers, A.D. dates by positive numbers.

Throughout prehistory, antiquity, and the early Middle Ages, it is often difficult to place exact dates. Therefore, most of the dates in this book up to the year A.D. 1000 should be considered approximate (with certain exceptions, such as May 28, 585 B.C., the precise date of a solar eclipse predicted by the Greek philosopher Thales). After A.D. 1000, dates can generally be considered exact unless marked with a *c.* for *circa*.

This chronology is primarily a record of action—experiments, achievements, discoveries, and assertions. To avoid clutter, birth and death dates of those performing the actions have mostly been left out of the timeline. However, the birth and death dates of many of the scientists and inventors named in the timeline are included in the appendix.

# B.C.

| | |
|---|---|
| −2,500,000 | In East Africa the hominid (humanlike) species *Homo habilis* makes the first stone tools. TECH |
| −1,000,000 | *Homo erectus*, and possibly other hominid species, begin to move out of Africa into Asia and Europe. Some groups of hominids may have left Africa for Asia even earlier, as early as 2,000,000 B.C. EARTH |
| −500,000 | *Homo erectus* discovers the use of fire. Evidence of firepits is found in China, Hungary, and France. *See also* 13,000 B.C., TECH. TECH |
| −400,000 | *Homo erectus* invents the spear. TECH |
| −100,000 | The earliest anatomically modern *Homo sapiens* populations evolve in Africa. (There is still debate about whether *Homo sapiens* moved out of Africa to spread across Asia and Europe, or simply evolved independently in different regions from local *Homo erectus* populations.) PALEO |
| −48,000 | By now, *Homo sapiens* has reached the continent of Australia from southeast Asia. EARTH |
| −38,000 | The Cro-Magnon people appear in Europe, the Middle East, and North Africa. These anatomically modern humans will replace the existing Neanderthal populations. Along with other modern human populations in places as far apart as Australia and southern Africa, they will invent art, develop specialized tools, and exhibit cultural differences over place and time. TECH |
| −33,000 | Early Europeans make body ornamentation such as beads and pendants, the first known form of art. TECH |
| −30,000– 28,000 B.C. | Europeans make paintings on cave walls. TECH |
| −28,000 | Europeans mark notches on bones and stones to tally numbers. MATH |
| −28,000 | Flutes, the earliest musical instruments, appear in Europe. TECH |
| −28,000 | Europeans make the first known sculptures from stone, bone, antler, and horn. Some are of animals, and others, called Venuses, are of exaggerated female shapes. TECH |
| −25,000 | In central Europe, in what will become the Czech Republic, weaving and fired-clay ceramics have been invented. TECH |

| −24,000 | Europeans invent sewing needles made of bone. | TECH |

| −24,000 | The fish hook and fishing line are in use in Europe. | TECH |

| −20,000 | The bow and arrow have been invented and are in use in Spain and North Africa. | TECH |

| −20,000 | Primitive oil lamps are invented. They are probably fueled by animal fat in hollowed-out stones, with wicks of plant fiber. | TECH |

| −13,000 | The spear thrower and harpoon are invented. | TECH |

| −13,000 | Huts built from mammoth bones and maps drawn on bone appear in Mezhirich in eastern Europe. | TECH |

| −13,000 | Humans begin making their own fires. Previously they had relied on "found" fires, which they carefully carried and maintained. *See also* 500,000 B.C., TECH. | TECH |

| −12,000 | Dogs, descendants of the Asian wolf, are domesticated in Mesopotamia (Iraq). | BIO |

| −10,000 | Herding begins with the domestication of goats in Persia (Iran). | BIO |

| −10,000 | By this time the ancestors of Native Americans have crossed the land bridge from Siberia to Alaska, entering the Western Hemisphere, though some archaeologists will place the first crossing as early as 33,000 B.C. | EARTH |

# Prehistoric Surgery

Long before modern brain surgery there was trepanation, the removal of bone by boring a small hole in the human skull. In Russia, Europe, and the Near East, trepanation was practiced as long ago as the Neolithic period, or New Stone Age (9000–6000 B.C.). In some places it was common as recently as the Middle Ages. Trepanned skulls have even been found associated with the Inca civilization, which flourished in Peru c. A.D. 1200–1533.

Trepanation may have been done sometimes as a religious ritual, but in many cases it appears to have been a medical treatment for a blow to the cranium and the resulting hematoma, or swelling filled with blood. Cranial drilling was intended to cure head injuries by allowing evil spirits to escape from the heads of the possessed. As late as the 19th century, trepanning was used to treat migraine headaches and epilepsy.

The procedure varied by place and time, with more than a dozen possible methods of scraping and grooving the skull to remove bone without damaging the underlying dura, a fibrous membrane that protects the brain. The Incas were trained to anaesthetize the patient with herbs and nerve pressure, and there is evidence that some cultures successfully used primitive antibiotics to stave off infection.

−9000     Mass extinctions of large animals in North and South America may be caused by the intrusion of human hunters or by the diseases they bring with them. BIO

−8000     Agriculture is invented in northern Mesopotamia (Iraq) with the farming of wheat and barley.                                                   BIO

−8000     Potatoes and beans are cultivated in Peru, rice in Indochina.              BIO

−8000     Polynesians in the East Indies and Australia begin to spread out over the islands of the South Pacific.                                          EARTH

−8000     The first cities appear in Mesopotamia (Iraq) and other sites in the Near East, including Jericho in Palestine.                                     TECH

−8000     In Mesopotamia (Iraq), clay tokens are used to tally shipments of grain and animals. This system will be the basis for the first system of numeration and writing. *See* 3500 B.C., TECH.                                         TECH

−7700     Sheep are domesticated in Persia (Iran).                                  BIO

−7000     From 7000 to 6000 B.C. the pig and the water buffalo are domesticated in China and East Asia, the chicken in South Asia.                           BIO

# The Last of the Wine

The earliest inventions—fire, bows and arrows, wheeled carts—are the most difficult to place in time. New excavations and methods of analysis can overthrow the hoariest of received opinions on who invented what when. A case in point is the first manufacture of wine.

For a long time, the earliest evidence of wine manufacture came from Egypt from about 3000 B.C. Then, in 1991, Canadian graduate student Virginia Badler made a new claim about a dirty fragment of pottery from a Sumerian site in western Persia (Iran) dating from about 3500 B.C. The interior of the pottery, housed at the Royal Ontario Museum, was stained red. Some archaeologists thought it was paint; Badler thought it was wine.

Chemists at the University of Pennsylvania put the issue to the test by analyzing the reddish residue with infrared spectroscopy, a method that distinguishes chemicals by the wavelengths of light they absorb. They found that the residue was rich in tannic acid, an organic substance found almost without exception in grapes. Badler was proved right and the date for the invention of wine was pushed back 500 years.

No sooner was the date settled than it was revised again. In 1996, University of Pennsylvania chemists discovered wine residue in another Iranian jar, this one dating from 5000 B.C.

| | |
|---|---|
| −7000 | New forms of wheat are cultivated in Syria and Turkey. Sugarcane is cultivated in New Guinea. Flax is grown in Southwest Asia. In Mexico, maize, squash, peppers, and beans are grown.     BIO |
| −7000 | Mortar is used with sun-dried brick in Jericho.     TECH |
| −6500 | Trepanning, the drilling of a hole in the skull as a treatment for head injuries, is practiced in Europe and Asia. In some regions it will continue through the Middle Ages and even into the 19th century.     PSYCH |
| −6400 | Cattle are domesticated in Turkey, probably from the long-horned wild ox called the auroch, or uru. Aurochs will become extinct in A.D. 1627.     BIO |
| −6000 | Modern-style wheat for bread is grown in Southwest Asia. Citrus fruit is domesticated in Indochina.     BIO |
| −6000 | Beer is brewed in Mesopotamia (Iraq).     TECH |
| −5000 | The llama and alpaca are domesticated in Peru.     BIO |
| −5000 | Wine is produced in Persia (Iran).     TECH |
| −5000 | Irrigation is invented in Mesopotamia (Iraq).     TECH |
| −5000 | Nuggets of metal, including gold, silver, and copper, are used as ornaments and for trade.     TECH |
| −4200 | Egyptians invent the first known calendar with a 365-day year broken into 12 30-day months plus 5 days of festivals. It will be the basis for the Roman and modern Gregorian calendars. The date of invention is uncertain and may be as late as 2700 B.C. *See also* 46 B.C., ASTRO.     ASTRO |
| −4000 | The horse is domesticated in Ukraine. The first known horse riders are the Ukrainian Sredny Stog culture.     BIO |
| −4000 | The first sail-propelled boats appear.     TECH |

## The Oldest Tree

In North America, about 2700 B.C., a bristlecone pine began growing in the White Mountains of what is now California. That tree, now known as Methuselah, is still alive, making it the oldest known living tree, at about 4,700 years of age. Bristlecone pines are believed to have a potential lifespan of 5,500 years and are rivaled in longevity only by the giant sequoias, which may live to 6,000. If no one cuts it down, Methuselah may still be alive in the 28th century A.D.

| | | |
|---|---|---|
| −4000 | The Egyptians mine and smelt copper ores. | TECH |
| −4000 | Bricks are fired in kilns in Mesopotamia (Iraq). | TECH |
| −3600 | Bronze, an alloy of 90 percent copper and 10 percent tin, is invented in the Middle East. Harder and more versatile than copper, this new metal ushers in the Bronze Age, the age of the Trojan War, Gilgamesh, and the Exodus. *See also* 1400 B.C., TECH. | TECH |
| −3500 | The Egyptians use papyrus boats to travel on the Nile. | TECH |
| −3500 | The plow is introduced in Mesopotamia (Iraq). | TECH |
| −3500 | The Sumerians develop cuneiform, the earliest known form of writing. This system, based on pictograms wedged into clay tablets, grows out of the earlier use of clay tokens. *See* 8000 B.C., TECH. | TECH |
| −3300 | Between now and 2500 B.C., speakers of proto-Indo-European begin to spread across a vast region from western Europe to central Asia. The homeland of this nomadic horse culture may have been the steppes of Ukraine and Russia. Their language will give rise to the Indo-European languages, including the branches called Germanic (English, German), Italic (Latin, French), Slavic (Russian), Indo-Iranian (Sanskrit), Baltic (Lithuanian), Celtic (Gaelic), and Greek, Albanian, Armenian, and Anatolian (Hittite). | SOC |
| −3300 | The wheel is invented in southern Mesopotamia (Iraq). It is put to use in hauling carts and making pottery. | TECH |
| −3100 | The Egyptians invent an early form of hieroglyphics. | TECH |
| −3000 | Hieroglyphic numerals are used in Egypt. | MATH |
| −3000 | The candle is introduced in Egypt and Crete. | TECH |
| −3000 | Cotton fabric is woven in India. | TECH |
| −3000 | Dyes for cloth are in use in China and Egypt. | TECH |
| −2900 | The Sumerians develop symbols for syllables, a key step in the evolution of writing. | TECH |
| −2700 | According to legend, Chinese emperor Shen Nung investigates and experiments with herbs and acupuncture. The *Pen Tsao* (*The Herbal*) is later attributed to him; he will be considered the founder of Chinese medicine. | MED |
| −2697 | Legendary Chinese emperor Huang-ti, known as the Yellow Emperor, will be said to have reigned from now until 2597 B.C. The medical text *Nei Ching*, later attributed to him, claims there are four steps to developing a medical diagno- | |

|         |                                                                                                                                                                                                                                                                                  |
|---------|--------------------------------------------------------------------------------------------------------------------------------------------------------------------------------------------------------------------------------------------------------------------------------|
|         | sis: observation, auscultation (listening to sounds that arise within organs), interrogation, and palpation (touching)—i.e., look, listen, ask, and feel.    MED                                                                                                              |
| –2680   | Imhotep, an Egyptian physician, architect, and counselor to King Zoser, flourishes between now and 2650 B.C. Often called the world's first scientist, he writes the first known medical manuscript and designs the step pyramid, or Pyramid of Zoser, the world's first large stone structure.    MISC |
| –2600   | In the first recorded seagoing voyage, Egyptians searching for cedarwood sail to Byblos in Phoenicia.    EARTH                                                                                                                                                                |
| –2600   | According to legend, silk manufacture begins in China.    TECH                                                                                                                                                                                                                |
| –2550   | Egyptian pharaoh Cheops, or Khufu, supervises the building of the Great Pyramid of Giza.    TECH                                                                                                                                                                              |
| –2500   | The construction of Stonehenge begins in southwestern England, near Salisbury, and is completed about 1700 B.C. The monument, with its concentric circles of stones, ditches, and holes, serves both religious and astronomical purposes. Some stones are aligned with the rising and setting of the sun and moon at the summer and winter solstices.    ASTRO |
| –2500   | Egyptian pharaoh Chefren, or Khafra, orders the building of the Great Sphinx at Giza.    TECH                                                                                                                                                                                 |
| –2500   | Glass ornaments appear in Egypt.    TECH                                                                                                                                                                                                                                      |
| –2500   | During the next few centuries the Sumerians develop a system of standard weights and measures, including such units as the shekel, the mina, the log, the homer, the cubit, and the foot.    TECH                                                                             |
| –2500   | The oldest written story, the Sumerian *Epic of Gilgamesh*, may have first appeared in written form around this time.    TECH                                                                                                                                                 |
| –2400   | Divination, the interpretation of omens perceived in natural phenomena, is used for medical purposes in Mesopotamia (Iraq). The body organs of sacrificed animals are thought to reveal a patient's fate.    MED                                                               |
| –2300   | Maps of lands and cities appear in Mesopotamia (Iraq).    TECH                                                                                                                                                                                                                |
| –2296   | The Chinese record the first known sighting of a comet.    ASTRO                                                                                                                                                                                                              |
| –2000   | The Park of Intelligence is founded as China's first zoo.    BIO                                                                                                                                                                                                              |
| –2000   | Horsedrawn battle chariots and metal riding bits are invented in the Near East.    TECH                                                                                                                                                                                       |
| –2000   | The palace of Minos in Crete has light and air shafts and interior bathrooms with their own water supply.    TECH                                                                                                                                                             |

| | |
|---|---|
| −2000 | Using wooden ships, the Minoans of Crete become the world's first sea power. <div align="right">TECH</div> |
| −1800 | Babylonian astronomers begin to compile records of celestial observations, including star catalogs. <div align="right">ASTRO</div> |
| −1800–<br>1600 B.C. | Babylonian mathematicians develop positional notation based on the number 60. They also develop squares and square roots, cubes and cube roots, quadratic equations, and multiplication tables. They discover what will become known as the Pythagorean theorem more than 1,000 years before Greek philosopher Pythagoras (*see* 530 B.C., MATH) and calculate an approximate value for pi. By this time, pi has also been calculated in Egypt. <div align="right">MATH</div> |
| −1800 | Leavened bread is invented in Egypt. <div align="right">TECH</div> |
| −1775 | Mesopotamian physicians are ruled by the Code of Hammurabi, which includes the earliest known system of medical ethics. <div align="right">MED</div> |
| −1700 | Rye is cultivated by eastern Europeans, whose growing season is not long enough to generate wheat. <div align="right">BIO</div> |
| −1700 | Egyptian mathematicians develop a system of geometry, a ciphered numeral system, and tables of values for fractions. <div align="right">MATH</div> |
| −1700 | The Babylonians use windmills for irrigation. <div align="right">TECH</div> |
| −1600 | Egyptians use castor oil as a laxative. <div align="right">MED</div> |
| −1600 | Egyptian medical remedies and procedures are documented in papyrus manuscripts, especially those discovered in modern times by Edwin Smith (*see* A.D. 1930, ARCH) and Georg Ebers (*see* A.D. 1872, ARCH). Topics include arthritis, hookworm infection, and surgery for head injuries. <div align="right">MED</div> |
| −1600 | The Phoenicians, or Canaanites, invent the world's first purely phonetic alphabet, based on symbols for sounds, not things or syllables. This alphabet is the ancestor of all modern Western alphabets. <div align="right">TECH</div> |
| −1523–<br>1027 B.C. | In China during the Shang dynasty, the earliest recorded locust control takes place, as flames are used to destroy locusts. <div align="right">BIO</div> |
| −1500–<br>1000 B.C. | The ancient Hindus derive reserpine, the first modern sedative-antihypertensive, from the root of the *Rauwolfia serpentina* plant. Hindus also become the first to perform successful skin grafting and plastic surgery of the nose. <div align="right">MED</div> |
| −1500 | The Egyptians use the shadow cast by a gnomon, or marker, to tell time. <div align="right">TECH</div> |
| −1500 | The Chinese under the Shang dynasty (1523–1027 B.C.) develop a system of writing. <div align="right">TECH</div> |

| | | |
|---|---|---|
| –1470 | The Aegean volcanic island of Thera (Santorini) explodes. The ashes and tidal wave bring an end to the powerful Minoan civilization of Crete, permitting the rise of the Mycenaean Greeks and the Phoenicians. | EARTH |
| –1400s | The shadow of the Needle of Cleopatra in Heliopolis, Egypt, is used to estimate the solstices, seasons, and time of day. | ASTRO |
| –1400 | Using oared ships and guided by the stars, the Phoenicians learn to navigate in the open sea. | TECH |
| –1400 | The Hittites of Asia Minor (Turkey) develop a practical method for smelting iron, ushering in the Iron Age. Bronze (*see* 3600 B.C., TECH) is gradually replaced as the dominant metal. The revolution reaches Europe by 1000 B.C. | TECH |
| –1350 | The Chinese develop decimal numerals. | MATH |
| –1300 | The Chinese devise a working calendar by this time, if not earlier. | ASTRO |
| –1200 | The dye known as Tyrian purple is invented by the Phoenicians. Obtained from a Mediterranean snail, it will be a favorite of the rich and powerful throughout antiquity. | TECH |
| –1200 | The Egyptians dig the first of several canals from the Nile River to the Red Sea. | TECH |
| –1200 | Linen is woven from flax stalks in Egypt. | TECH |
| –1200 | The Olmec civilization of Mesoamerica (Mexico and Central America) raises pyramids and massive stone monuments. | TECH |
| –1000 | The Chinese develop the counting board, the forerunner of the abacus. | MATH |
| –1000 | Between now and c. 900 B.C., Aesculapius, the Greek (later Roman) god of medicine, becomes known as a deity. Aesculapian cults build temples where the sick congregate to be healed. | MED |
| –1000 | The Chinese burn coal for fuel. | TECH |
| –1000 | The Chinese store ice to use for refrigeration. | TECH |
| –800 | Egyptian physicians give drugs, along with magic spells, to heal the sick. The healers begin to notice the association between cause and effect, the basis of empirical medicine. | MED |
| –800 | By now, Egyptians are using sundials with six time divisions to tell time. The sundial is introduced in Greece by the sixth century B.C. | TECH |
| –763 | The Babylonians are the first to record a solar eclipse. | ASTRO |

| | | |
|---|---|---|
| −750 | The arch is in use among the Etruscans in Italy. | TECH |

−700   The Chinese begin keeping records of comets, meteors, and meteorites. ASTRO

−700   During the Chou dynasty (c. 1027–256 B.C.), two doctrines evolve to form the basis of Chinese medicine. One is the doctrine of yin and yang, the two principles of masculinity, light, and heaven (yang) and femininity, darkness, and earth (yin). The other is that of the five elements or phases: metal, wood, water, fire, and earth. It is believed that humans require equilibrium among the two principles and five elements to remain in good health.                     MED

−700   The Assyrians introduce the aqueduct.                                          TECH

−668   The reign of Ashurbanipal, king of Assyria, begins (ends 627 B.C.). He establishes a library in his capital at Nineveh, which is destroyed c. 612 B.C.       MISC

−650   The Lydians of Asia Minor (Turkey) introduce the first standard coinage.   TECH

−625   Thales of Miletus, a Greek philosopher, astronomer, and mathematician, is born (*d.* 547 B.C.) in Asia Minor (Turkey). Among other things, he will theorize correctly that solar eclipses are the result of the moon's passing in front of the sun. *See also* 585 B.C., ASTRO.                                                       MISC

−600   The Japanese practice massage and acupuncture, adopted from the Chinese as healing therapies.                                                               MED

−600   The Zapotecs of Mesoamerica (Mexico and Central America) invent a system of hieroglyphics, the earliest known writing in the Americas.                   TECH

−585   A solar eclipse, predicted by Greek mathematician Thales of Miletus, occurs on May 28, during a battle between the Lydians and Medes. The warring parties take it as an omen and make peace.                                             ASTRO

−585   Thales develops deductive geometry. The theorem of Thales, that an angle inscribed in a semicircle is a right angle, is attributed to him, though he may have learned it from the Babylonians.                                         MATH

−585   Thales studies magnetism.                                                    PHYS

−580   Thales theorizes that water is the fundamental element of which all other substances are made.                                                            CHEM

−570   The diluvial doctrine, a theory proposing that the earth's surface was transformed by great floods, begins to develop and spread. The concept can be traced to the early Greek thinkers Xenophanes and Anaximander. Diluvialism will flourish in the 17th and 18th centuries A.D.                               EARTH

-547        Greek philosopher and astronomer Anaximander dies. He introduced the idea
           of evolution, claiming that life begins in marshy slime conditions and slowly
           evolves onto drier areas.                                                    BIO

-530        Greek philosopher and mathematician Pythagoras argues that Earth is a
           sphere and that the sun, the moon, the stars, and the five visible planets
           (Mercury, Venus, Mars, Jupiter, and Saturn) revolve around Earth in eight con-
           centric spheres, rotating independently. The friction between the spheres, he
           claims, generates harmonious, virtually inaudible sounds called the music of
           the spheres.                                                                 ASTRO

-530        Pythagoras is the first Greek to learn that the morning star and evening star are
           the same object. He names this planet Aphrodite, after the goddess of love
           known to the Romans as Venus, hence the planet's modern name.              ASTRO

-530        Pythagoras proves what will become known as the Pythagorean theorem, that
           the square of the hypotenuse of a right triangle is equal to the sum of the squares
           of the other two sides. *See also* 1800–1600 B.C., MATH.                      MATH

-530        The idea that the brain is the center of higher activity will be credited to
           Pythagoras.                                                                  MED

-510        Greek traveler Hecataeus draws the first recognizable map of the Mediterranean
           world.                                                                       EARTH

-500s       The sundial is in use in Greece. *See* 800 B.C., TECH.                        TECH

-500        Hanno of Carthage navigates down the west African coast and describes,
           among other things, the gorilla.                                             BIO

-500        Phoenician navigators are believed to have reached the Atlantic Ocean, sailing as
           far as Cornwall, England, to the north, where they established tin mines, and cir-
           cumnavigating Africa to the south.                                           EARTH

-500        By now the abacus, the first significant calculating device, is known in Egypt.
                                                                                        MATH

-500        The work known as the *Sulvasutras* ("Rules of the Cord") summarizes Indian geom-
           etry.                                                                        MATH

-500        Chinese philosopher Confucius is one of the first to discuss human nature and
           how it can be modified.                                                      PSYCH

-470        The Greek Alcmaeon becomes the first known physician to dissect human
           bodies. Because of objections to human dissection, anatomy studies will
           decline until the works of Mondino de' Luzzi are written in A.D. 1316.        MED

-460        Greek physician Hippocrates is born on Cos (an island off Turkey, known later
           as Kos). He will set medicine free of the shackles of philosophy and religion

by being the first to record case histories, practice bedside observations, and provide physicians with moral inspiration and ethical standards. The Hippocratic oath, administered to new physicians, will become the best known of the Hippocratic writings, but its original authorship will remain uncertain.                                                                                      MED

−450        Greek philosopher Leucippus epitomizes the study of rationalism by stating that every event has a natural cause, ruling out supernatural intervention as an explanation.                                                                                    MISC

−440        Greek philosopher Democritus theorizes that the Milky Way is made up of many stars, that matter is composed of invisible particles called atoms, and that the moon is similar to Earth.                                                           MISC

−430        Greek philosopher Empedocles of Acragas (Agrigentum) speculates that the world is made up of four elements: earth, air, water, and fire. *See also* 350 B.C., EARTH.                                                                                        CHEM

−428        Greek natural philosopher and mathematician Anaxagoras dies. The author of *On Nature*, he was imprisoned for suggesting that the sun is a big, hot stone rather than a deity and that the moon is an inhabited body that borrows light from the sun.                                                                               ASTRO

---

## Proving the Earth Is Round

Most people today accept that the earth is round, but many would be hard put to defend this claim. After all, as seen from a casual stroll or even from an airplane, the earth looks flat. Without resorting to pictures from space, how can one prove that the earth is round?

This question was answered more than 2,000 years ago by the Greek philosopher Aristotle (384–322 B.C.), who argued that one can see the earth's shape clearly during a lunar eclipse. As the moon passes under the earth's shadow, the shape of the shadow is always round. This effect might sometimes be produced by a flat disk, but not always. For example, the sun would sometimes strike the disk edge on, producing a shadow in the shape of a straight line. Only a sphere will always produce a round shadow.

For the unconvinced, Aristotle pointed out that travelers going north or south saw new stars appear over the horizon ahead, while stars that had been visible disappeared under the horizon in the rear. Ships going out to sea disappeared hull first, whichever direction they took. These effects could be explained only by a round earth.

As it turns out, the earth is not a perfect sphere but an oblate spheroid, a sphere slightly flattened at its poles and slightly bulging at the equator. English physicist Isaac Newton (A.D. 1642–1727) showed that this effect would result from the earth's rotation. However, Aristotle's conclusion, still roughly correct, has been held by educated people ever since.

−420    Greek mathematician Hippias introduces the first curve beyond the circle and the straight line, the trisectrix or quadratrix.                                                          MATH

−420    Greek mathematicians discover incommensurable line segments.                          MATH

−420    Greek physician Hippocrates believes, incorrectly, that only women suffer hysteria, claiming it is caused by a "wandering uterus." (The term hysteria, which is derived from the Greek word *hystera*, or womb, is rooted in this notion.) He does present accurate descriptions of mania, phobias, paranoia, and melancholia.                                                                                              PSYCH

−414    Greek mathematician Theaetetus is born (*d.* 369 B.C.). He will study the five regular solids and develop the theorem that there are five—and only five—regular polyhedra.                                                                                            MATH

−408    Greek philosopher Eudoxus of Cnidus is born (*d.* 355 B.C.). He will develop a model of celestial motion involving a complex combination of rotating spheres. He will also establish the geometric theory of irrational numbers. MATH

−400    By now the Babylonians have established the zodiac circle of constellations, the band in the sky that includes the apparent paths of the sun, the moon, and the planets. Horoscopes become available describing the presumed influence of the sun, the moon, and the planets given their positions in the zodiac at the time of one's birth.                                                                              ASTRO

−400    Philolaus, a member of the Pythagorean school, argues that Earth, the sun, the moon, the planets, and the stars are all in motion around a "central fire." This is the earliest known theory of an orbiting Earth.                                          ASTRO

−400    By now the Greeks have formulated three famous problems that will puzzle mathematicians for centuries: squaring the circle, duplicating the cube (*see* 360 B.C., MATH) and trisecting the angle. By the 19th century it will be shown that the three problems are unsolvable using the straightedge and compass alone. MATH

−400    Greek philosopher Democritus argues that objects in the external world radiate beams that induce perceptions in the human mind.                                          PHYS

−400    Greeks working for Dionysius of Syracuse, Sicily, invent the catapult, the first artillery weapon.                                                                                      TECH

−390    Greek astronomer Heracleides is born in Pontus, later part of Turkey (*d.* c. 320 B.C.). He will be the first to argue that Venus and Mercury orbit the sun.  ASTRO

−387    Plato founds the Academy in Athens, often considered the world's first university.                                                                                                  MISC

−384    Greek philosopher Aristotle is born (*d.* 322 B.C.). After studying under Plato, he will write on logic, ethics, poetics, rhetoric, metaphysics, politics, and nature. His teachings on biology, medicine, and the physical world will be transmitted to Europe, mainly through Arab scholars, during the Middle Ages. These views will be considered authoritative until the scientific revolution (c. A.D. 1550 to 1700) calls them into question.     MISC

−372     Greek botanist Theophrastus is born (*d.* c. 287 B.C.). He studies under Plato and Aristotle, carrying on the tradition of biology and concentrating on the plant world. He will be considered the founder of botany.     BIO

−370     Between now and 350 B.C., Diocles of Carystos writes the first Greek herbal. MED

−360     Greek mathematician Eudoxus of Cnidus introduces a new theory of proportion: a definition of equal ratios that deals with the problem of comparing ratios of incommensurable magnitudes. He also develops the axiom of continuity that serves as the basis for the method of exhaustion.     MATH

−360     Greek mathematician Menaechmus discovers conic sections, the curves later known as the ellipse, parabola, and hyperbola. He uses conics to provide a solution to the problem of duplicating the cube. *See* 400 B.C., MATH.     MATH

−360     Greek mathematician Dinostratus uses the trisectrix or quadratrix of Hippias (*see* 420 B.C., MATH) to provide a solution to the problem of squaring the circle (*see* 400 B.C., MATH).     MATH

−352     The Chinese report the first recorded supernova.     ASTRO

−350     Chinese astronomer Shin Shen prepares a star catalog with about 800 entries.     ASTRO

−350     Aristotle classifies animals. He will be considered the founder of classical biology and zoology.     BIO

−350     In *De caelo* (*On the Heavens*), Aristotle defines chemical elements as constituents of bodies that cannot be broken down into other parts.     CHEM

−350     Aristotle theorizes that the universe is arranged in concentric shells with the earth dominating at the center, then water, air, and fire. A fifth shell, the site of the heavenly bodies, is unchanging and incorruptible, composed of a fifth element, ether.     EARTH

−350     Aristotle writes on disease, comparative anatomy, embryology, and psychology. His approach to disease is grounded in the theory of four humors (blood, phlegm, yellow bile, and black bile), four qualities (hot, dry, moist, and wet), and four elements (earth, fire, air, and water). Despite evidence to the contrary, Aristotle's ideas will dominate medicine for centuries.     MED

−350     In his *Organon* Aristotle systematically outlines the rules of logic.     MISC

−350     Aristotle claims that memory is based on three principles of association: similarity, contrast, and contiguity. He argues that arousing violent emotions through drama has a cathartic effect on human audiences, allowing viewers to purge themselves of aggressive impulses.                                    PSYCH

-340        Greek physicist Strato is born (d. 270 B.C.). Like Aristotle, he will detect the acceleration of falling bodies but state incorrectly that heavier bodies fall faster than lighter ones.                                    PHYS

-335        Aristotle founds the university called the Lyceum in Athens. His lectures there will be collected into about 150 volumes, 50 of which will survive to modern times.                                    MISC

-330        Athenian female physician Agnodike challenges a law prohibiting women from practicing medicine on other women. As a result of her successful efforts at assisting women in childbirth, the law is changed and Athenian women are allowed to practice medicine.                                    MED

-320        The Greek philosopher Theophrastus writes the first systematic book on botany, describing over 500 plant species.                                    BIO

-320        Mathematicians Aristaeus and Euclid write on conics, the curves formed by a plane intersecting a cone.                                    MATH

---

## The Appian Way

The world's first all-weather road system was built to facilitate modern warfare. Following their defeat in the Samnite Wars, particularly their humiliation at the Battle of the Caudine Forks along the rocky Apennines in 321 B.C., the Roman military began to develop more effective attack formations and better transportation routes through uneven terrain. The formation, the legion, allowed troops to scatter when facing troublesome roads, then reunite easily when conditions improved. The improved transportation route was the Via Appia, or Appian Way.

The Roman censor Appius Claudius Caecus ordered construction of the Appian Way, a paved road uniting Rome and Capua, to be usable by troops in all weather. Begun in 312 B.C., the road was built of multiple layers of durable materials, the top layer composed of a mixture of concrete, rubble, and stones set in mortar. The road was instrumental in facilitating Roman victory in future wars with the Samnites.

Equally important were the Appian Way's various political uses. It was crucial to building commercial interests and sustaining cultural links with and political control over the provinces. Over time, several roads were built to link Rome with other cities and colonies, including the Via Flaminia (Flaminian Way), which headed north to link Italy with the Latin colony of Ariminum. In all, the Roman road system covered more than 50,000 miles and crossed through 30 countries. Only remnants of the roads still exist.

−314        Theophrastus of Eresus writes *Peri Lithon* (*On Stone*), which catalogs the min-
            eral substances then found in Athenian trade. This short treatise is the first known
            geology text.                                                                    EARTH

−312        The Roman censor Appius Claudius Caecus begins building the Appian Way,
            which will stretch 132 miles from Rome to Capua and later be extended to
            Brundisium (Brindisi). Initially covered by gravel and later by stone, it will be
            the best road yet built.                                                         TECH

−300s       By now the Babylonians have developed a symbol for zero.                         MATH

−300        Greek adventurer Pytheas sails into the Atlantic, voyaging as far as Scandinavia
            and the Baltic Sea. En route he observes and describes tides, a phenomenon
            little known in the Mediterranean.                                               EARTH

−300        Euclid, a Greek in Alexandria, Egypt, writes his *Elements*, a textbook summariz-
            ing and systematizing Greek mathematics, including plane and solid geometry
            and the theory of numbers. It will be accepted in the West as a basic reference
            until the modern age.                                                            MATH

−300        A classic work of Chinese mathematics, the *Chou pei suan ching*, is written.    MATH

−300        Chinese mathematicians use a system of "rod" symbols for numerals (some-
            times written, sometimes represented by physical rods) to carry out calculations of
            large numbers.                                                                   MATH

−300        The third century B.C. is a golden age of Greek mathematics, due largely to the
            work of Euclid, Apollonius of Perga, and Archimedes at Alexandria, Egypt. MATH

−300        Indian mathematicians develop the Brahmi numerals, a decimal system of
            numeration without a place-value notation.                                       MATH

−300        Alexandria's Greek school of medicine is founded.                                MED

−300        Ptolemy I, ruler of Egypt, founds the university in Alexandria called the
            Museum. Its library will be the largest yet known.                               MISC

−295        Greek physician Praxagoras distinguishes between veins and arteries. (The
            word *arteries* is derived from this physician's mistaken belief that arteries carry
            air.)                                                                            MED

−287        Greek scientist, mathematician, and inventor Archimedes is born (*d.* 212 B.C.).
            He will discover the law of specific gravity and study the mathematics of the
            lever. Among his inventions will be the Archimedean screw, a device to lift
            water and loose materials such as sand.                                          MISC

−280        Greek anatomist Herophilus divides nerves into sensory and motor, names the
            first section of the small bowel the duodenum, and names the prostate gland.
            After studying the function of the arteries and veins, Herophilus advocates

bloodletting, which will be used as a therapeutic for more than 2,000 years.

MED

−280    The Colossus of Rhodes, a 105-foot-high statue of the sun god, is completed.

TECH

−280    Sostratus of Cnidus builds a 300-foot lighthouse on Pharos near Alexandria, Egypt. Projecting light from a series of concave mirrors, it will become one of the seven wonders of the ancient world.    TECH

−270    Greek scientist Ctesibius invents a popular new version of the ancient water clock, a device that tells time according to the steady accumulation of water in a chamber.    TECH

−260    Greek mathematician and scientist Archimedes calculates the value of pi. MATH

−260    The Roman numeral system is at an advanced stage. It will survive in Europe until the Middle Ages, when it will be gradually replaced by Arabic numerals. *See* A.D. 1202, MATH.    MATH

−260    While sitting in a public bath Archimedes discovers the law of specific gravity now known as Archimedes' principle: a body dropped into a fluid displaces a volume of fluid equal to its own volume.    PHYS

−260    Greek scientist Archimedes works out the mathematics of the lever.    TECH

−250    The *Chui-chang suan-shu* (*Nine Chapters on the Mathematical Art*) is among the most influential Chinese books of mathematics. It contains more than 200 problems on engineering, surveying, calculation, agriculture, and right triangles as well as solutions to problems in simultaneous linear equations using positive and negative numbers.    MATH

−250    From now until 48 B.C., the Alexandrian medical school enjoys its greatest prominence. It is the only center in the ancient world where human dissection is regularly practiced for scientific reasons.    MED

−250    The Chinese book *Mo Ching*, written by followers of the philosopher Mo-tzu, contains a statement of the first law of motion that will be stated in A.D. 1687 by Isaac Newton in his *Principia*: a body continues in a state of rest or in uniform motion unless acted upon by outside forces.    PHYS

−240    Chinese astronomers make the first known observation of Halley's comet. *See* A.D. 1705, ASTRO.    ASTRO

−240    Eratosthenes of Cyrene, librarian at Alexandria, Egypt, correctly calculates the diameter of the earth as about 8,000 miles and the circumference as about 25,000 miles. *See also* 100 B.C. and A.D. 1684, EARTH.    EARTH

| | |
|---|---|
| −225 | Greek mathematician Apollonius of Perga, known in antiquity as the Great Geometer, publishes his *Conics*, which makes several important advances in the study of these curves.    MATH |
| −214 | The Great Wall of China is begun by Emperor Shih Huang Ti, founder of the Ch'in dynasty (221−206 B.C.). The Great Wall will eventually extend 1,500 miles from the Pacific Ocean to central Asia.    TECH |
| −202−220 A.D. | In China during the Han dynasty, Wang Zhong (Wang Chung) writes about using open ditches as a trap for immature locusts.    BIO |
| −200s | Greek astronomer Aristarchus of Samos is the first person known to argue that Earth revolves around the sun. He also proposes that day and night are caused by Earth's rotation and makes estimates of the sun's distance and size that are several orders of magnitude too small. *See also* A.D. 1650, ASTRO.    ASTRO |
| −200 | The Greeks invent the astrolabe, a device for measuring the positions of heavenly bodies.    ASTRO |
| −170 | Scholars working for Eumenes II of Pergamum, in Asia Minor (Turkey), invent parchment, a writing material made from hides. It will compete effectively with the more ancient writing vehicle, papyrus.    TECH |
| −165 | The Chinese make the first recorded observations of sunspots.    ASTRO |
| −150 | Greek astronomer Hipparchus of Nicaea correctly calculates the distance of the moon from Earth as about 240,000 miles.    ASTRO |
| −140 | Chinese philosopher Han Ying makes the first known reference to the hexagonal structure of snowflakes.    CHEM |
| −140 | Hipparchus compiles the first trigonometric table. His table of chords helps to introduce the systematic use of the 360° circle.    MATH |
| −101 | The Romans become the first to employ water power to mill flour.    TECH |
| −100s | Hipparchus compiles a star catalog and discovers the precession of the equinoxes, an apparent shift of the stars. He argues that Earth is motionless at the center of the universe, a view that will dominate European thinking until the time of Copernicus. *See also* A.D. 140 and A.D. 1543, ASTRO.    ASTRO |
| −100s | Hipparchus invents a system of magnitude for measuring the brightness of stars, the basis for the modern system. *See also* A.D. 1856, ASTRO.    ASTRO |

| | |
|---|---|
| −100–<br>1 B.C. | The tradition of environmental conservation in China has its roots in philosophical statements of the responsibilities of humanity in relation to nature. BIO |
| −100 | Greek astronomer Posidonius of Apamea erroneously calculates that the earth's circumference is 18,000 miles. This value will be accepted as true through the Middle Ages, while the correct value (deduced by Eratosthenes in 240 B.C.) will be forgotten. EARTH |
| −100 | Chinese mathematicians use negative numbers. MATH |
| −99 | The Roman physician Asclepiades opposes the theory of humors put forth by Hippocrates. Asclepiades teaches that the body is composed of disconnected atoms, separated by pores, and that orderly motion of the atoms must be maintained. He attempts to cure disease through exercise, bathing, and varying the diet. This theory is revived in different forms far into the 18th century. MED |
| −63 | A primitive system of shorthand is developed by former slave Marcus Tullius Tiro. TECH |
| −60 | In the poem *On the Nature of Things*, the Roman philosopher Lucretius speculates, as had Democritus (*see* 440 B.C., MISC), that matter is made of atoms. PHYS |
| −52 | Chinese astronomer Ken Shou-ch'ang builds a stellar observation device called an armillary ring, which consists of a metal circle representing the equator. ASTRO |
| −50 | Mayan written records begin in Mesoamerica (Mexico and Central America). Though writing in the New World did not begin with the Maya (*see* 600 B.C., TECH), they were to give it its greatest refinement, using a mix of ideographic and phonetic elements. The Mayan Classic period will last from A.D. 250 to 900. TECH |
| −50 | In Palestine, the process of glassblowing is invented. TECH |
| −46 | Following the advice of Greek astronomer Sosigenes, Julius Caesar institutes the Julian calendar, a reform of the Roman calendar based on estimates that the year is 365¼ days rather than 365 days. This calendar alternates three regular years of 365 days with one leap year of 366 days. With the reforms proclaimed by Pope Gregory XIII in A.D. 1582, this calendar will become the basis for the one used throughout most of the world. ASTRO |
| −44 | In May and June, Roman and Chinese observers report a red comet visible in daylight. Many Romans believe it to be the departed spirit of Julius Caesar, exalted to divine status after his assassination on March 15. The red color is probably due to dust in the air from recent eruptions of Mount Etna in Sicily. ASTRO |
| −44 | From March to May, Mount Etna in Sicily undergoes a series of eruptions. Volcanic dust darkens the skies. Three years of crop failures are reported by the Chinese. EARTH |

-40s   The Tower of Winds in Athens is built by Andronikos of Kyrrhestes. Its time-keeping device combines a water clock and eight solar clocks.   TECH

-28   From now until A.D. 1638, Chinese astronomers keep continuous records of sunspot activity.   ASTRO

-27   Construction of the Roman Pantheon, an early domed building, is begun this year.   TECH

# A.D.

20   Greek historian and philosopher Strabo summarizes the geographical knowledge of his day in *Geographia*.   EARTH

43   In *De situ orbis* ("A Description of the World"), Roman geographer Pomponius Mela divides the earth into five climatic zones: North Frigid, North Temperate, Torrid (equatorial), South Temperate, and South Frigid.   EARTH

50   Roman philosopher Seneca speculates that there is change and imperfection beyond the moon, contrary to prevailing belief in the unchanging heavens.   ASTRO

50   Roman poet Lucius Junius Moderatus Columella suggests that grain and legume crops be rotated and fields be "dunged" to preserve the earth's fertility. Previously Romans have used a mixture of blood and bone as fertilizer.   BIO

50   Roman encyclopedist Aulus Cornelius Celsus writes the first organized medical history, which describes the four cardinal symptoms of inflammation: redness, swelling, heat, and pain.   MED

50   Greek engineer and mathematician Hero, or Heron, of Alexandria invents a primitive steam engine, but it is never put to productive use.   TECH

50   Windows of colored glass are in use in the Roman world. Clear glass will not be invented until 1291.   TECH

60   Greek physician Pedanius Dioscorides compiles the first systematic pharmacopeia. This famous herbal, *De materia medica*, describes more than 500 plants and 35 animal products. Ninety of the plants he mentions will still be in use in the 20th century. *See also* 1544, BIO.   BIO

70   Roman natural philosopher Pliny the Elder, who will die of asphyxiation while witnessing the eruption of Mount Vesuvius in 79, publishes *Natural History*, a 37-volume work on zoology, botany, astronomy, and geography. It will become as famous for its errors as for the facts it propagates throughout the Middle Ages.   BIO

70          In China the religion of Buddhism (imported from India, where it was found-
            ed in the sixth century B.C.), introduces faith healing, hypnotism, autosugges-
            tion, and meditation arts as components of medical practice.          MED

75          Greek engineer and mathematician Hero, or Heron, of Alexandria publishes
            the *Metrica*, in which he demonstrates what will become known as Heron's
            formula for the area of a triangle.          MATH

79          Mount Vesuvius erupts near Naples, burying the towns of Pompeii and
            Herculaneum. The buried towns, rediscovered 15 centuries later, in 1592, will
            serve as a spur to early archaeologists. *See also* 1592, ARCH.          ARCH

90          In Rome, multiple aqueducts provide 250 gallons of water per day to the citizenry.
            TECH

100s        The Chinese note that a magnetic sliver, allowed to turn freely, always points
            north–south. *See also* 1180, PHYS.          PHYS

100         In China it is discovered that dried chrysanthemum flowers can kill insects.
            The Chinese proceed to develop a powder from these flowers and invent the first
            insecticide.          BIO

100         An alchemist known as Mary the Jewess, living in the first or second century,
            invents or at least elaborates such types of chemical apparatus as a three-armed
            still, a hot-ash bath, the water bath (later named the *bain-marie* in her honor),

---

## The Eruption of Mount Vesuvius
### August 24, 79

Meanwhile on Mount Vesuvius broad sheets of fire and leaping flames blazed at several points, their bright glare emphasized by the darkness of night....

They [his uncle's household] debated whether to stay indoors or take their chance in the
open, for the buildings were now shaking with violent shocks, and seemed to be swaying
to and fro as if they were torn from their foundations. Outside, on the other hand, there
was the danger of falling pumice stones, even though these were light and porous....

We also saw the sea sucked away and apparently forced back by the earthquake; at any
rate it receded from the shore so that quantities of sea creatures were left stranded on
dry sand....

At last the darkness thinned and dispersed like smoke or cloud; then there was genuine
daylight, and the sun actually shone out, but yellowish as it is during an eclipse. We
were terrified to see everything changed, buried deep in ashes like snowdrifts.

> —*Eyewitness impressions of the disaster that destroyed the towns of Pompeii and
> Herculaneum, from Pliny the Younger, nephew of Pliny the Elder, in Letters (trans-
> lated by Betty Radice, 1969)*

and the dung bed. Her writings combine practical techniques, mystical imagery, and theoretical ideas. CHEM

100 Alexandrian mathematician Menelaus writes on spherical geometry in his *Spherics*. MATH

100 Mathematician Nicomachus of Gerasa, which is near Jerusalem, writes the *Introductio arithmeticae*, which uses mathematics in the service of neoplatonic philosophy. MATH

100 Indian physician Charaka presents ethical standards to be required of those caring for the sick, including purity, cleverness, kindness, good behavior, and competence in cooking. MED

105 Chinese inventor Tsai Lun devises paper, a writing surface that can be produced cheaply from wood, rags, or other substances containing cellulose (as opposed to papyrus, made from an Egyptian reed, or parchment, made from hides). It will not reach Europe until 1320. TECH

106 In central Asia, traders from China meet to barter Chinese silk and spices with Roman traders of gems, precious metals, glassware, pottery, and wine. MISC

117 Between now and 138, Greek physician Soranus of Ephesus serves as a respected authority on gynecology, obstetrics, and infant diseases. His treatise on pediatrics contains the earliest description of rickets. MED

122 Hadrian's Wall is built in Britain to defend against northern tribesmen, including the Picts. The wall, built mainly of stone, runs 72 miles, from the Tyne to the Solway. TECH

125 Zhang Heng of China refines the armillary ring, first introduced in 52 B.C. and used for observing the stars. ASTRO

126 Between now and 145, during the Shun Ti reign, the Taoist religious leader Chang Tao-ling composes a guide to charms and incantations intended to cure disease. MED

128 Imported wheat from Egypt and North Africa lowers grain prices and decreases the number of Roman farmers, who cannot compete with foreign prices. BIO

132 The Chinese invent the wind vane. EARTH

140 In the *Megalé syntaxis tēs astronomias*, later known as *The Almagest*, the Alexandrian astronomer, geographer, and mathematician Claudius Ptolemaeus (Ptolemy) synthesizes the geocentric Ptolemaic system that will dominate Western cosmology until the Copernican revolution of 1543. In Ptolemy's system all the heavenly bodies revolve around a fixed Earth. It will come to western Europe by way of Arabic translation in 827. ASTRO

140        Ptolemy introduces the concept of epicycles, hypothetical small circles on which each planet moves. Epicycles are used to account for apparent anomalies in planetary motion that will not be correctly explained until Kepler's works in 1609.       ASTRO

140        Ptolemy's *Almagest* includes a table of chords and writings on trigonometry. MATH

160        Greek physician and anatomist Claudius Galen dissects animals, applying the results (sometimes mistakenly) to humans. He shows the importance of the spinal cord, uses the pulse as a diagnostic tool, and describes the flow of urine to the bladder. He also describes respiration and proves that the arteries carry blood (*see* 295 B.C., MED) but he incorrectly explains the passage of blood through the heart. Right and wrong, his pronouncements will carry medical authority for the next 17 centuries.       MED

160        Galen establishes the doctrine of vitalism, which claims that a force that is neither chemical nor mechanical is responsible for the processes of life. He specifically identifies animal spirits in the brain, vital spirits in the heart, and natural spirits in the liver.       MED

180        The first known alchemy manuscripts appear in Egypt.       CHEM

185        The Chinese observe a supernova in the constellation Centaurus that remains visible for 20 months.       ASTRO

190        The Chinese calculate pi to five decimal places: 3.14159.       MATH

250        Greek mathematician Diophantus devises solutions to problems that represent the beginnings of algebra. His problems include ones that must be solved with whole numbers, known as Diophantine equations.       MATH

250        The Mayan Classic period begins in Mesoamerica (Mexico and Central America). Lasting until 900, the Classic Mayan civilization will make great advances in agriculture, astronomy, mathematics, writing, and architecture. MISC

270        In China, Wu dynasty alchemists manufacture gunpowder by combining sulfur and saltpeter.       TECH

284        The Coptic calendar is introduced on August 29 in Egypt and Ethiopia.    TECH

300        Zosimos of Panopolis, Egypt, writes a summary of alchemy.       CHEM

300        The Maya use sweat baths for medicinal purposes.       MED

304        Integrated pest management begins in China when Hsi Han records how to use specific types of ants to control other insect pests attacking mandarin oranges.       BIO

| | |
|---|---|
| 350 | The Chinese invent an early form of printing using blocks of raised, reversed symbols smeared with ink. TECH |
| 390 | Roman matron St. Fabiola is influential in the founding of the first general public hospital in western Europe. MED |
| 400 | Alexandrian philosophers coin the term *chemistry* to denote the process of change in material substances. CHEM |
| 400 | Indian texts called the *Siddhantas* contain the first trigonometric use of half-chords, the predecessor of the modern sine function. MATH |
| 400 | Indian physician Susruta describes plastic surgery operations for earlobe deformity, skin grafting, and rhinoplasty (nasal reconstruction). MED |
| 400 | The Chinese invent the wheelbarrow. TECH |
| 406 | Rye, oats, and spelt (wheat for animal feed) are brought to Europe by such invaders as the Alans, Sciri, and Vandals. BIO |
| 410 | Alexandrian mathematician Proclus is born (*d.* 485). He will preserve information on Greek mathematics before Euclid, particularly in his summary of the lost work of Eudemus, *History of Geometry* (c. 335 B.C.). MATH |

## Hindu Numerals

In 529 the Byzantine emperor Justinian closed the philosophical schools of Athens, including the Academy, founded by Plato some nine centuries earlier, in 387 B.C. Now regarded as an archetypal victory of ignorance over knowledge, this move was made to defend the state religion, Christianity, from what were then perceived as pagan influences.

Some of the Academy's scholars moved to Syria, where they founded centers of Greek learning. However, they regarded their new home as an academic backwater and expressed disdain for the level of knowledge of non-Greeks. On hearing of this, Syrian bishop Severus Sebokht was moved, writing in 662, to let the Greeks know that "there are also others who know something." Noting that the Hindus in particular had made great advances in astronomy and mathematics, Sebokht praised their "valuable methods of calculation, and their computing that surpasses description. I wish only to say that this computation is done by means of nine signs." This is now considered the first explicit mention anywhere of the Hindu numeral system, based on 10 but at the time still lacking the zero. This system would later (820) be adopted by the Arabs and still later by the entire Western world.

Thus, the first mention of Hindu numerals was an indirect result of Justinian's distaste for Greek learning and the Greeks' distaste for everyone else's.

433 St. Patrick spreads Christianity throughout Ireland. By 795 Irish monks, following the path of the navigator St. Brendan, are believed to have reached Iceland. EARTH

476 The last Roman emperor, Romulus Augustulus, is deposed on September 4 by the Goths under Odoacer. This date traditionally marks the end of the western Roman empire and of classical antiquity, with its heritage of art and learning. The Middle Ages that follow in Europe are considered to last roughly until the mid-15th century, when the Renaissance (15th through 17th centuries) will revive and extend classical art, scholarship, and science. See 1453, MISC. MISC

499 Indian mathematician Aryabhata writes the *Aryabhatiya*, a summary of astronomical and mathematical knowledge. Among other things, he recalculates the Ptolemaic measurements of celestial motion and suggests that Earth rotates. ASTRO

500 Polynesians begin settling the islands of Hawaii. EARTH

529 St. Benedict founds the monastic order of Benedictines, under whom care of the sick becomes a part of monastery life. The era is marked by the belief that saints and miracles can heal the sick and dying. MED

529 Byzantine emperor Justinian closes the Academy and the Lyceum, two universities in Athens, founded respectively by Plato in 387 B.C. and Aristotle in 335 B.C. The closing is motivated by the Christian church's distrust of pagan learning. MISC

537 The church of Hagia Sophia in Constantinople takes full advantage of the art of dome design, with a large dome placed on a square support and pierced with many windows. TECH

552 Silkworms are smuggled from the Far East at the order of Byzantine emperor Justinian, thus introducing silk production to Constantinople. BIO

560 Eutocius's commentaries on Archimedes and Apollonius preserve many of their mathematical ideas. MATH

595 Hindu numerals make their first appearance on a plate where the date 346 is written in decimal-place value notation. At this time the Hindu system includes only 9 numerals—the 10th, the zero, will not appear until 876. Hindu numerals will be adopted by the Arabs and later by the Europeans. See 820, MATH. MATH

600 Chinese mathematicians Zu Chong-zhi and his son Zu Geng-shi calculate pi to seven decimal places. See 190, MATH. MATH

600 Eastern and northern European agricultural production improves with the refinement by the Slavs of the coulter (a blade that cuts vertically into the ground in advance of the plowshare) and moldboard plow. TECH

| | |
|---|---|
| 602 | Priests from Korea introduce the Chinese calendar and astronomy to Japan. ASTRO |
| 620 | Indian mathematician Brahmagupta uses negative numbers. His *Brahma-sphuta-siddhānta* treats trigonometry, algebra, and mensuration.                    MATH |
| 622 | The Arab prophet Muhammad (c. 570–632), founder of the religion of Islam, flees on September 20 from his native city of Mecca to Medina. Muslims will date their calendars from this event, called the Hegira, or flight. After his death, Muhammad's followers will establish an Arab empire stretching from Spain, North Africa, and the Middle East to central Asia. In the following centuries the Arabs will sponsor a revival of science and learning that will greatly influence European civilization.                    MISC |
| 628 | Sugar is introduced to Constantinople by soldiers returning through India from fighting in Persia (Iran).                    TECH |
| 635 | The Chinese observe that a comet's tail always points away from the sun. *See also* 1540, ASTRO.                    ASTRO |
| 642 | The Arabs conquer Alexandria, Egypt, the ancient site of Hellenistic learning. Although they destroy its museum and library, the Arabs will preserve, translate, and extend the scientific and mathematical knowledge of the Greeks. MISC |
| 662 | Syrian bishop Severus Sebokht makes the earliest specific reference to Hindu numerals. *See* 595, MATH.                    MATH |
| 673 | Callinicus, an alchemist in Constantinople, invents Greek fire, a chemical mixture (perhaps naphtha, potassium nitrate, and calcium oxide) that burns on water and is useful in naval battles.                    TECH |
| 675 | The first sundial in England is erected in Newcastle.                    ASTRO |
| 700s | Independently from the Hindus, the Maya of Mesoamerica (Mexico and Central America) have developed a system of positional notation based on 20 and using zero as a place holder. With this system they are able to calculate enormous numbers.                    MATH |
| 700s | The Arabs introduce spices from Indonesia to the Mediterranean world. This commercially successful innovation will one day provide an incentive to European explorers seeking easy access to the East.                    MISC |
| 700 | The Chinese invent porcelain, a form of pottery that eventually reaches Europe under the popular name *china*.                    TECH |
| 700 | Windmills are invented in Persia (Iran). The Crusaders will bring the idea to Europe in the 12th century.                    TECH |

708    Safer than local water and said by some to have medicinal powers, tea becomes a commercially popular drink in China, though it has been known there since prehistoric times. It will come to Europe in the 17th century.    TECH

721    Arab alchemist Jabir ibn Hayyan, or Geber, is born (*d.* c. 815). He will draw on ancient Greek alchemy to try to turn base metals into gold and discover the elixir of life or a panacea, a substance that can cure any disease. In the course of these fruitless efforts he will make important discoveries in chemistry, including aluminum chloride, acetic acid, nitric acid, and white lead.    CHEM

750    Geber distills acetic acid from vinegar to create the first pure acid known.    CHEM

751    After learning papermaking techniques from Chinese prisoners, Muslim engineers construct the first paper mill in Muslim territory. A Chinese-influenced paper mill will open in Baghdad, Mesopotamia (Iraq), in 793.    TECH

765    Three-field crop rotation, which allows land to be productive for two out of three years, is first mentioned in European texts. It will later be popularized by Carolingian king Charlemagne.    BIO

770    Hindu works on mathematics are translated into Arabic as the Arabs begin to synthesize, then extend the discoveries of Greek and Indian mathematicians.    MATH

770    Iron horseshoes come into widespread use in Europe.    TECH

780    Arab mathematician Muḥammad ibn Mūsā al-Khwārizmī is born (*d.* 850) in Khwarizm (Uzbekistan). His translated works will introduce Hindu-Arabic notation to Europe. (His name, al-Khwārizmī, is the source of the modern word *algorithm.*)    MATH

800    Chinese mathematicians solve equations with the method of finite differences.    MATH

800    French ecclesiastic St. Bernard forbids Cistercian monks from studying medical books, declaring that prayer is the only remedy allowed to treat the sick.    MED

800    Persians elevate the professional standards of their physicians by requiring examinations before licensing.    MED

808    Islamic translator Hunayn ibn Ishaq is born (*d.* 873). He will translate many volumes of Greek natural science, including works by Plato, Aristotle, Hippocrates, and Galen.    MISC

820    Having borrowed Hindu numerals (now known as Arabic numerals) from India, Arab mathematician Muḥammad ibn Mūsā al-Khwārizmī outlines rules for computing with these numerals. In *Al-jabr wa'l muqabalah*, known as *Algebra* in Europe, he features these numerals and the system of positional notation. This

work also shows how to solve all equations of the first and second degree with positive roots. *See also* 1202, MATH.                                                                    MATH

827     Ptolemy's synthesis of Greek astronomy, *Megalé syntaxis tēs astronomias* (*see* 140, ASTRO), is translated into Arabic as *Al magiste,* or *The Greatest.* It becomes known to history as *The Almagest.*                                                                ASTRO

836     Arab mathematician Thabit ibn Qurra is born in Haran, Turkey (*d.* 901). He will translate Greek works into Arabic and work on solving the problem of Euclid's fifth postulate.                                                                           MATH

850     The astrolabe, used for astronomical observations, is refined by Arab scientists.
                                                                                        ASTRO

858     Arab astronomer Abū `Abd Allāh Muhammad ibn Jābir ibn Sinān al-Battānī al-Harrānī as-Sabt (Albatemius) is born in Haran, Turkey (*d.* 929). He will refine Ptolemy's system, introduce trigonometry as an astronomical tool, recalculate the length of the year, and improve measurements of the precession of the equinoxes.                                                                     ASTRO

863     The Cyrillic alphabet is developed by Macedonian missionary Cyril and his brother Methodius. Eventually it is used by the Russians and various other peoples.                                                                                 TECH

868     The first printed book, *The Diamond Sutra,* is manufactured in China.       TECH

870     Ottar, a Viking, sails more than 100 miles north of the Arctic Circle to become the first human known to cross the Arctic Circle by sea.                                    EARTH

874     Ingolfur Arnarson, a Viking, lands in Iceland, leading to the first permanent settlement there.                                                                             EARTH

---

# Before Gutenberg

In 1454, German printer Johannes Gutenberg founded the modern publishing industry when he set the Latin Bible in lead-alloy movable type and printed it on a wood printing press. But Gutenberg was not the first to invent printing or movable type. The first experimenters with the printing process were the Sumerians, who cut into cylindrical stones symbols representing a signature, and pressed the image into clay that was then baked. The technique of reversed characters being inked onto paper was first developed by the Chinese, who engraved images onto wood blocks as early as the 8th century; the first such book, *The Diamond Sutra,* was printed in 868. By the 11th century, both Chinese and Korean printers were using clay, wood, bronze, and iron to develop movable type. The Chinese printer Pi Sheng was a leader in the field of setting individual ideograms in clay type.

| | |
|---|---|
| 876 | The symbol for zero appears in India. MATH |
| 880 | Arab alchemists and physicians produce concentrated alcohol by distilling wine. CHEM |
| 880 | Persian physician and alchemist Ar-Razī, or Rhazes, uses what will be called plaster of Paris to form cast material for holding broken bones in place. Later he will be the first to give authentic descriptions of smallpox and measles. He will also divide all substances into animal, vegetable, or mineral. His Greco-Arabic medical encyclopedia, *Continens*, translated into Latin in 1279, will be a major source of therapeutic knowledge in Europe for three centuries thereafter. MED |
| 900s | By now, coffee is being cultivated in Ethiopia. *See* 1400s, TECH. TECH |
| 900 | The Classic Mayan civilization (begun c. 250) collapses. MISC |
| 900 | The horse collar is in use in Europe. This innovation completes the transformation of the horse into a powerful farm animal, allowing for still greater food production in northern Europe. *See also* 600, TECH. TECH |
| 940 | With the Dunhuang star map Chinese astronomers invent a map-projection technique of the kind later called the Mercator projection. *See* 1568, EARTH. ASTRO |
| 960–1279 | During the Song period in China, the rain gauge is invented. EARTH |
| 980 | Persian physician Avicenna is born (*d.* 1037). His medical writings will be considered some of the most important textbooks in medical education from the 12th to the 17th centuries. MED |
| 982 | Icelander Erik Thorvaldson, or Erik the Red, discovers Greenland. A Viking colony is begun there in 986. *See also* 1576, EARTH. EARTH |
| 1000 | The Maoris, a Polynesian people, colonize New Zealand. Within a few hundred years they will exterminate most of the island's unique animal species, including the large flightless birds called moas. EARTH |
| 1000 | By now the Malagasy people, originally from Indonesia, have colonized the island of Madagascar off eastern Africa. They exterminate much of that island's unique fauna, including gorilla-sized lemurs and flightless elephant birds. EARTH |
| 1000 | Bjarne Herjulfson, a Viking, becomes the first European to see the Americas when he sails west past Greenland to Newfoundland, Canada. Two years later, Leif Eriksson will travel there to found the short-lived colony of Vinland. EARTH |
| 1000 | Eskimos arrive at Greenland, and competition begins to develop between their group and the Vikings. In 1415 the Viking colony will end and Greenland will be left to the Eskimos. EARTH |

| 1000 | By this time Polynesians have traversed 14 million square kilometers (5.6 million square miles) of ocean and occupy a triangle from New Zealand north to Hawaii and east to Easter Island. EARTH |

1000    By this time Polynesians have traversed 14 million square kilometers (5.6 million square miles) of ocean and occupy a triangle from New Zealand north to Hawaii and east to Easter Island. EARTH

1000    In China the pivoting needle on a magnetic compass is discovered and becomes an important aid to navigators. By 1100 this knowledge is picked up by Arab traders through contacts with the Indonesian islands. EARTH

1000    Persian or other Asian travelers introduce the concept of the seven-day week to the Chinese. Before this, ten-day weeks were common in China. TECH

1000    The Bridge of the Ten Thousand Acres in Foochow, China, is constructed. TECH

1006    A bright "guest star" or supernova, visible for a number of years, is observed in China, Japan, various Islamic lands, and Europe. ASTRO

c. 1010    Arab astronomer Ibn Yunus compiles two centuries of observations in *The Large Astronomical Tables of al-Hakim*, used by future Arab astronomers. ASTRO

c. 1025    Arab physicist Abū `Alī al-Hasan ibn al-Haytham (Alhazen) is one of the first scientists to study optics. He analyzes lenses, develops parabolic mirrors, and theorizes that vision is the result of light falling on the eye, not light emanating from the eye, as had been previously thought. *See also* c. 1270, PHYS. PHYS

1027    As acupuncture becomes more systematic, China's reigning emperor requests that copper models of the human body be made to illustrate the principles of this form of medical therapy. MED

1041    Movable type of Chinese ideograms, fashioned from clay blocks, is used in China by printer Pi Sheng. TECH

c. 1050    Arab poet, scientist, and mathematician Omar Khayyám is born (*d.* 1123). His *Algebra* will go beyond al-Khwārizmī's (*see* 820, MATH) to include equations of the third degree. MATH

c. 1050    The crossbow, the first mechanized hand weapon, is introduced into France. It uses a device such as a two-handed crank to increase the tension of the bow and therefore the force of the bolt or arrow. TECH

c. 1050    More efficient iron plows are used in northern Europe instead of wooden plows. TECH

1054    On July 4, Chinese, Japanese, and Arab astronomers report a supernova that remains visible for 22 months. Its residue will form what will become known as the Crab Nebula. ASTRO

| | |
|---|---|
| c. 1065 | Jewish rabbi, physician, and philosopher Moses Maimonides becomes known for his medical teachings. In the 20th century the invocation known as the Prayer of Maimonides will be used at some medical school graduation ceremonies.    MED |
| 1066 | During William the Conqueror's invasion of England, the bright phenomenon that will later become known as Halley's comet is sighted.    ASTRO |
| 1067 | To prohibit international development of its invention of gunpowder, China outlaws the exportation of sulfur and saltpeter.    TECH |
| 1071 | Forks as eating utensils are introduced to Venice and western Europe from the Byzantine Empire, but they are slow to gain general acceptance.    TECH |
| c. 1075 | The Arab astronomer Arzachel theorizes correctly that the orbits of planets are elliptical, not circular.    ASTRO |
| 1086 | The magnetized compass is popularized by Shen Cha, an official working with the Chinese water systems.    TECH |
| c. 1100s | Latin translators such as Adelard of Bath, Michael Scot, and Gerard of Cremona introduce Arabic works on astronomy and mathematics to medieval Europe. These works, based ultimately on ancient Greek learning, eventually become the foundation of European science.    MISC |
| c. 1100 | The Chinese demonstrate the cause of solar and lunar eclipses.    ASTRO |
| 1100 | Arab physician Ibn Zuhr (Avenzoar) is the first to describe the parasite, an itch mite, causing the highly contagious skin disease scabies.    MED |
| c. 1100 | A history of science by Arab writer Abu'l Fath al-Chuzini includes tables of specific densities and observations on gravity.    PHYS |
| 1120 | The use of latitude and longitude measurements, in degrees, minutes, and seconds, is introduced by an Anglo-Saxon scientist known as Welcher of Malvern.    TECH |
| 1137 | The Abbey of St. Denis near Paris, designed by Abbé Suger, becomes the first major building to make use of flying buttresses, a system of architectural support that permits the building of gigantic cathedrals. The new architecture is known as Gothic.    TECH |
| 1142 | Adelard of Bath translates Euclid's *Elements* into Latin.    MATH |
| 1148 | Soldiers returning from the Crusades in the Middle East introduce sugar to Europe.    TECH |
| c. 1159 | Between now and 1173, Benjamin of Tudela travels east through Islamic lands, keeping a written account of his journey. Though he is the first western |

European to do so, his status as a Jew prevents his account from being influential in the Christian world.                                                                                EARTH

1175    Gerard of Cremona, Italy, translates Ptolemy's astronomical compendium *The Almagest* into Latin, along with other Greek and Arab works.                                ASTRO

1176    In England, rabbits are introduced as local livestock.                                        BIO

1180    English scholar Alexander Neckam is the first European to note that a magnetic needle always points north–south (*see also* 100s, PHYS). Three centuries later, this discovery contributes to the navigation feats of the age of exploration.                                                                                                   PHYS

1184    The cathedral at Sens, France, becomes one of the earliest examples of Gothic architecture and technology. It is designed by architect William of Sens. *See also* 1137, TECH.                                                                                                            TECH

1189    In France, the first paper mill in Christian Europe is opened. *See also* 1276 and 1494, TECH.                                                                                                TECH

1193    Indigo is exported from India for use in dyeing fabrics. *See also* 1741, BIO, and 1870, CHEM.                                                                                            TECH

c. 1194 Viking explorers reach Spitzberg, 450 kilometers north of the Arctic Circle. This is the farthest north the Vikings go.                                                         EARTH

1200s   Coal, known as a fuel in China since 1000 B.C., is mined in England by early in this century. *See* 1233, TECH.                                                              TECH

c. 1200 Medical instruction becomes more theoretical and scholarly, spreading to medical schools at Montpellier, Paris, Oxford, and Bologna.                            MED

1202    Italian mathematician Leonardo Fibonacci publishes his *Book of the Abacus*, which introduces Arabic numerals and positional notation to Europe, though these are not fully adopted for 300 years. *See also* 820, MATH.                            MATH

c. 1220 Scottish naturalist Michael Scot translates Aristotle's classifications of animals into Latin.                                                                                        BIO

c. 1220 Jordanus Nemorarius publishes *Mechanica*, describing a law of the lever and the law of the composition of movements.                                               PHYS

1225    Cotton is first manufactured in Spain.                                                          TECH

1233    Coal is first mined in Newcastle, England. The mine is so successful that it generates the phrase "carrying coals to Newcastle," connoting an unnecessary activity.                                                                                                       TECH

| | |
|---|---|
| 1237 | Chinese physician Chen Tzu-ning publishes his *Fu Jen Liang Fang*, the first Asian monograph on the diseases of women. MED |
| 1240 | European shipbuilders construct vessels with rudders, an innovation borrowed from the Arabs. TECH |
| c. 1245–1247 | Franciscan friar Giovanni da Pian del Carpini leads a conversion mission into Mongol lands. His account of these travels, *Liber tartarorum*, becomes the first opportunity for Westerners to read an accurate description of central Asia. EARTH |
| 1249 | English scholar and scientist Roger Bacon notes that lenses can be used for improving eyesight. Eyeglasses appear in China and Europe at about the same time; it is not clear where they were invented first. TECH |
| 1249 | Gunpowder, developed in China in 270, is mentioned for the first time in European writings by English scholar and scientist Roger Bacon. TECH |
| c. 1250 | German scientist Albertus Magnus introduces Aristotle's ideas on botany and biology to Europe. His *De vegetabilibus* classifies plants and vegetables and describes the function and structure of various plant parts. BIO |
| c. 1250 | Albertus Magnus discovers the element arsenic. CHEM |
| 1250 | Crusaders returning to Europe from Arab lands help spread acceptance of Arabic numbering and decimal systems. MATH |
| 1259 | Construction begins on an observatory at Maragha, Persia (Iran). ASTRO |
| c. 1260 | An observatory is built at Beijing, China. ASTRO |
| 1264 | Bakers' marks, the forerunner to trademarks, are used for the first time in England. Through them bakers identify their wares with individualized icons slashed into the bread. TECH |
| c. 1265 | During the Kamakura period (1185–1333), Japanese swordsmiths reach their technical apex. Their *tachi* (slashing swords) are sharp enough to behead an enemy with one stroke. TECH |
| 1267 | The Council of Venice forbids Jews to practice medicine among Christians. MED |
| 1269 | William of Moerbeke translates the major scientific and mathematical treatises of Archimedes into Latin. MATH |
| 1269 | Tolls are charged on some roads in England. MISC |
| 1269 | French scholar Pèlerin de Maricourt (Petrus Peregrinus) performs early experiments with magnets, describing magnetic poles and refining the use of a magnet as a compass. TECH |

c. 1270    The Polish scientist Witelo writes *Perspectiva*, which will be combined with works by Arab physicist Abū `Alī al-Hasan ibn al-Haytham (Alhazen; *see* c. 1025, PHYS) in *Opticae Thesaurus* (1572), the most influential treatise on optics until the 17th century.                                                                    PHYS

1270       English scholar and scientist Roger Bacon researches optics and refraction, the bending of a light ray as it passes from one medium into another. *See* 1249, TECH.                                                                                                             PHYS

1272       The Alphonsine tables, planetary charts whose compilation was ordered in 1250 by Alphonso X of Castile, are completed. They will remain in use until the 1500s.                                                                                                            ASTRO

c. 1276    Italian scientist Giles of Rome writes a treatise, *De formatione corporis humani in utero*, on the development of the human fetus. It includes a discussion of the timing of the soul's entry into the fetus and the biological importance of each of the two parents.                                                                                                 BIO

1276       Papermaking begins in Italy, in the city of Montefano.                               TECH

c. 1280    Arab physician Alquarashi is the first to identify the pulmonary transit of blood, from the right to the left ventricle via the lungs.                                          MED

1288       The first known gun, a small cannon, is made in China.                              TECH

# When Glass Lost Its Color

For thousands of years after sand was fired into glass objects in the Near East around 2500 B.C., this now commonly clear material was usually produced only in color. Impurities lent color to glass, and a workable decolorization process had yet to be invented.

By the end of the 13th century, the technique of adding clarifying substances to make glass clear was perfected in Venice, which had become, and to this day remains, a world center for exquisite glass production. One way Venetian glass manufacturers retained their monopoly on certain glassmaking processes was to move in 1291 to an isolated island where materials were hoarded, techniques kept secret, and penalties assessed on trespassers and talkative employees.

To infuse glass with color, impurities can still be useful. Here are some common compounds added to glass to produce popular colors:

| | |
|---|---|
| Blue | Cobalt and copper |
| Bottle Green | Oxidized iron |
| Brown | Iron and sulfur |
| Purple | Chromates |
| Red | Copper or selenium |
| Ruby | Gold |

1289        Block printing is used for the first time in Europe. *See also* 1041, TECH.        TECH

c. 1290     French surgeon Henri de Mondeville advises doctors to cleanse wounds and let them dry without salves or wine-soaked dressings. He also recommends applying pressure to stop bleeding and advocates the use of sutures.        MED

1291        In Venice, glassmakers learn to produce clear glass (only colored glass has been available since glass ornaments first appeared in Egypt about 2500 B.C.). The colorless glass will make modern mirrors and windowpanes possible.        TECH

c. 1292     A new type of vessel, the great galley, is developed in Venice. These long, shallow boats are driven by multiple rows of oarsmen and can carry a great deal of cargo.        TECH

1295        French physician Lanfranchi becomes the first to describe a brain concussion and the symptoms of a skull fracture.        MED

1298        Venetian merchant Marco Polo publishes *Divisament dou monde*, describing his travels (1275–1295) in China. The book will inspire future explorers.        EARTH

1300s       Mechanical clocks, driven by the force of gravity on weights, are invented in Europe.        TECH

1300        The False Geber, an anonymous alchemist writing under the name Geber, discovers sulfuric acid, the most powerful acid yet known.        CHEM

1300        Spanish alchemist Arnau de Villanova produces brandy from wine for the first time.        TECH

1303        Chinese mathematician Chu-Shi-kié (Chu Shih-chieh) writes the *Precious Mirror of the Four Elements*, which marks the apex of Chinese algebra. It contains simultaneous equations, equations of degrees up to 14, the Horner transformation method, and the arithmetic triangle later called the Pascal triangle.        MATH

1304        Theodoric of Freibourg, Germany, accurately explains several aspects of the formation of rainbows.        PHYS

1312        Europeans reach the Canary Islands off Morocco for the first time.        EARTH

1316        Italian anatomist Mondino de' Luzzi writes *Anatomia*, the first book in history devoted entirely to anatomy.        MED

1320        By this time in Europe, paper, a Chinese invention (*see* 105, TECH), has largely replaced vellum, a parchment made from the skins of animals such as the calf or lamb.        TECH

1328        English philosopher and mathematician Thomas Bradwardine publishes his *Tractatus de proportionibus*, in which he broadens the theory of proportions and proposes an alternative to Aristotle's (incorrect) law of motion.        MATH

| 1333 | In Venice the first botanical garden since antiquity is established. | BIO |

| 1346 | English king Edward III uses longbows and cannons loaded with gunpowder at the battle of Crécy, France, on August 26. The longbows are more important to Edward's victory, but gunpowder is the weapon of the future. | TECH |

| c. 1348 | Bubonic plague, called the black death, begins to sweep Europe after devastating Asia and North Africa. As much as three quarters of the Old World's population will die from the plague within 20 years. | MED |

| c. 1350 | French philosopher Jean Buridan proposes the concept of impetus, which is similar to the concept of inertia as articulated in Newton's first law of motion. *See* 1687, PHYS. | PHYS |

| c. 1360 | French mathematician Nicole d'Oresme generalizes Thomas Bradwardine's theory of proportions (*see* 1328, MATH), suggests notations for fractional powers, argues that irrational powers are possible, and develops the graphic representation of functions known as the latitude of forms. | MATH |

| 1360–1644 | In China during the Ming dynasty, texts and diagrams detail the locust's behavorial characteristics. | BIO |

---

# The Plague and Anti-Semitism

Bubonic plague, known in medieval times as the black death, is known today to be caused by the bacterium *Yersinia pestis*, a microbe transmitted to humans by the bite of fleas from infected hosts, especially rats. However, in the mid-14th century, when the black death moved like a wave across Central Asia, the Middle East, North Africa, and Europe, its cause was a mystery. Its sufferers experienced chills, high fever, vomiting, diarrhea, and the formation of buboes, or painful inflammations of lymph nodes, most commonly in the groin. Black hemorrhages might form and the disease turn into pneumonic plague, in which the patient's lungs became infected, leading swiftly to death. The pneumonic form was particularly contagious, transmitted from one person to another by water droplets in the air.

Unchecked by medical knowledge, this lethal epidemic decimated the population. Sufferers grappled for a way to explain its destruction. Many Christians, considering it divine punishment for moral wrongdoing, turned to prayer and acts of penance. Others took a course as virulent as the disease itself: blaming the Jews.

In communities throughout Provence, Catalonia, Aragon, Switzerland, southern Germany, and the Rhineland, Jews were accused of poisoning Christian water sources. Violence erupted against Jews. In some areas, governments resisted the clamor for retribution, but in others they abetted the Christian protests, destroying hundreds of Jewish communities. Jews in western Europe were burned, tortured, imprisoned, and exiled in numbers large enough to shift the center of European Jewry permanently eastward.

| | |
|---|---|
| 1370 | John of Arderne, the earliest known English surgeon, writes extensively on modern surgery, the use of irrigation (cleansing by flushing with water), and the repair of anal fistulas. **MED** |
| 1391 | English writer Geoffrey Chaucer's *Treatise on the Astrolabe* explains how to construct and use this instrument for measuring the position of stars. **ASTRO** |
| 1400s | Coffee, first cultivated in the 900s but rarely drunk until now, spreads from the Yemen region of southern Arabia to become popular in the Arab world. It will reach most of Europe by the 17th century. **TECH** |
| c. 1400 | Italian architect Filippo Brunelleschi begins the first archaeological digs in Rome. **ARCH** |
| 1400 | English monks Johann Sprenger and Heinrich Kraemer publish *The Witches' Hammer*, claiming in it that witches are possessed by the devil and should be killed. Although this book has a long, influential history in Europe, by the 20th century researchers will theorize that most "possessed" people in the 1400s were in fact mentally ill. **PSYCH** |
| c. 1402 | Emperor Yung-lo of China encourages overseas conquest. Admiral Cheng-ho's fleet sails as far west as the Red Sea, visiting Mecca and Egypt as well as Indonesia, Malaya, and Sri Lanka. **EARTH** |
| 1418 | Prince Henry the Navigator of Portugal opens an observatory and school for navigation at Sagres on Cape St. Vincent, Portugal. His goal is to find a way around Africa to reach the trade riches of China. **EARTH** |
| 1418 | Portuguese navigators discover and claim Madeira. **EARTH** |
| 1427 | The Portuguese navigator Diogo de Sevilha discovers the Azores, an island chain more than 700 miles west of Portugal. **EARTH** |
| 1428 | Mongol astronomer Ulūgh Beg builds an observatory at Samarkand featuring a quadrant, a device to measure stellar positions, that is 180 feet high. **ASTRO** |
| 1430s | Ulūgh Beg publishes a new star map and tables of star positions, improving on Ptolemy's work. *See* 140, **ASTRO**. **ASTRO** |
| 1436 | Italian artist Leon Battista Alberti writes on using mathematical principles to achieve perspective in art. **TECH** |
| 1439 | The military forces of French king Charles VII become the first to make systematic use of gunpowder artillery. **TECH** |
| 1440 | German scholar Nicholas of Cusa argues that space is infinite and that the stars are suns, each with its own inhabited planets. **ASTRO** |

| | |
|---|---|
| c. 1450 | It is discovered that transporting soil up from the valleys in the Andes mountains makes the higher regions arable.                                                    BIO |
| c. 1450 | The idea of bodily humors (*see* 350 B.C., MED) remains central to European medicine as the Middle Ages draw to a close. Purging, cupping, bloodletting, and leeching are all in ordinary use to control the body's balance of blood, phlegm, yellow bile, and black bile.                                                    MED |
| 1450 | The Dutch arquebus (also harquebus) becomes the first firearm small enough to be carried and fired by a single person.                                                    TECH |
| 1450 | The Chinese are printing pages using movable wooden type, an innovation that will soon (*see* 1454, TECH) spread to Europe.                                                    TECH |
| 1451 | Nicholas of Cusa, a German scholar, introduces the idea of using concave lenses to amend nearsightedness. Previously, only convex lenses for farsightedness were used.                                                    TECH |
| c. 1452 | Between now and 1519, Italian scientist and artist Leonardo da Vinci makes notes proposing the marine origin of fossils. Knowing it would be thought heretical, he does not publicly reveal this observation.                                                    EARTH |
| 1453 | Constantinople falls to the Ottoman Turks on May 30, marking the end of the Byzantine, or Eastern Roman, Empire and the traditional end of the Middle Ages. The city will become known as Istanbul. Christian scholars fleeing the Muslims help bring the knowledge of classical learning to western Europe, contributing to the flourishing of arts, letters, and sciences during the Renaissance (15th to 17th centuries).                                                    MISC |
| 1454 | With a printing press of his own invention, using movable metal type, Johannes Gutenberg of Mainz, Germany, prints 300 copies of the Bible in Latin. This book, the first to be printed in Europe (*see* 868, TECH), will become known as the Gutenberg Bible.                                                    TECH |
| 1456 | Sugar is made more widely available in England.                                                    TECH |
| 1457 | The first medical publication printed, a calendar, advises physicians as to when purging should be most therapeutic.                                                    MED |
| c. 1460 | Prince Henry the Navigator dies (*b.* 1394). By this time Portuguese explorers have sailed down the west coast of Africa as far as the country that will become known as Gambia.                                                    EARTH |
| 1470 | The mainspring, a spiral string whose gradual unwinding powers a clock, is invented.                                                    TECH |
| 1472 | German mathematician and astronomer Regiomontanus (Johann Müller) makes the first scientific study of the comet that will later become known as Halley's comet (*see* 1758, ASTRO).                                                    ASTRO |

1473     Astronomer Nicolaus Copernicus is born on February 19 in Torun, Poland (*d.* 1543). His *De revolutionibus orbium coelestium* (1543), proposing that Earth and the other planets revolve around the sun, will revolutionize the field of astronomy.     ASTRO

1473     The first medical dictionary is printed.     MED

1473     The atomic theory of Democritus (*see* 440 B.C., MISC) becomes known to Western scholars with the translation into Latin of Lucretius's *On the Nature of Things* (60 B.C.).     PHYS

1474     Regiomontanus publishes his *Ephemerides astronomicae*, a compendium of celestial coordinates that will prove useful to navigators. *See* 1504, ASTRO.     ASTRO

1474     After 425 years under construction, Winchester Cathedral in England is completed.     TECH

c. 1475     Aristotle's *Meteorologia* and Ptolemy's *Geographia* are translated into Latin. EARTH

1476     In Venice, Aristotle's fourth-century B.C. work on animal structure, function, reproduction, physiology, and development is published under the title *De animalibus*. It is the first zoological compilation.     BIO

1479     The first book set from metal type in England is printed by William Caxton: *The Game and Playe of Chesse*, a Latin work translated into French.     TECH

1481     German cleric Konrad von Megenberg's *Buch der Natur*, the first printed book to contain animal figures, is published.     BIO

1483     Greek botanist and philosopher Theophrastus (372–287 B.C.) is posthumously honored when his treatise on botany *De historia et causis plantarum* is published in Treviso. This work is considered the earliest on scientific botany.     BIO

1487     Portuguese navigator Bartolomeu Dias discovers the southernmost tip of Africa, the Cape of Good Hope.     EARTH

1487     John II of Portugal, Prince Henry the Navigator's grand-nephew, organizes an expedition through the Mediterranean and the Red Sea under the leadership of Pero da Covilhão.     EARTH

1490     The *Tabulae directionum* of German mathematician and astronomer Regiomontanus is published posthumously. In this work and *De triangulus*, not published until 1533, Regiomontanus first organizes trigonometry as a discipline, making numerous advances in the field. He also applies algebraic methods to geometric problems.     MATH

1490     Italian scientist and artist Leonardo da Vinci describes capillary action, the rise of fluids in a small-diameter (capillary) tube.     PHYS

1492          On October 12, Italian mariner Christopher Columbus (1451–1506), leading
              a fleet of three ships from Spain, becomes the first European since the Vikings
              (*see* 1000, EARTH) to reach the Americas when he disembarks on an island in the
              Bahamas. Unlike the Vikings, Columbus will open the Western Hemisphere to
              wholesale colonization by Europe.                                          EARTH

1492          Italian mathematician Francesco Pellos introduces the decimal point.       MATH

1492          Through the voyages of Christopher Columbus several foods, including all-
              spice, peppers, plantain, and pineapples, are made known to Europe.        TECH

1493          On September 25, Columbus's second voyage sets sail from Cadiz, Spain. This 17-
              ship expedition will explore Dominica, Puerto Rico, Cuba, and Jamaica.     EARTH

---

# Leonardo's Science

Italian Renaissance artist Leonardo da Vinci is well known. Leonardo the scientific
investigator has been less well appreciated. Yet his notebooks, particularly the two previously unknown ones found in Madrid in 1965, reveal the Italian artist and thinker as
a man who merged the worlds of science and art to better understand the mechanics
of life and the possibilities of technology.

A brief sampling of Leonardo's scientific achievements shows not only his ability to
intertwine the disciplines of the arts and sciences but to balance theoretical pursuits
and the activity of practical tasks:

- In 1494, while serving as artist and scientist to the court of Lodovico Sforza in
  Milan (one year before completing the painting *The Last Supper*), he devised plans
  to harness the waters of the Arno River.
- As military engineer for Cesare Borgia in 1502–1503, Leonardo explored problems of swamp reclamation.
- While examining mathematical theory in Florence, he studied anatomy at the
  city's hospital of Santa Maria Nuova.
- During a time of great artistic achievement (his *Mona Lisa* was created in 1503), he
  furthered his study of anatomy by analyzing the movements of birds in flight and
  carrying out a variety of cadaver dissections that culminated in a book of life drawings, *Anatomy* (1508, unpublished).
- While architect and engineer to French king Louis XII from 1506 to 1513,
  Leonardo also undertook scientific studies of botany, geology, and hydraulic
  power.

His other areas of exploration included rock stratification, the making of eyeglass
lenses, and inquiries into flying machines. Whether theoretical, artistic, or practical,
Leonardo's explorations were experiments in vision. He believed that the key to understanding the world was *saper vedere*, "to know how to see."

| 1493 | A town crier is directed by Paris officials to order all those who have the "greater pox" (syphilis) to leave the city—or be thrown into the Seine. By 1496 syphilis will be epidemic in Europe. MED |
|---|---|

1494　Italian mathematician Luca Pacioli publishes his *Summa de arithmetica, geometria, proportioni et proportionalita*. This highly influential work includes not only the first printed material on algebra but also useful information on double-entry bookkeeping. MATH

1494　The first English paper mill opens. TECH

c. 1497　Italian navigator Amerigo Vespucci explores the coast of the continental area south of the islands discovered by Columbus. Vespucci maintains that this land is not Asia but a "new world." EARTH

1497　Italian mariner Giovanni Caboto (John Cabot), sailing for the English, reaches Newfoundland and Nova Scotia. He becomes the first European since the Vikings (*see* 1000, EARTH) to reach the mainland of North America. EARTH

1497　On November 22, Portuguese navigator Vasco da Gama becomes the first European to round the southernmost part of Africa, the Cape of Good Hope, and reach Asia, arriving in India in 1498. EARTH

1498　In May Columbus departs from Spain on his third voyage, commanding a fleet of six ships. After stopping at Trinidad, the fleet reaches the Paria Peninsula on the coast of Venezuela on August 5, marking Columbus's first visit to the mainland of the Americas. EARTH

1500　German engineer Ulrich Rulein von Kalbe writes *Bergbuchlein*, the first known mining manual. EARTH

c. 1500　Italian scientist and artist Leonardo da Vinci dissects human bodies and records his anatomical findings in accurate, detailed drawings. MED

1501　German botanist Leonhard Fuchs is born in Wemding, Bavaria (*d.* 1566). *Fuchsia*, a genus of tropical shrubs and trees, will be named for him, as will the color fuchsia after the purplish-red of the shrub's flowers. BIO

1501　From now to 1587, a pandemic outbreak spreads through Europe of a disease characterized by fever, headache, sweating, and a black tongue. It is initially called Mobus Hungaricus (the Hungarian disease) but later will be regarded as an outbreak of typhus. MED

1502　On May 9, Columbus departs from Cadiz, Spain, on his fourth and final voyage to the Americas. On this voyage, Columbus will visit Santo Domingo, the Cayman Islands, Jamaica, Cuba, and Honduras. He will then sail south along the coasts of Nicaragua and Costa Rica looking in vain for a sea passage to India. After being marooned in Jamaica for a year, he will arrive back in Spain on November 7, 1504, and will die two years later, on May 20, 1506. EARTH

1503        The properties of rubber are noted for the first time by Europeans during Columbus's fourth voyage. This substance, made of latex from the plant *Hevea brasiliensis* or *Parthenium argentatus*, first came to the explorers' attention in the form of a ball used in various games by the Native Americans. *See* 1615, TECH.

        TECH

1504        Using German mathematician and astronomer Regiomontanus's *Ephemerides astronomicae* (1474), Columbus correctly predicts a total lunar eclipse on February 29, to the astonishment of the local Native Americans.    ASTRO

1504        German inventor Peter Henlein devises the first watch, a clock small enough to fit in a pocket.    TECH

1506        The Laocöon sculpture (first century B.C.) is discovered near Santa Maria Maggiore, Italy. It will influence Michelangelo and other Renaissance sculptors.    ARCH

1507        The name *America* appears on a map for the first time. German cartographer Martin Waldseemüller names the newly discovered lands for Amerigo Vespucci, an Italian explorer reputed to have been the first to recognize that these areas were not part of Asia. *See* c. 1497, EARTH.    EARTH

1510        French barber and surgeon Ambroise Paré (*d.* 1590) is born near Laval, Mayenne. He will be considered the father of modern surgery for his common-sense treatments of injury and disease. At a time when other surgeons are treating gunshot wounds with boiling oil, Paré treats them with salves and cleanliness. And when colleagues are cauterizing bleeding arteries—without anaesthesia—Paré learns to tie off arteries to stop blood loss.    MED

1510        Italian scientist and artist Leonardo da Vinci designs a horizontal water wheel, a forerunner of the water turbine.    TECH

1512        Hieronymus Brunschwygk publishes his *Big Book* (an expansion of his *Little Book* of 1500), on chemical apparatus and techniques such as stills, furnaces, and distillation.    CHEM

1513        Exploring Florida, Juan Ponce de León of Spain is the first European to reach the portion of North America that will become the United States.    EARTH

1513        On September 25, Spanish explorer Vasco Nuñez de Balboa, traveling through Panama, becomes the first European to see the Pacific Ocean from the Americas.    EARTH

1513        Orange and lemon trees are introduced to Florida by Spanish explorer Ponce de León.    TECH

1514        Copernicus writes his first account of his heliocentric theory, that Earth and the other planets revolve around the sun. His writings circulate quietly for years, but he does not publish a complete account until 1543.    ASTRO

1514      Plus (+) and minus (–) signs are first used in equations for addition and subtraction.      MATH

1514      French physician Pierre Brissot opposes the current method of bloodletting, in which physicians drain blood from the veins farthest from the pathogenic lesion. Brissot claims that blood withdrawal from a surgically opened vein should be near the lesion to be effective. For this heresy he is banished by the French Parliament.      MED

1514      In Brussels, Flemish anatomist Andreas Vesalius is born on December 31 (*d.* 1564). He will create the first accurate illustrated book on the structure of the human body, marking the beginnings of the modern study of anatomy. He will oppose Galen's theories on anatomy (*see* 160, MED), proving them incorrect. Vesalius will also oppose Aristotle's view that the heart is the seat of life, claiming that role instead for the brain and nervous system.      MED

1517      Spanish explorer Francisco Fernández de Córdoba becomes the first European to discover the Yucatán Peninsula and find the remains of Mayan civilization.      EARTH

1517      Henry VIII's physician, Thomas Linacre, writes a new Latin translation of Galen's medical treatises (*see* 160, MED). Physicians all over Europe now realize they had been placing their full confidence in distorted, secondhand versions of the famous physician's work.      MED

1518      England's Royal College of Physicians is founded on September 23.      MED

1519      Chocolate, peanuts, sweet potatoes, and vanilla, among other foods, are found in use in Mayan communities.      BIO

1519      Spanish explorer Hernando Cortés sails from Cuba to Mexico, where he encounters Aztec civilization, with its capital at Tenochtitlán, the site of what will become Mexico City. In time the Spanish destroy the Aztec civilization.      EARTH

1519      Ferdinand Magellan, a Portuguese navigator wounded while fighting for his country against Morocco but denied a pension, offers his services instead to Spain. Charles V decides to sponsor Magellan on a westward voyage to circumnavigate the globe. Magellan departs on September 20 with five ships.      EARTH

1520      The rifle is developed by German gunmaker August Kotter.      TECH

1521      Fighting with natives in an area that will come to be known as the Philippines, Magellan dies on April 27.      EARTH

1522      In September, a ship from Ferdinand Magellan's fleet returns to Spain, the first to circumnavigate the globe. Magellan himself and four of his five ships failed to complete the trip. The lone surviving ship is led by Juan Sebastián de Elcano.                                                                    EARTH

1524        Commissioned by France to search for a northwest passage, Italian navigator Giovanni da Verrazano becomes the first European to enter New York harbor.                                                                    EARTH

1524        The Hospital of the Immaculate Conception in Mexico City is built by Spanish explorer Hernando Cortés. It is the first hospital on the American continent.                                                                    MED

1525        Christoff Rudolf introduces the square root symbol ($\sqrt{\phantom{x}}$) and makes use of decimal fractions in *Die Cass*.                                                                    MATH

1525        In Rome the first Latin transcription of the works of Hippocrates (*see* 460 B.C., MED) is published.                                                                    MED

1526        Spanish historian Gonzalo Fernández de Oviedo publishes *Summary of the Natural History of the West Indies*; he will later expand the work into an encyclopedic tome covering all Spanish possessions in the Americas. He is considered the first ethnographer of the New World.                                                                    SOC

1527        Swiss-born chemist and physician Philippus Aureolus Theophrastus Bombast von Hohenheím (Paracelsus) publicly burns the works of Galen and Avicenna, whose theories of humors he rejects, claiming that the purpose of alchemy is to make medicines that will cure and treat disease. Although he is ridiculed for his belief in astrology and an elixir of life, Paracelsus stresses the importance of minerals, especially zinc, in treating disease.                                                                    MED

1529        Sweet rather than bitter oranges are introduced from Asia to Europe by the Portuguese. Bitter oranges had been available for centuries.                                                                    BIO

1530        Italian physician and poet Girolamo Fracastoro writes *Syphilis sive morbus Gallicus*, giving this sexually transmitted disease its modern name and recognizing its venereal cause.                                                                    MED

1530        Spanish explorer Gonzalo Jiménez de Quesada becomes the first European to learn about the potato, from Native Americans in Colombia. It will become a staple of Old World as well as New World cooking.                                                                    TECH

1530        Dutch mathematician Gemma Frisius suggests that the local time of a prime meridian should be the standard time in determining longitude.                                                                    TECH

1531        Tobacco is commercially grown in the Spanish West Indian colonies.                                                                    BIO

1532–        Spanish explorer Francisco Pizarro conquers Peru, destroying the Inca civiliza-
1533        tion as Cortés destroyed the Aztec one. *See* 1519, EARTH.                                                                    EARTH

| | |
|---|---|
| 1532 | Bills of mortality, showing that the population can be estimated from birth and death rates, are introduced in England. They are the first attempt at vital statistics. MED |
| 1535 | Italian mathematician Niccolò Tartaglia works out a method for solving cubic equations. The solution is later published by mathematician Geronimo Cardano. *See* 1545, MATH. MATH |
| c. 1536 | Sarsaparilla is promoted as an antisyphilitic drug. MED |
| 1540 | In *Astronomicum caesareum*, German astronomer Petrus Apianus (Peter Bennewitz) observes that the tails of comets always point away from the sun, as the Chinese had previously discovered. *See* 635, ASTRO. ASTRO |
| 1540 | In his *Narratio prima de libris revolutionum*, German mathematician Georg Joachim Iserin von Lauchen (Rhäticus or Rheticus) offers a summary of the heliocentric system developed but not yet published by Copernicus. *See* 1514, ASTRO. ASTRO |
| 1540 | *On Pyrotechnics*, by Italian mine supervisor Vannoccio Biringuccio, discusses ore processing, smelting, distillation, and other such topics. This posthumous opus is considered the first important work on metallurgy. CHEM |
| 1540 | German instrument maker Georg Hartman discovers magnetic inclination or dip and is believed to be the first to measure magnetic declination on land. EARTH |
| 1540 | During an expedition in 1540–1542, Spanish explorer Francisco Vásquez de Coronado becomes the first European to see the Grand Canyon. EARTH |
| 1540 | Swiss-born chemist and physician Paracelsus is the first to use tincture of opium, which he calls laudanum, for medical purposes. MED |
| 1540 | Prussian physician Valerius Cordus (1515–1544) discovers sulfuric ether, although it will not become widely used as an anesthetic until the 19th century. MED |
| 1540 | Italian physician Pietro Andrea Mattioli uses mercury to treat syphilis, but this popular cure is painful and often kills the patient. Mattioli also advocates oil of scorpions to treat the plague. MED |
| 1541 | During an expedition in 1539–1542 led by Spanish explorer Hernando de Soto, Europeans see the Mississippi River for the first time. EARTH |
| 1541 | During an expedition in 1541–1542, Spanish explorer Francisco de Orellana becomes the first European to see the Amazon River and cross South America from ocean to ocean. EARTH |
| 1542 | In Leipzig, Germany, a botanical garden is established. BIO |

1542      German botanist Leonhard Fuchs (*see also* 1501, BIO) describes peppers, pumpkins, and corn (maize) from the New World in his botanical masterpiece *De historia stirpium*. His work makes no attempt to classify plants, instead emphasizing firsthand observation of plant habits, locales, and characteristics.    BIO

1543      Shortly before the death this year of Polish astronomer Nicolaus Copernicus (*b.* 1473), his *De revolutionibus orbium coelestium* (*On the Revolutions of Celestial Bodies*) is published. This work argues that Earth and the other planets travel around the sun, in contradiction to the prevailing geocentric world view codified by Ptolemy (*see* 140, ASTRO) and accepted ever since. Like Ptolemy, though, Copernicus argues that planetary motion is basically circular, with epicycles to account for observed anomalies. The publication of this work marks the beginning of what will become known as the Copernican revolution in astronomy.    ASTRO

1543      On June 1, Flemish anatomist Andreas Vesalius publishes *De corporis humani fabrica* (*On the Structure of the Human Body*), the first accurate human anatomy text.    MED

1544      Italian physician and botanist Pietro Andrea Mattioli publishes an Italian version of the classic botany text *De materia medica* by Dioscorides, a Greek physician and herbalist. *See* 60, BIO.    BIO

1545      Italian mathematician Geronimo Cardano publishes his *Ars magna*, a landmark book on algebra often considered to mark the beginning of modern mathematics. It includes Tartaglia's method for solving cubic equations (*see* 1535, MATH) and Ludovico Ferrari's method for solving quartic equations. It also includes methods for working with negative numbers.    MATH

c. 1545      French surgeon Ambroise Paré devises artificial limbs for war-injured soldiers. These "hands" include individually moving fingers and a holder for a quill pen.    MED

1546      Danish astronomer Tycho Brahe (*d.* 1601) is born at Skøane, now part of Denmark and later part of Sweden. He will set a new standard for precision in astronomical measurements. He will also make important observations of the supernova of 1572 and the comet of 1577.    ASTRO

1546      Italian physician Girolamo Fracastoro formulates the first theory suggesting that the tiny, autonomous living entities later known as bacteria are what cause disease. *See also* 1632, BIO.    MED

1546      In *De natura fossilium*, German metallurgist Georgius Agricola (Georg Bauer) coins the word *fossil* for anything dug from the earth, including rocks in the shape of bones and shells.    PALEO

1547      In London the St. Mary's of Bethlehem Hospital establishes a separate asylum for the insane. It will become known as Bedlam, a term that will become synonymous with a place or state of uproar and confusion.    MED

| | |
|---|---|
| 1550 | Italian scientist Geronimo Cardano publishes a book on natural history that implies a belief in evolutionary change. **BIO** |
| c. 1550 | In Yucatán, Mexico, during the postclassic period, Mayan Indians roast green corn, then leave it to produce mold, which is used to treat wounds, cuts, ulcers, and intestinal infections. **MED** |
| c. 1550– 1700 | This period in Europe will become known as the scientific revolution. During this time the works of such scientists as Copernicus, Galileo, Harvey, Pascal, and Newton and such philosophers as Francis Bacon transform the modern world's approach to understanding nature. Traditional deference to classical authorities on nature, such as Aristotle, is replaced by reliance on the empirical methods of science. **MISC** |
| 1551 | German astronomer Erasmus Reinhold publishes *Tabulae prutenicae*, astronomical tables based on Copernicus's heliocentric theory that improve on the 13th-century Alphonsine tables. *See* 1272, **ASTRO**. **ASTRO** |
| 1551 | German-Swiss physician and naturalist Konrad von Gesner begins publication of what will be considered the most authoritative zoological study since Aristotle's, his *Historiae animalium* (1551–1558, 1587). It includes lists and descriptions of each known animal species and their physical appearance, emotions, habits, locale, diseases, and uses for humankind. **BIO** |
| 1551 | German mathematician Georg Joachim Iserin von Lauchen (Rhäticus or Rheticus) produces detailed trigonometric tables. **MATH** |
| 1551 | Italian physician Bartolommeo Maggi proves that gunshot wounds are not poisonous. **MED** |
| 1551 | Italian anatomist Gabriel Fallopius describes the tubes that carry the human ovum from the ovary to the uterus. These passages, which will become known as the Fallopian tubes, are where fertilization takes place. **MED** |
| 1552 | The convex lens is developed by Italian physicist Giambattista della Porta and used to refine the camera obscura, an artist's tool for tracing that had been invented by Roger Bacon three centuries earlier. **TECH** |
| 1553 | English mariner Richard Chancellor opens a northeastern sea route to Russia, encouraging trade between the two countries. **EARTH** |
| 1554 | Italian naturalist Ulisse Aldrovandi publishes his systematic study of plant classification, *Herbarium*. **BIO** |
| 1554 | French physician Jean-François Fernel publishes his work called *Medicina*, the first modern medical textbook. **MED** |
| 1555 | French naturalist Pierre Belon describes the homologies (basic similarities) in the body plans of vertebrates. **BIO** |

1556    Tobacco is introduced to continental Europe by Franciscan monk André Thevet, who brings seeds of the plant to Spain from a trip to Rio de Janeiro.    BIO

1556    *De re metallica* (*Concerning Metallic Things*), by German physician Georgius Agricola (Georg Bauer), now published posthumously, is the first important work on mineralogy.    EARTH

1556    The deadliest earthquake in recorded history devastates Shansi, China, on January 24, with a death toll in the hundreds of thousands.    EARTH

1556    French surgeon Pierre Franco is the first to perform a suprapubic lithotomy, or incision into the bladder to remove stones.    MED

1557    English mathematician Robert Recorde makes the first known use of the modern equals (=) sign.    MATH

1559    Ice cream is developed in Italy, by a freezing process using ice and salt.    TECH

1560    Italian physicist Giambattista della Porta founds the first scientific association designed for the exchange of information, the Academia Secretorum Naturae, or Academy of the Mysteries of Nature.    MISC

1564    Galileo Galilei is born in Pisa, Italy on February 15 (*d.* 1642). Often considered the founder of the experimental method, he will become known for his achievements in astronomy, physics, and mathematics as well as a clash with the Roman Catholic church over his support for the heliocentric theory (that Earth and the other planets revolve around the sun).    MISC

1565    Swiss scientist Konrad von Gesner's *De rerum fossilium* (*On Things Disinterred from the Earth*) contains the first illustrations of fossils.    BIO

1565    Tobacco is introduced from Florida to England by explorer John Hawkins. *See also* 1556, BIO.    BIO

1565    Muskets are in use in Europe.    TECH

1565    The lead pencil (or lead) is seen for the first time in a woodcut from a book about fossils by Swiss-German naturalist Conrad von Gesner.    TECH

1567    Bologna's botanical garden is founded.    BIO

1568    The first map to use the Mercator projection appears. Designed by Flemish cartographer Gerardus Mercator (Gerhard Kremer), this map employs a cylindrical projection that distorts the sizes of areas in order to preserve their shapes.    EARTH

1568    Ambroise Paré recognizes the difference between syphilis (called the greater pox) and smallpox (the lesser pox).    MED

1570      Geographer Abraham Ortelius of Antwerp publishes his *Theatrum orbis terrarum*, which contains 70 maps and is the first comprehensive atlas of the world.      EARTH

1572      A supernova as bright as Venus is observed in the constellation Cassiopeia by Chinese astronomers and Danish astronomer Tycho Brahe. This stellar explosion remains visible to the naked eye for 16 months. Tycho calls it a *nova* in his *De nova stella* (*On the New Star*) the following year. His hypothesis that this supernova is farther away than the moon contradicts traditional belief by indicating that change can happen in the celestial sphere.      ASTRO

1572      Pigeons are used to transport messages in Haarlem, the Netherlands.      BIO

1572      In Rafael Bombelli's *Algebra*, complex numbers are applied to solve equations for the first time. Bombelli also uses continued fractions to approximate roots.      MATH

c. 1574      Italian philosopher Giordano Bruno is accused of heresy and forced to leave the Dominican order. Among Bruno's heretical beliefs is his defense of Copernicus's heliocentric theory on metaphysical grounds.      ASTRO

1574      Danish astronomer Tycho Brahe, under the patronage of Danish king Frederick II, establishes an observatory on the Danish island of Ven. For 20 years he and his assistants carry out accurate, detailed naked-eye observations of the stars and planets.      ASTRO

1575      Spanish physician Juan Huarte recommends that ability tests be given and vocational counseling be used to match people with their occupations. He also claims that intelligence and higher culture are possible only in moderate climatic zones.      PSYCH

1575      Porcelain dinnerware is produced for the first time in Europe by Tuscan grand duke Francesco Maria de Medici.      TECH

1576      English explorer Martin Frobisher fails to find a northwest passage from Europe to Asia but does discover Baffin Island and, in 1578, rediscovers Greenland. *See* 982, EARTH.      EARTH

1577      Using parallax theory, Danish astronomer Tycho Brahe proves that a bright comet he is observing is at least three times as far away as the moon, contradicting the conventional belief that comets are luminous vapors in the atmosphere.      ASTRO

1577      During an expedition in 1577–1580, English mariner Francis Drake circumnavigates the globe for the first time since the voyage of Magellan in 1522. On the way he discovers the Drake Passage or Drake Strait south of Tierra del Fuego and sails up the California coast as far as San Francisco Bay.      EARTH

1578    French physician Guillaume de Baillou is the first to describe whooping cough and coins the term *rheumatism* for soreness, stiffness, and inflammation of the joints and muscles.                    MED

1578    The first medical school on the North American continent is established, at the University of Mexico.                    MED

1578    English physician William Harvey is born on April 1 (*d.* 1657) in Folkestone, Kent, England. He will be known for establishing that the heart is a muscle and that blood circulates. He will bring about the end of unquestioning acceptance of Galen (*see* 160, MED) and Greek medicine. Some will consider him the founder of modern physiology.                    MED

1580    Italian scientist Prospero Alpini discovers that plants have two sexes.                    BIO

1581    In Siberia, Russian explorer Yermak Timofievich conquers the Mongol kingdom of Sibir.                    EARTH

1581    While watching hanging lamps during a service in the cathedral of Pisa, the 16-year-old Galileo notes that the duration of a pendulum's swing seems to be determined solely by its length, not by the width of its swing. This observation will lead to the manufacture of accurate pendulum clocks by the late 1600s.                    PHYS

1582    Following the advice of Bavarian astronomer Cristoph Clavius, Pope Gregory XIII reforms the calendar. In what would become known as the Gregorian, or New Style, calendar, century years not divisible by 400 are not leap years. This change from the old Julian calendar, which had been in use in Europe since 46 B.C., results in 10 days being dropped. Thus, the day after October 4, 1582 is proclaimed to be October 15, 1582. The Gregorian calendar will gradually be adopted throughout Europe and the Western world and in parts of Asia.                    ASTRO

1583    Italian botanist Andrea Cesalpino proposes a plant classification system in his treatise *De plantis*. He classifies plants according to their roots and fruit organs, putting lichens and mushrooms at the bottom of the plant hierarchy.                    BIO

1583    Dutch engineer and mathematician Simon Stevin founds the science of hydrostatics with discoveries about factors determining the pressure of liquids on surfaces.                    PHYS

1584    English explorer Sir Walter Raleigh brings the plant extract curare to England from South America.                    MED

1586    Potatoes imported from the Americas are planted in Ireland by English explorer Sir Francis Drake.                    BIO

1586     Dutch engineer and mathematician Simon Stevin works out a system of decimal fractions that allows fractions to be included in positional notation. He also discovers rules for locating the roots of equations.     MATH

1586     On dropping two objects of different weight, Dutch engineer and mathematician Simon Stevin notes that they hit the ground at the same time, disproving Aristotle's long-held proposition that heavier bodies fall faster than lighter ones. *See also* 1590, PHYS.     PHYS

1588     Italian botanist Giambattista della Porta tries to draw parallels between the medicinal properties of a plant and its external shape, arguing that plants resembling human organs can be useful in healing diseases of those organs.     BIO

c. 1589     William Lee of Cambridge, England, invents the stocking frame, the first knitting machine. It slowly gains popularity during the 17th century.     TECH

1590s     Italian scientist Galileo privately accepts the Copernican heliocentric explanation of the solar system.     ASTRO

1590     Galileo publishes his *De motu* (*On Motion*), showing how his experiments with falling bodies (similar to those of Simon Stevin in 1586) refute Aristotle's physics.     PHYS

1590     Rudolf Goeckel publishes a book of essays by different authors on human nature and the soul. This is the first book to have the word *psychology* (*Psychologia*) in its title.     PSYCH

1590     Dutch spectacle maker Zacharias Janssen invents the compound microscope. *See also* 1609, TECH.     TECH

1591     French mathematician François Viète introduces algebraic sign language using consonants for known quantities and vowels for unknown ones.     MATH

1592     Italian engineer Domenico Fontana discovers the ruins of the cities of Pompeii and Herculaneum, buried by the eruption of Mount Vesuvius in 79. Deliberate excavation will not occur until 1738, but this discovery marks the beginning of the science of archaeology.     ARCH

c. 1592     Galileo invents the thermoscope, a primitive thermometer.     TECH

1593     A shortage of lumber and firewood in England encourages the expansion of coal mining.     TECH

1594     Flemish cartographer Gerardus Mercator dies (*b.* 1512). His son publishes Mercator's great work, *Atlas sive cosmographicae*, posthumously.     EARTH

1595     The word *trigonometry* appears in print for what may be the first time, in a work by German mathematician Bartholomaeus Pitiscus.     MATH

1596          Dutch astronomer David Fabricius reports for the first time on the irregular variation of the star Omicron Ceti, or Mira. *See also* 1638, ASTRO.          ASTRO

1596          English botanist John Gerard's (1545–1612) *Herbal*, the greatest survey of botanical knowledge to date, is published.          BIO

1596          Dutch mathematician Ludolph van Ceulen calculates pi to 20 decimal places, later extending it to 35 places.          MATH

1596          Philosopher, mathematician, and scientist René Descartes is born in La Haye, France (*d.* 1650). Considered the founder of modern philosophy and analytic geometry, he will originate Cartesian coordinates and Cartesian curves. His contributions to science will include work in physiology, optics, and psychology. MISC

1596          Korean admiral Visunsin builds the first ironclad warship.          TECH

1596          The water closet, meant to replace the chamber pot and privy, is developed by English poet Sir John Harington.          TECH

1597          After 20 years of research, Danish astronomer Tycho Brahe is forced to leave his observatory at Ven when the new king of Denmark, Christian IV, cuts off his support. He then goes to Prague as court astronomer for Holy Roman Emperor Rudolph II.          ASTRO

1597          German alchemist Andreas Libau (Libavius) publishes his *Alchymia*, a landmark text in chemistry. Among other things, it explains how to prepare hydrochloric acid and ammonium sulfate.          CHEM

1597          Italian physician Gaspare Tagliacozzi publishes his studies on the reconstruction of the nose. He thus becomes established as the first modern plastic surgeon. MED

1598          Italian aristocrat Carlo Ruini illustrates a work called *Dell' anatomia et dell' infirmita del cavallo*, the first comprehensive monograph on an animal's anatomy, in this case the horse.          BIO

1599          Italian naturalist Ulisse Aldrovandi publishes his classic studies in ornithology, the branch of zoology dealing with birds.          BIO

1600          Young German mathematician Johannes Kepler becomes Danish astronomer Tycho Brahe's assistant at Prague.          ASTRO

1600          On February 17 the Italian philosopher Giordano Bruno (*b.* 1548) is burned at the stake in Rome for heresy, including his support for Copernican theory.          ASTRO

c. 1600          Johann Thölde, writing as Basil Valentine, is the probable discoverer of the elements antimony and bismuth.          CHEM

1600      English physician and physicist William Gilbert (1540–1603) publishes *De magnete* (*On Magnetism*), the first work of physical science based completely on experimentation. In it he argues that the earth acts like a giant magnet with poles near the geographic poles.      EARTH

1600      English physician and physicist William Gilbert is named president of the Royal College of Physicians and personal physician to Queen Elizabeth I of England (1533–1603).      MED

1601      Danish astronomer Tycho Brahe (*b.* 1546) dies on October 24 at Benatky, near Prague. German mathematician Johannes Kepler succeeds him as court astronomer to Holy Roman Emperor Rudolph II.      ASTRO

1601      French mathematician Pierre de Fermat is born (*d.* 1665). Not a professional mathematician, he publishes virtually nothing during his lifetime, but will eventually be regarded as the founder of number theory, a codiscoverer of analytic geometry, and a codiscoverer also of differential calculus. He will also be renowned for the theorems he scribbled in the margins of books, particularly one called Fermat's Last Theorem. *See* 1637 and c. 1810s, MATH.      MATH

1601      Fifty-three stations with overnight inns called *ryokans* and horse-changing stops are built in Edo and Osaka, Japan. Developed by a man named Ieyasa, regent of Tokugawa, these inns ease the burden of long-distance travel.      TECH

1601      Coffee, introduced to England by traveler Anthony Shirley, is sold for five pounds per ounce.      TECH

1601      Pepper is imported in large quantities to England by the East India Company, a trading concern.      TECH

1602      German astronomer Johann Bayer's celestial atlas *Uranometria* introduces a new system for naming and describing the locations of stars that will still be in use in the 20th century. In this system a star is named by a Greek letter and its constellation.      ASTRO

1602      Hugh Platt discovers coke, the residue left after distillation of coal, which later becomes an important fuel.      TECH

1604      Korean and Chinese astronomers, independently from Johannes Kepler at Prague, observe a supernova in the constellation Ophiuchus that lasts 12 months. Kepler's observations are published as *De stella nova* (*On the New Star*) in 1606.      ASTRO

1604      Italian scientist Galileo observes that a falling body increases its distance as the square of time.      PHYS

1605      English essayist Francis Bacon publishes his treatise *The Advancement of Learning*, which promotes experimentation and observation as the basis for knowledge.      MISC

| | |
|---|---|
| 1606 | Spanish navigator Luis Vaez de Torres sails completely around New Guinea, showing it to be an island.       **EARTH** |
| 1608 | Dutch spectacle maker Hans Lippershey invents the first telescope to attract public notice. Military telescopes had been used secretly by the Dutch for about 20 years.       **ASTRO** |
| 1609 | Italian scientist Galileo builds his own telescope with three-power magnification, eventually making one with a magnification of 30.       **ASTRO** |
| 1609 | German mathematician and astronomer Johannes Kepler's *Astronomia nova* (*The New Astronomy*), published this year, contains both his first law (that the |

## *Hamlet* and Probability

About the year 1601, English playwright William Shakespeare wrote the tragedy *Hamlet*, a masterpiece that sparked centuries of theatrical and literary interpretation. As an exemplar of the unique, unrepeatable nature of great art, the play also sparked a question that has become part of popular scientific lore: How long would it take for a monkey sitting at a typewriter to write *Hamlet* by randomly pounding on the keyboard?

This question has been raised in many contexts, sometimes by nihilists seeking to suggest the meaninglessness of human works, at other times by creationists arguing that random events could not have led to the evolution of intelligence. (The latter application is, however, misleading, since the natural selection of useful characteristics is, by definition, not random but selective.) Whatever its significance, the question itself is easily answered, not by setting a monkey in front of a typewriter and waiting, but with techniques drawn from probability theory. This area of mathematics was founded by French mathematicians Blaise Pascal and Pierre de Fermat around 1654, not long after *Hamlet* was written.

Assume that the number of symbols or characters in *Hamlet*—including all letters, punctuation, and spaces—equals 200,000. (Whether it is more or less, the basic idea will be the same.) There are 46 characters on a typewriter keyboard, not counting the shift key and such subtleties as tabs and returns. Thus each time the monkey hits a key, it has a one in 46 ($\frac{1}{46}$) chance of hitting the right one, as for example the initial *t* in "to be or not to be." When it hits the next key, it again has a one in 46 chance of hitting the right key for the *o* in "to." The odds of its typing the complete word "to" are, then, $\frac{1}{46} \times \frac{1}{46}$, or $(\frac{1}{46})^2$, $\frac{1}{2116}$. Similarly, the odds of its typing all 18 characters in "to be or not to be" in the correct sequence are $(\frac{1}{46})^{18}$, equivalent to $\frac{1}{46}$ multiplied by itself 18 times. The odds of the monkey's typing all 200,000 characters therefore are $(\frac{1}{46})^{200,000}$. This probability is so small as to be virtually zero, or, to put it another way, the time it would take is virtually infinite. The only known way for monkeys to write *Hamlet* is for them to evolve to be as intelligent as Shakespeare.

orbit of a planet around the sun is an ellipse) and second law (that these orbits sweep out equal areas in space in equal periods of time).          ASTRO

1609          English mathematician Thomas Harriot uses a telescope to sketch the moon.
                                                                                    ASTRO

# The Moons of Jupiter

In 1609 Italian scientist Galileo Galilei received word that a Dutchman, Hans Lippershey, had invented a device "by the aid of which visible objects, although at a great distance from the eye of the observer, were seen distinctly as if near." Wasting no time, Galileo learned how to build his own telescope and constructed several models. In 1610 he discovered a startling fact about the planet Jupiter. In *The Sidereal Messenger*, published later that year, he reported his find.

> On the 7th day of January in the present year, 1610, in the first hour of the following night, when I was viewing the constellations of the heavens through a telescope, the planet Jupiter presented itself to my view, and as I had prepared for myself a very excellent instrument, I noticed a circumstance which I had never been able to notice before, owing to want of power in my other telescope, namely, that three little stars, small but very bright, were near the planet....
>
> I scarcely troubled at all about the distance between them and Jupiter, for, as I have said, at first I believed them to be fixed stars; but when on January 8th, led by some fatality, I turned again to look at the same part of the heavens, I found a very different state of things, for there were three little stars all west of Jupiter [where previously two had been east and one west], and nearer together than on the previous night....
>
> [After more days of observation:]
>
> I, therefore, concluded, and decided unhesitatingly, that there are three stars in the heavens moving about Jupiter, as Venus and Mercury around the Sun; which at length was established as clear as daylight by numerous other subsequent observations. These observations also established that there are not only three, but four, erratic sidereal bodies performing their revolutions round Jupiter....
>
> [Galileo concludes:]
>
> [W]e have a notable and splendid argument to remove the scruples of those who can tolerate the revolution of the planets round the Sun in the Copernican system, yet are so disturbed by the motion of one Moon about the Earth, while both accomplish an orbit of a year's length about the Sun, that they consider that this theory of the universe must be upset as impossible; for now we have not one planet only revolving about another, while both traverse a vast orbit about the Sun, but our sense of sight presents to us four satellites circling about Jupiter, like our Moon about the Earth, while the whole system travels over a mighty orbit about the Sun in the space of twelve years.

1609        Italian anatomist Giulio Casserio finishes a series of five books on the hearing, sight, smell, taste, and touch organs of humans. This work, *Pentaestheseion*, will be noted for its literary style and illustrative plates.                    BIO

1609        Independently from Zacharias Janssen (*see* 1590, TECH) Dutch scientist Hans Lippershey invents the compound microscope, a central tube with lenses attached to both ends.                    TECH

1609        In northern Germany, the world's first regular newspapers are published.    TECH

1610        Italian scientist Galileo becomes the first to make significant astronomical observations using the telescope. He sees four of Jupiter's moons, the phases of Venus, and the individual stars of the Milky Way. He notes that Saturn has an odd appearance, later found to be rings. He publishes his discoveries in a work called *The Sidereal Messenger*.                    ASTRO

c. 1610     Italian scientist Galileo uses the microscope to study insect anatomy.    BIO

1610        Jean Beguin of France publishes his *Tyrocinium chymicum*, the first textbook on chemistry rather than alchemy.                    CHEM

1610        English navigator Henry Hudson, attempting to find the northwest passage, instead finds and enters the Canadian waterway later known as the Hudson Strait, leading to a large, landlocked body of water that will come to be called Hudson Bay.                    EARTH

1611        Italian scientist Galileo, English mathematician Thomas Harriot, Dutch astronomer Johannes Fabricius, and German astronomer Father Christoph Scheiner independently discover sunspots, which were first recorded by the Chinese in 165 B.C.                    ASTRO

1611        English physician John Woodall recommends citrus fruit for protection against scurvy on long sea voyages.                    MED

1612        German astronomer Simon Marius (Simon Mayr) is the first to study the Andromeda galaxy.                    ASTRO

1612        Through advice from Native Americans, American colonists learn to grow and cure tobacco on a large scale and use it as a prime export commodity to England. Over 1,500 pounds per acre are grown in Virginia's James River Valley.                    BIO

1613        In *The Sunspot Letters* Galileo supports the Copernican theory of heliocentrism in print for the first time.                    ASTRO

1613        Italian mathematician Pietro Cataldi develops techniques for handling continued fractions.                    MATH

1614    Italian chemist Angelo Sala discovers that light darkens the white compound silver nitrate, a phenomenon relevant to the invention of photography two centuries later. *See* 1802, TECH.                                                    CHEM

1614    Scottish mathematician John Napier publishes a table of logarithms based on powers of 2.                                                    MATH

1614    Italian physician Sanctorius (Santorio Santorio) publishes his studies on body weight, food, and excreta—the first metabolic balance studies.                                                    MED

1615    Italian scientist Galileo travels to Rome to defend the Copernican theory. ASTRO

1615    French explorer Samuel de Champlain reaches the eastward extension of Lake Huron, called Georgian Bay. He thus becomes the first European to sight the Great Lakes.                                                    EARTH

1615    Rubber is introduced to Europe from South America, but its uses will not be fully developed for centuries. *See also* 1503, TECH.                                                    TECH

1616    In a rebuff to Galileo, the Roman Catholic church issues a decree stating that the Copernican doctrine is "false and absurd" and should not be held or defended. Copernicus's *De revolutionibus* (1543) is placed on the church's *Index of Prohibited Books*, where it will remain until the 19th century.                                                    ASTRO

1616    English explorer William Baffin reaches what will become known as Baffin Bay and travels to within 800 miles of the North Pole, a record held until the latter half of the 19th century.                                                    EARTH

1616    English physician William Harvey lectures to the Royal College of Physicians about the circulation of the blood.                                                    MED

1617    English mathematician Henry Briggs's *Logarithmorum chilias prima* (*Logarithms of Numbers from 1 to 10*) introduces logarithms based on powers of 10, or common logarithms.                                                    MATH

1617    Dutch mathematician Willebrord Snell develops a technique for finding distances by trigonometric triangulation.                                                    MATH

1617    French clergyman St. Vincent de Paul (1581–1660) organizes the Dames de Charité, women who visit the sick and dying, administering nursing services. With this organization St. Vincent introduces the modern principles of home health care and the idea that poverty should not keep people from giving or receiving medical care.                                                    MED

1617    In London, King James I grants a charter to a newly formed society of pharmacists, allowing pharmacists to emerge as a distinct group of craftsmen separate from grocers.                                                    MED

1619          In his *Harmonice mundi* German mathematician and astronomer Johannes
              Kepler propounds his third law, that the squares of the times of revolution of
              any two planets are proportional to the cubes of their distances from the sun.
                                                                                    ASTRO

1619          German mathematician and astronomer Johannes Kepler publishes *Epitome
              astronomiae copernicae* (*Epitome of the Copernican Astronomy*), a defense of the
              Copernican doctrine. The Roman Catholic church places it on its *Index of
              Prohibited Books*.                                                    ASTRO

1619          From Ingulstadt, Germany, comes the first report of telescopic observations of a
              comet.                                                                ASTRO

1620          English philosopher Francis Bacon points out that the outlines of Africa and
              South America generally mesh, an observation that will become important in
              the development of plate tectonics.                                   EARTH

1620          Francis Bacon publishes his *Novum organum*, a treatise outlining the scientific
              method based on the principles of experimentation and induction.      MISC

c. 1620       In London, Dutch inventor Cornelis Drebbel constructs the first submarine,
              using greased leather over a wooden form. Powered by rowers, the vessel cruis-
              es beneath the surface of the Thames.                                 TECH

## Bacon's Philosophy

English philosopher, essayist, and statesman Francis Bacon (1561–1626) was not an
expert in any one science but dabbled in many, doing things like stuffing a chicken with
snow to test the idea that cold can preserve meat from decay. Much more valuable than
such experiments was his development of a philosophy of knowledge that is one of the
principal sources of the scientific method, and therefore of modern science.

In his works *The Advancement of Learning* (1605) and *Novum organum* (1620), Bacon
laid out his belief that the Aristotelian method of deducing truth from a priori assump-
tions was not a valid way of uncovering truths about nature. He proposed instead what he
called the inductive method: making numerous observations and experiments in order to
build to general conclusions. Bacon perhaps overstated the case, since modern science uses
both deduction (to frame theories, hypotheses, and predictions) and induction (to test
predictions against the real world and provide evidence for the framing or reframing of
theories). But he did provide philosophical underpinnings for science that served to
increase the pace and rigor of scientific discovery.

As for Bacon's chicken-and-snow experiment, the outcome was never recorded.
Bacon caught a severe chill while collecting the snow and died a few days later, a martyr, if an unglamorous one, to his own method.

1621      In his treatise *On the Formation of Eggs and Chickens*, Italian scientist Girolamo Fabrici gives detailed, sequential illustrations of chick embryo development. *See also* 1673, BIO.      BIO

c. 1621      English mathematician William Oughtred invents the slide rule.      MATH

1621      Dutch mathematician Willebrord Snell formulates what will become known as Snell's law, which concerns refraction of light. It states that the ratio of the sine of the angle of incidence to the sine of the angle of refraction is equal to the ratio of the refracting medium's index of refraction to the original medium's index of refraction.      PHYS

1622      Italian anatomist Gasparo Aselli discovers the lacteal vessels, an important intestinal lymph vessel.      MED

1623      Botanist Gaspard Bauhin of Switzerland classifies some 6,000 plants and introduces the practice of using two names—one for the genus, another for the species—to classify living things.      BIO

1623      French mathematician, scientist, and religious philosopher Blaise Pascal is born (*d*. 1662). He will become the founder of probability theory, discover many properties of the cycloid, and lay the groundwork for the hydraulic press.      MISC

1624      Flemish physician Jan Baptista van Helmont studies the metabolism of the willow tree.      BIO

1624      Physician Thomas Sydenham is born on September 10 in Wynford Eagle, England (*d*. 1689). Known as the English Hippocrates, he will be the first to describe measles and scarlet fever and will recommend such remedies as opium for pain and iron for anemia.      MED

1624      Jan Baptista van Helmont coins the word *gas* (from the Flemish word for *chaos*) to describe substances like air. One of the gases he identifies is carbon dioxide, which he calls *gas sylvestre* (wood gas).      PHYS

1624      French philosopher Pierre Gassendi contributes to sensory psychology by being the first to measure sound velocity.      PSYCH

1625      German chemist Johann Glauber finds that hydrochloric acid can be formed with sulfuric acid and sodium chloride. The residue of this compound will gain popularity as the laxative known as sodium sulfate.      MED

1626      English philosopher Francis Bacon experiments with refrigerating food by placing snow in the cavities of chickens.      TECH

1627      German mathematician and astronomer Johannes Kepler publishes his *Rudolphine Tables*, planetary tables based on his theory of elliptical orbits.      ASTRO

| | |
|---|---|
| 1627 | The auroch, or uru, the long-horned wild ox believed to be the ancestor of domestic cattle (*see* 6400 B.C., BIO), becomes extinct when the last specimen dies in Poland.     BIO |
| 1628 | British physician William Harvey publishes a description of the circulation of blood that is largely correct, in contrast to the many erroneous ideas extant on blood movement. *See also* 1660, MED.     MED |
| 1628 | A rudimentary version of the steam engine is developed by English engineer Edward Somerset.     TECH |
| c. 1629 | French mathematician Pierre de Fermat discovers a method of finding maximum and minimum values for functions which represents the genesis of differential calculus.     MATH |
| 1630 | Pierre Vernier, a French military engineer, invents the Vernier scale, which measures angles and small distances with great precision. Though its original applications will be in navigation and astronomy, the Vernier scale will come into general use near the end of the 17th century.     EARTH |
| 1630 | Englishman Peter Chamberlen devises the first obstetrical forceps.     MED |
| 1630 | Italian natural philosopher Niccolò Cabeo observes that electrically charged bodies first attract, then repel each other. He is the first to use the term *lines of force* to describe the curves assumed by iron filings on a sheet of paper above a magnet.     PHYS |
| 1631 | Following German mathematician and astronomer Johannes Kepler's predictions, Pierre Gassendi observes the transit of Mercury across the sun.     ASTRO |
| 1631 | English mathematician Thomas Harriot's posthumously published work, *Artis analyticae praxis*, introduces a raised centered dot for multiplication and the symbols > and < for "greater than" and "less than."     MATH |
| 1631 | English mathematician William Oughtred introduces the × sign for multiplication.     MATH |
| 1632 | Galileo's *Dialogue on the Two Great World Systems* is published in Italian, not Latin, to reach a general audience. Using the conceit of a hypothetical debate among three philosophers, it makes a solid case for the Copernican theory. The Roman Catholic church promptly places a ban on it that will not be lifted until 1822.     ASTRO |
| 1632 | In Delft, the Netherlands, microscopist and zoologist Antoni van Leeuwenhoek is born (*d.* 1723). Although he will receive little scientific education, Leeuwenhoek will make pioneering discoveries regarding microbes, red blood cells, capillary systems, and insects' life cycles. After his invention of a double-convex microscope, he will discover protozoa and be the first to |

observe bacteria, three types of which he will describe—bacilli, cocci, and spirilla. *See* c. 1660, MED, and c. 1677, BIO. BIO

1632    English philosopher John Locke is born (*d.* 1704). He will found British empiricism, the philosophical doctrine that all knowledge is derived from experience. Locke's views will be associated with the rise of experimental science in the 18th century. *See also* 1690, PSYCH, and 1690, SOC. MISC

1633    Galileo, at age 69, is called before the Inquisition in Rome on charges of heresy. Pleading guilty, he recants his views and is sentenced to house arrest for the remainder of his life. ASTRO

1633    English physician Stephen Bradwell writes *Helps in Sudden Accidents*, the first book on first aid. MED

1635    English astronomer Henry Gellibrand presents evidence that the earth's magnetic poles shift position over time. EARTH

1635    Italian mathematician Francesco Bonaventura Cavalieri publishes the influential *Geometria indivisibilibus continuorum*, in which he develops the theory of indivisibles (infinite processes), an important stage in the development of the calculus. MATH

1635    Inland mail delivery by wheeled coaches is inaugurated between London and Edinburgh. TECH

1636    Sugar is first grown in Barbados. Introduced by the Dutch, it will become a mainstay crop in Barbados and other islands in the Caribbean. TECH

1637    King Christian IV of Denmark establishes a permanent astronomical observatory in Copenhagen. ASTRO

1637    In an appendix to his *Discourse on Method*, French mathematician René Descartes introduces analytic geometry, a branch of geometry in which all points are represented with respect to a coordinate system. (Pierre de Fermat developed analytic geometry independently in 1636 or earlier, but his work will not be published until 1670.) Descartes' work also introduces a system of exponential notation. MATH

1637    French mathematician Pierre de Fermat formulates, but does not prove, Fermat's Last Theorem, so called because for centuries it will remain the last proposition of Fermat's to go unproven. It states that the expression $x^n + y^n = z^n$ has no positive integral solutions if $n$ is an integer greater than 2. In 1993 British mathematician Andrew Wiles will finally prove it. *See also* c. 1810s, MATH. MATH

1637    French mathematician René Descartes explains the process of accommodation, in *Dioptrics*, his work on ophthalmology. MED

1637    Descartes publishes his *Discourse on Method*, which begins from the premise of universal doubt, describes a mechanistic physical world divorced from the mind, and promotes the use of deduction in science.                    MISC

1638    Dutch astronomer Phoclides Holwarda identifies the first known variable star, Omicron Ceti, or Mira, initially observed by David Fabricius in 1596.                    ASTRO

1638    Italian scientist Galileo publishes *Mathematical Discourses and Demonstrations on Two New Sciences* in which he discusses the laws of motion and friction, refuting Aristotle on several points.                    PHYS

c. 1639    English astronomer William Gascoigne invents the micrometer, a device placed in a telescope to measure the angular distance between stars.                    ASTRO

1639    In his work known as *Brouillon project* (*Rough Draft*), French mathematician Girard Desargues develops projective geometry.                    MATH

c. 1639    French mathematician René Descartes claims that the human body functions as a machine, a system of mechanical devices. He mistakenly believes the pineal gland to be the center of the human mind and soul.                    MED

1640    French mathematician Pierre de Fermat develops modern number theory. MATH

c. 1640    Italian matron Jeanne Biscot develops the abilities of the patients she cares for during the Thirty Years' War (1618–1648). Her hospitals become workshops where patients enrich their personal skills, marking the beginning of occupational therapy.                    MED

1640    In *De motu gravium*, Evangelista Torricelli of Italy applies Galileo's laws of motion to fluids, thus founding the science of hydrodynamics.                    PHYS

1640    Stagecoaches to transport people are introduced in England.                    TECH

1641    The first living chimpanzee to be brought out of the wild is introduced to the Netherlands.                    BIO

1641    Using his father's theories, Galileo's son designs a clock with a pendulum. TECH

1642    Sailing from Dutch outposts in Indonesia, mariner Abel Tasman becomes the first European to reach what will become known as the island of Tasmania and the southern island of New Zealand.                    EARTH

1642    French scientist Blaise Pascal invents the adding machine and also contributes to the development of differential calculus.                    MATH

1642    On Christmas Day of the same year that Italian scientist Galileo Galilei dies (*b.* 1564), English mathematician and physicist Isaac Newton is born (*d.* 1727). He will found the field of celestial mechanics, co-invent calculus, and make revolutionary breakthroughs in the studies of optics, gravitation, and motion.

In his most famous work, known as the *Principia* (1687), he will present his three laws of motion and the law of universal gravitation. MISC

1643    Italian mathematician Evangelista Torricelli invents the barometer. TECH

1644    Dutch explorer Abel Tasman becomes the first European to discover the continent of Australia. EARTH

1644    Evangelista Torricelli publishes discoveries about the cycloid, including methods of finding its area and constructing the tangent. Earlier and independently, French mathematician Gilles Personne de Roberval made similar discoveries but did not publish them. MATH

1644    French mathematician René Descartes explains reflex action, the involuntary response to a stimulus. MED

1644    Descartes publishes his *Principles of Philosophy*, which builds on the *Discourse on Method* (1637) in describing natural phenomena in mechanistic terms. MISC

1645    German physicist and engineer Otto von Guericke invents the first practical air pump. Vacuums created by this device allow von Guericke to carry out experiments revealing that in vacuums sound does not travel, fire is extinguished, and animals stop breathing. Von Guericke will also make the first measurement of the density of air. PHYS

1647    *Selenographia*, by German-Polish astronomer Johannes Hevelius, is the first map of the visible side of the moon. ASTRO

1647    In Italy, physician Georg Wirsung discovers the pancreatic duct, which passes pancreatic juice into the small bowel, where the juice assists with food breakdown. MED

1648    Flemish physician Jan Baptista van Helmont claims that baby mice are produced either by spontaneous generation or through the joining of adult male and female mice. BIO

1648    Using barometers, French scientist Blaise Pascal demonstrates that air pressure decreases with altitude and shows that air has a finite height. EARTH

1648    Pascal formulates what will become known as Pascal's law, that in a confined fluid, externally applied pressure is transmitted uniformly in all directions and pushes at right angles to any surface in or surrounding the fluid. This principle will be the basis for the hydraulic press. PHYS

1648    The Taj Mahal, a massive marble and sandstone mausoleum, is built in Agra, India, by Mughal leader Shāh Jahān. TECH

1648    In the Massachusetts colony, iron production, developed by John Winthrop Jr., son of the colony's first governor, flourishes. TECH

| | |
|---|---|
| 1650 | Belgian astronomer Godefroy Wendelin calculates the sun's distance from Earth as being 240 times the moon's distance. Though this is less than the actual value (400 times), it is still more accurate than Aristarchus's old value of 20 times. *See* 200s B.C., ASTRO.    ASTRO |
| 1650 | Italian astronomer Giambattista Riccioli discovers the first known double star, Mizar.    ASTRO |
| c. 1650 | Anglican bishop James Ussher calculates from biblical genealogies that the creation of the world took place in 4004 B.C., a finding that will later be contradicted by geology.    EARTH |
| 1650 | Between now and 1677, English anatomist and physiologist Francis Glisson is the first to prove that muscles contract when performing activity and to describe childhood rickets. He also lays the foundation for modern knowledge of the anatomy of the liver.    MED |
| 1650 | French mathematician Pierre de Fermat articulates the principle of least time to describe the behavior of light, which he says travels from one point to another in such a way that the travel time is as short as possible.    PHYS |
| 1651 | English physician William Harvey denounces what he sees as erroneous conceptions of animal generation, including the belief that embryos are miniature versions of adult organisms. Harvey claims instead that embryo growth involves the successive development of structures. He also rejects the currently popular idea that the primary generative agent of reproduction is the male, arguing rather that *"ex ovo omina"* (all creatures come from an egg).    BIO |
| 1651 | The practice of whaling is popularized in Massachusetts by New Bedford developer Joseph Russell.    BIO |
| 1651 | English physician Nathaniel Highmore discovers the maxillary sinus.    MED |
| 1651 | English philosopher Thomas Hobbes publishes his *Leviathan*, which provides a rationalistic explanation for the existence of governments. In it he argues that humans voluntarily submit to absolute authority in order to protect themselves from each other's violent tendencies. Hobbes will be considered the father of political science. *See also* 1690, SOC.    SOC |
| 1654 | French mathematicians Blaise Pascal and Pierre de Fermat found probability theory, developing methods for judging the likelihood of outcomes in games of dice.    MATH |
| 1654 | German physicist Otto von Guericke demonstrates the force of air pressure through experiments in which the muscle power of humans or animals competes with air pressure.    PHYS |

| | |
|---|---|
| 1655 | English mathematician John Wallis publishes his *Arithmetica infinitorum*, in which he applies algebra to the treatment of infinite processes. *See* 1635, MATH. <br>                   MATH |
| 1655 | A book by Isaac de la Peyrère is publicly burned for hypothesizing that unusually chipped stones found in France were made by humans before the time of Adam.           PALEO |
| 1656 | Dutch scientist Christiaan Huygens discovers the rings of Saturn, explaining the oddness of the planet's shape identified by Galileo in 1610.      ASTRO |
| 1656 | German anatomist Werner Rolfink demonstrates that a cataract is a clouding of the eye lens, using executed criminals for his dissections.     MED |
| 1656 | English physician Sir Christopher Wren is the first to successfully inject drugs into veins.              MED |
| 1656 | Dutch scientist Christiaan Huygens invents an accurate pendulum clock, ushering in a new era of precision in timekeeping. *See also* 1673, PHYS.    TECH |
| 1657 | English physicist Robert Hooke demonstrates that all bodies fall at equal rates in a vacuum.              PHYS |
| 1658 | Johann Rudolf Glauber publishes *Opera omnia chymica*, an important early work of chemistry.              CHEM |
| c. 1658 | Further experimenting with pendulum clocks, Christiaan Huygens discovers that the cycloid curve is a tautochrone, a curve along which a mass point in a gravitational field reaches its lowest point in a time independent of the starting point. He also studies involutes and evolutes of cycloids and other curves.    MATH |
| 1658 | Dutch mathematician Jan de Witt devises kinematic and planimetric definitions of conic sections.              MATH |
| 1658 | Dutch naturalist Jan Swammerdam announces the discovery of the oxygen-carrying element of the blood, red blood corpuscles (red blood cells or erythrocytes).              MED |
| 1658 | Moravian educator Jan Ámos Komenský (Comenius) stresses the presentation of educational material in accordance with a child's developmental stages. He also introduces the planned school year and group instruction formats. PSYCH |
| 1659 | The modern sign for division (÷) is introduced by German mathematician Johann Heinrich Rahn.              MATH |
| c. 1660 | Dutch biologist Antoni van Leeuwenhoek develops the single-lens microscope, able to magnify nearly 200 times. Although not the first to invent microscopes, Leeuwenhoek does more with microscopy than many other sci- |

entists. In 1677, for instance, he will discover the one-celled animals called protozoa as well as sperm cells. *See also* 1632, BIO.                                                         MED

1660 | Italian physician Marcello Malpighi discovers capillary circulation, an important missing link in William Harvey's 1628 discovery of blood circulation.                 MED

1660 | Otto von Guericke designs a rotating sulfur globe that can be electrified by rubbing, demonstrating the large-scale existence of static electricity.                 PHYS

1661 | Irish chemist Robert Boyle publishes *The Skeptical Chymist*, a work distinguishing scientific chemistry from medieval alchemy and defining elements as substances that cannot be converted into anything simpler.                 CHEM

1662 | British mathematicians John Graunt and William Petty compile the first statistical study of population.                 MATH

1662 | British king Charles II charters the scientific association known as the Royal Society.                 MISC

1662 | Irish chemist Robert Boyle discovers what will become known as Boyle's law, that the volume of a mass of gas at a constant temperature is inversely proportional to its pressure. This argument will also be called Mariotte's law, after its independent discovery by French physicist Edme Mariotte in 1676. This principle supports the hypothesis that gases, and perhaps all other forms of matter, are made up of atoms that can be pressed closer together.                 PHYS

1663 | Girolamo Cardano's *Book on Games of Chance* is the first known work on probability theory.                 MATH

1664 | Italian astronomer Giovanni Alfonso Borelli discovers that a comet's orbit is a parabola.                 ASTRO

1665 | Italian-born French astronomer Giovanni Cassini determines the rotation rates of Mars and Jupiter.                 ASTRO

1665 | In his classic landmark book *Micrographia*, English biologist Robert Hooke publishes the first drawings of cells and is the first to use the word *cell* to describe the living fibers he sees under a compound microscope.                 BIO

1665 | Blaise Pascal's *Treatise on Figurative Numbers*, published posthumously this year, widely disseminates the process of mathematical induction.                 MATH

c. 1665 | English mathematician and physicist Isaac Newton discovers the general binomial theorem.                 MATH

1665–1666 | English mathematician and physicist Isaac Newton has his two most fertile years of discovery. During this time he discovers the general method of the calculus, which he calls the "theory of fluxions," and achieves his most important insights into gravitation and the composition of light (*see* 1666, PHYS).

Publication of his discoveries will await the *Principia* (1687) and later works of 1704 and 1736.                                                                                    MATH

1665   Dutch anatomist Fredrick Ruyson is the first to demonstrate the existence of valves in the lymphatics.                                                                    MED

1665   In a posthumous publication, Italian physicist Francesco Maria Grimaldi reveals his discovery of the diffraction of light, the bending of light waves as they pass through an aperture or around a barrier. *See also* 1801, PHYS.    PHYS

1665   Parisian government official Pierre Petit invents the filar micrometer, a device for measuring very small distances, angles, or objects.                                TECH

1666   Italian-French astronomer Giovanni Cassini discovers the polar ice caps of Mars.                                                                                          ASTRO

1666   English physician Thomas Sydenham uses Jesuits' bark, containing quinine, to treat malaria. *See also* 1819, MED.                                                        MED

1666   While working in his garden at Woolsthorpe, England, mathematician and physicist Isaac Newton observes an apple falling from a tree and begins the train of thought that will lead to his theory of universal gravitation, as expounded two decades later in the *Principia* (1687).                          PHYS

1666   Experimenting with a prism, Newton discovers that color is a property of light and that white light is composed of a spectrum of colors.                                PHYS

1667   English naturalist John Ray classifies plants by the number of their seed leaves, establishing the categories of monocots and dicots.                                    BIO

1667   Scottish mathematician James Gregory makes important advances in infinitesimal analysis, including the extension of the Archimedean algorithm to the quadrature of hyperbolas and ellipses.                                          MATH

1667   English physician and microscopist Robert Hooke demonstrates the function of the lungs by exhibiting the process of artificial respiration.                             MED

1667   On June 15 French physician Jean-Baptiste Denis carries out the first modern blood transfusion by infusing 12 ounces of lamb's blood into a 15-year-old boy. The boy's health improves after the procedure.                         MED

1668   English mathematician and physicist Isaac Newton invents the reflecting telescope.                                                                                      ASTRO

1668   Italian naturalist and physician Francesco Redi experiments with meat, both uncovered and covered in jars, and disproves that maggots come from decaying tissue. Although his results refute the theory of spontaneous generation, they go unacknowledged by his peers.                                    BIO

| | |
|---|---|
| 1668 | James Gregory discovers what is later called Gregory's series, the series for arctan $x$.  MATH |
| 1668 | German mathematician Nicolaus Mercator publishes the logarithmic approximation formula now known as Mercator's series.  MATH |
| 1668 | English mathematician John Wallis articulates the law of conservation of momentum, stating that the total momentum (mass times velocity) of a closed system remains unchanged.  PHYS |
| 1669 | Italian anatomist Marcello Malpighi publishes his microscopic findings on the anatomy of a silkworm. He shows the insect to have no lungs, only a tracheal system distributing air through its body through holes in its sides.  BIO |
| 1669 | German chemist Hennig Brand discovers the element phosphorus.  CHEM |
| 1669 | English mathematician and physicist Isaac Newton writes *De analysi per aequationes numero terminorum infinitas*, first published in 1711. It contains his infinite analysis and the first systematic account of the calculus. Also for the first time, an area is found through the inverse of what is now called differentiation.  MATH |
| 1669 | English physician Thomas Sydenham advocates fresh air for sickrooms, activity for tuberculosis sufferers, and simplified prescriptions.  MED |
| 1669 | Danish geologist Nicolaus Steno proposes that fossils are the petrified remains of ancient creatures, a view that is eventually accepted.  PALEO |
| 1669 | While studying a crystal of Icelandic feldspar, Danish physician Erasmus Bartholin observes the phenomenon of double refraction, the apparent doubling of images when viewed through the crystal. *See also* 1808, PHYS.  PHYS |
| 1670s | Using a pendulum, French astronomer Jean Richer concludes that the diameter of the earth is greater around the equator than from pole to pole.  EARTH |
| 1670 | English mathematician Isaac Barrow develops a method of tangents quite similar to that used in the differential calculus.  MATH |
| 1670 | Clocks are built with minute hands for the first time.  TECH |
| 1670 | A decimal-based system for measurement is developed by French cleric Gabriel Mouton.  TECH |
| 1671 | The Paris Observatory, begun in 1667, is completed, under director Giovanni Cassini, an Italian-French astronomer.  ASTRO |
| 1671 | Cassini discovers Iapetus, a satellite of Saturn. He will later discover other satellites of Saturn: Rhea (1672) and Tethys and Dione (1684).  ASTRO |

1671            Rice is cultivated in the colony of South Carolina.                    BIO

c. 1671         English mathematician and physicist Isaac Newton writes a second account of the calculus, titled *Method of Fluxions*. In it he also proposes eight new types of coordinate systems, including what are now called bipolar coordinates, and suggests what will become known as Newton's method for approximate solutions of equations. This work will not be published until 1736.                    MATH

1671            German mathematician Gottfried Wilhelm Leibniz invents a calculating machine that multiplies and divides.                    MATH

1672            French scientist N. Cassegrain invents the Cassegrain type of reflecting telescope.                    ASTRO

1673            Astronomer Giovanni Cassini of the Paris Observatory, assisted by Jean Richer, determines the distance of Mars from Earth and uses it to calculate the scale of the solar system. His figure of 86 million miles for the sun's distance from Earth is only 7 percent off.                    ASTRO

1673            Italian Marcello Malpighi describes a chick embryo's development, contributing to the early science of embryology. *See also* 1621, BIO.                    BIO

1673            In 1673–1676, working independently from Isaac Newton, German mathematician Gottfried Wilhelm Leibniz begins to develop the calculus.                    MATH

1673            Dutch scientist Christiaan Huygens publishes *Horologium oscillatorium*, a work on pendulum clocks (*see* 1656, TECH) that contains several important laws of mechanics, including the law of centripetal force, Huygens's law for pendular motion, and the principle of conservation of kinetic energy.                    PHYS

1674            British chemist John Mayow identifies the action of oxygen in burning, or oxidizing, metals and in respiration. Oxygen itself will not be isolated until Joseph Priestley's work in 1774.                    CHEM

1674            Pierre Perrault, a lawyer and government official in Paris, solves the mystery of springs' origins by proving that rainfall is more than sufficient to supply the flow of springs and rivers. This analysis probably marks the beginning of the study of the hydrologic cycle.                    EARTH

1674            The tourniquet is invented to arrest hemorrhage.                    MED

1674            Between now and 1675, French philosopher and physicist Nicholas de Malebranche expresses the belief that the human soul has two kinds of faculties: the understanding and the will. The understanding is passive, including sensory impressions, imagination, and memory. The will consists of attitudes and inclinations.                    PSYCH

1675            The Royal Observatory at Greenwich, England, is completed. Its director and first Astronomer Royal is John Flamsteed.                    ASTRO

| | |
|---|---|
| 1675 | Italian-French astronomer Giovanni Cassini discovers what will become known as the Cassini division, a gap in Saturn's rings.                    **ASTRO** |
| 1675 | Danish astronomer Olaus Roemer makes the first reasonable estimate of the speed of light: 141,000 miles per second, about three-fourths of the actual value.                    **PHYS** |
| 1676 | English mathematician and physicist Isaac Newton writes a third account of the calculus, *De quadratura curvarum*, in which he introduces the concept of prime and ultimate ratios. This work will not be published for nearly 30 years. *See* 1704, **PHYS**.                    **MATH** |
| 1676 | English physicist Robert Hooke articulates what will become known as Hooke's law, saying that within the limit of elasticity the stress applied to a material is proportional to the strain that results in its change in dimension or stretch.                    **PHYS** |
| c. 1677 | With the microscope he invented c. 1660, Dutch scientist Antoni van Leeuwenhoek discovers microscopic organisms.                    **BIO** |
| 1677 | Leeuwenhoek describes spermatozoa.                    **BIO** |
| 1678 | After two years' work on the island of St. Helena in the South Atlantic, English astronomer Edmund Halley publishes the first catalog of the southern stars. **ASTRO** |
| 1678 | Dutch astronomer Christiaan Huygens maintains that light consists of waves, whereas Newton contends that light is made up of particles. Huygens will publish his wave theory of light in 1690, in his *Treatise on Light*.                    **PHYS** |
| 1678 | Brick, formerly used only for ovens and fireplaces, is used for an entire house in Boston.                    **TECH** |
| 1679 | Italian naturalist Marcello Malpighi observes the detailed structure of plant cells and publishes his findings in *Anatomes plantarum*. At the same time, English naturalist Nehemiah Grew provides detailed, definitive descriptions of the sexual reproduction of plant cells in *The Anatomy of Plants* (1682). Their work sets a long-lasting standard for discussing cells as structural units and also shows the connection between microscopic exploration and cell study.                    **BIO** |
| 1679 | English physicist Robert Hooke becomes the first to formulate the movement of planets as a mechanical problem, though his theory that gravitational attraction varies inversely with distance from the sun is later proven incorrect. *See* 1687, **PHYS**.                    **PHYS** |
| 1679 | French physicist Denis Papin invents the pressure cooker, the first practical application of steam power.                    **TECH** |

| | |
|---|---|
| 1680s | Swiss mathematicians Jakob (Jacques) and Johann (Jean) Bernoulli make a number of contributions to differential and integral calculus as well as to the integration of many ordinary differential equations. Jakob contributes the logarithmic spiral, the lemniscate, and the isochrone. He also contributes to the study of polar coordinates, the catenary, and isoperimetric figures. The two brothers and Gottfried Wilhelm Leibniz devise and solve the Bernoulli equation.    MATH |
| 1680 | The last dodo dies on the island of Mauritius in the Indian Ocean. This flightless bird was exterminated by Dutch settlers, who arrived in 1598.    BIO |
| 1680 | German biochemist Johann Joachim Becher proves that fermentation cannot happen without the presence of essential sugars.    BIO |
| 1680 | American entomologist John Banister classifies 52 American insect species. BIO |
| 1682 | English astronomer Edmund Halley observes the comet that will be named for him after his correct prediction, in 1705, that it will return in 1758.    ASTRO |
| 1683 | Dutch scientist Antoni van Leeuwenhoek is the first to discover bacteria in the human mouth.    BIO |
| 1684 | In a posthumous publication, French astronomer Jean Picard (1620–1682) reports a fairly accurate figure for the circumference and diameter of the earth. These figures are the first real improvement on those proposed by Eratosthenes in 240 B.C.    EARTH |
| 1684 | German mathematician Gottfried Wilhelm Leibniz publishes his first paper on the calculus, with a second following in 1686. He coins the terms *differential calculus* and, with Jakob (Jacques) Bernoulli, *integral calculus*.    MATH |
| 1685 | English mathematician John Wallis devises methods for working with imaginary numbers.    MATH |
| 1686 | English astronomer Edmund Halley produces the first meteorological world map showing the prevailing tropical winds, which are monsoons and trade winds. EARTH |
| 1686 | English physicist Thomas Sydenham is first to describe chorea (dancing mania), a severe nerve disorder caused by the streptococcus responsible for rheumatic fever.    MED |
| 1687 | English mathematician and physicist Isaac Newton theorizes that the earth is an oblate spheroid slightly flattened at the poles and bulging at the equator.    EARTH |
| 1687 | In September Newton publishes his greatest work, *Philosophiae naturalis principia mathematica* (*Mathematical Principles of Natural Philosophy*), known as the *Principia*. This work outlines the law of universal gravitation and the three laws of motion. It also includes the first published documentation of Newton's dis- |

covery of the calculus (*see* 1665–1666, MATH) and several new theorems on conics. *See also* 250 B.C., PHYS.                                                                 **PHYS**

1688          In France, large sheets of glass are being made for mirrors and windows, an innovation that will lead to the commonplace use of panes of glass.      **TECH**

c. 1690s      Swiss mathematician Johann (Jean) Bernoulli, often considered the inventor of the exponential calculus, studies exponential curves.      **MATH**

---

## The *Principia*

A scientific work that contributes a small advance in knowledge may be considered worthwhile; a work that contributes a momentous advance may be considered great; but a work that contributes several momentous advances is in another league altogether. Such a work is Isaac Newton's *Principia*. Published in Latin late in 1687, it laid the basis for the science of mechanics (the study of the interaction between matter and forces) for over two centuries by defining the unifying principles known as the laws of motion and the law of gravitation.

The three laws of motion govern the way bodies move in almost all situations, from the fall of a leaf to the collision of automobiles, from the spinning of dancers to the orbits of planets. Although Albert Einstein showed in 1905 that Newton's laws of motion do not apply when objects move at relative speeds close to that of light, Newtonian mechanics still holds in most human-scale circumstances.

The three laws of motion are:

1. A body at rest remains at rest, and a moving body continues to move in a straight line at the same velocity unless acted upon by external forces.
2. When a force acts upon a moving body, the rate at which its momentum (mass times velocity) changes is proportional to, and in the same direction as, the force applied.
3. When a force acts upon a body, the body exerts an equal force, or reaction, in the opposite direction.

The *Principia* also presented Newton's law of gravitation, which states that the gravitational force between two bodies is proportional to their mass and inversely proportional to the square of the distance between them. Einstein's general theory of relativity (1916) brought new understanding to the study of gravitation, but Newton's law still holds well enough to serve as the basis for such tasks as analyzing orbits and planning space missions. In mathematical terms, Newton's law of gravitation states:

$$F = G \, \frac{m_1 m_2}{d^2}$$

Where *F* is the force of gravitation between two bodies, *G* the gravitational constant (a quantity equal to about $6.67 \times 10^{-11} \, N \, m^2 \, kg^{-2}$), $m_1 m_2$ the masses of the two bodies, and $d^2$ the square of the distance between them.

1690     The Mathematische Gesellschaft, the oldest mathematical society still in existence 300 years later, is founded in Germany.      MATH

1690     In his *Essay Concerning Human Understanding*, English philosopher John Locke (*see* 1632, MISC) opposes the belief in innate ideas, arguing that the mind at birth is a blank slate that acquires all its knowledge from experience. Locke's ideas will influence philosophers David Hume (*see* 1739, PSYCH) and George Berkeley (*see* 1709, PSYCH) and will form the underlying basis for the 20th-century American psychological school of behaviorism.      PSYCH

1690     English philosopher John Locke publishes *Two Treatises on Civil Government*, in which he offers an alternative to Thomas Hobbes's view of the origin of governments (*see* 1651, SOC). In it Locke argues that human nature is good, that people are born equal, free, and with certain inalienable rights, and that people form a "social contract" to guarantee those rights. Locke's views will influence centuries of political theory and practice, notably in the U.S. Declaration of Independence and Constitution.      SOC

1691     English naturalist John Ray publishes a major classification of organic life, *The Wisdom of God Manifested in the Works of Creation*, bringing together information on the vast number of new plants and animals discovered around the world over the last few centuries.      BIO

1691     Like Nicolaus Steno in Denmark (*see* 1669, PALEO), English naturalist John Ray argues that fossils are the remains of ancient creatures.      PALEO

1693     English naturalist John Ray continues to classify animals, sorting them on the basis of hoofs, toes, and teeth.      BIO

1693     In unpublished letters, German mathematician Gottfried Wilhelm Leibniz makes the first Western reference to the method of determinants.      MATH

1693     English astronomer Edmund Halley prepares the first mortality tables, statistically relating human death rates to age.      MATH

1694     Botany professor Joseph Pitton de Tournefort, considered at this time to be the leader of French botanical thought, publishes *Elemens de botanique*. It inventories and describes more than 8,000 plants and devises an artificial classification system that will be accepted until the work of Carolus Linnaeus in 1735. In 1719, this work will be republished in English as *The Complete Herbal*.      BIO

| | |
|---|---|
| 1695 | English chemist Nehemiah Grew isolates magnesium sulfate, popularly known as epsom salts.     CHEM |
| 1696 | Swiss mathematician Johann (Jean) Bernoulli discovers L'Hôpital's rule on indeterminate forms, named for French mathematician Guillaume François Antoine de L'Hôpital, who publishes it this year in the first printed textbook on the differential calculus, *Analyse des infiniment petits*.     MATH |
| c. 1697 | Swiss mathematicians Jakob (Jacques) and Johann (Jean) Bernoulli solve the problem of the brachistochrone, the curve of quickest descent for a mass point moving between two points in a gravitational field, proving that this curve is a cycloid. As a result, they are often considered the inventors of the calculus of variations.     MATH |
| 1698 | English astronomer Edmund Halley undertakes the first ocean voyage for a purely scientific purpose, to measure and map magnetic declinations all over the world, such as the distance between the direction of the compass needle and true north.     EARTH |
| 1698 | English engineer Thomas Savery invents a steam-driven pump, the "miner's friend," for use in coal mining.     TECH |
| 1698 | The London Stock Exchange opens. The New York Stock Exchange will follow in 1792.     TECH |
| 1698 | Champagne is invented by French cellarer Dom Pierre Pérignon in the abbey of d'Hautvilliers.     TECH |
| 1699 | French physicist Guillaume Amontons invents an air thermometer that measures temperature by the change in gas pressure. Using this device he shows that the volume of a fixed quantity of a gas increases as the temperature rises and decreases as the temperature falls, and that the rate of change in volume is the same for all gases. His breakthrough is forgotten, however, until rediscovered by Jacques-Alexandre-César Charles in 1787. *See also* 1802, PHYS.     PHYS |
| 1700s | Swiss naturalist and entomologist Charles Bonnet is the first to use the term *evolution* to describe the concept that periodic catastrophes result in increasingly higher life forms.     BIO |
| 1700s | Eastern Native American tribes are observed to keep wounds clean. In clashes with European colonists this "primitive" treatment is often found to be effective. Native Americans also isolate the wounded for treatment, while white soldiers are kept together in infirmaries, where they often die of what will become known as iatrogenic, or hospital-acquired, infections.     MED |
| 1700s | "Idiot cages" are used to confine and display mentally ill people, usually as a source of public entertainment.     PSYCH |
| 1700s | In England the straitjacket is invented to restrain agitated asylum patients.     PSYCH |

1700        Ole Rømer of Denmark invents the meridian telescope.      ASTRO

c. 1700     German chemist George Ernst Stahl proposes that objects burn or rust because they lose a combustible substance called phlogiston, a theory disproved by Antoine-Laurent Lavoisier in 1772.      CHEM

1700        German mathematician Gottfried Wilhelm Leibniz explains that any base such as 10, 12, or 2 can be used for positional notation and that a system with a base of 2, the binary system, is particularly useful. This system, consisting of the symbols 1 and 0, will one day become the basis for digital computers.      MATH

c. 1700     Native American women have long been using quinine, sassafras, datura, ipecac, cascara, and witch hazel to treat minor illnesses and injuries. All of these substances will become part of modern pharmacopeia.      MED

1700        Italian physician Bernardino Ramazzini publishes a work describing 40 occupational diseases. He is the first to write on the subject.      MED

1700        French physicist Joseph Sauveur coins the term *acoustics* for describing the relations of musical tones.      PHYS

1701        English farmer Jethro Tull invents a multirow machine drill for planting three lines of seeds simultaneously, which decreases waste and increases productivity. TECH

1702        Dutch chemist William Homberg discovers boric acid.      CHEM

1702        David Gregory, Savilian Professor of Astronomy at Oxford, publishes *Astronomiae physicae & geometricae elementa*, the first textbook of astronomy based on Newtonian principles.      EARTH

1702        The *Daily Courant*, the first daily newspaper in England, begins publication in London. Three years later the first regular newspaper in the American colonies, the weekly *News-Letter*, is published in Boston.      TECH

1703        English mathematician and physicist Isaac Newton, a member of the Royal Society since 1672, is elected its president, a post in which he will serve until his death in 1727.      MISC

1704        Newton publishes *De quadratura curvarum*, written in 1676. This is the first clear published account of Newton's version of the calculus, though hints of it had appeared in the *Principia* (1687).      MATH

1704        Newton publishes *Enumeration of Curves of [the] Third Degree*, the first work devoted solely to graphs of higher plane curves in algebra, written about 1676.      MATH

1704        In his *Optics* Newton theorizes that light is made up of particles called corpuscles, a view that conflicts with Christiaan Huygens's wave theory of light (*see* 1678, PHYS). Newton also argues that white light is made up of the colors of the spectrum.                                                                    PHYS

1704    German mathematician Gottfried Wilhelm Leibniz finishes his *New Essays on Understanding*, in which he disputes John Locke's theory that empirical knowledge is humans' only source of truth. Leibniz says the human mind has innate intelligence, inborn ideas, truths, dispositions, habits, and potentials. This book, with its nativist views, will not be published until 1765.                PSYCH

1705    English astronomer Edmund Halley theorizes that the comet of 1682, observed since antiquity (*see* 240 B.C., ASTRO), is a regular visitor, and he correctly predicts it will return in December 1758.                                               ASTRO

1705    The Royal Observatory of Berlin is established.                                ASTRO

1705    Queen Anne knights English mathematician and physicist Isaac Newton.  MISC

1705    Experimenting with a clock inside a vacuum, English physicist Francis Hauksbee shows that sound can travel only in a fluid medium such as air.        PHYS

1706    The Greek letter $\pi$ (pi) is first used as the symbol for the ratio of a circle's circumference to its diameter.                                                    MATH

1706    Francis Hauksbee invents a crank-operated glass sphere that can produce an intense charge of static electricity.                                              PHYS

1707    Swiss mathematician Leonhard Euler, one of the most productive mathematicians in history, is born (*d.* 1783). He will publish 560 books and papers during his lifetime and hundreds more will be published afterward. He will contribute to every mathematical field of his day as well as to such related areas as astronomy, hydraulics, artillery, shipbuilding, and optics. His systems of notation will remain standard.                                                          MATH

1708    The United East India Company is formed in England from smaller trading companies, making it the premier shipping firm for importing fabrics, foods, and military materials like saltpeter.                                          TECH

1709    Anglo-Irish empirical philosopher George Berkeley publishes his *Essay towards a New Theory of Vision*, in which he agrees with John Locke that all knowledge comes from experience and depends on human perception.                   PSYCH

1709    English ironworks master Abraham Darby shows that coke, a derivative of coal, can be used instead of wood-based charcoal to smelt iron. This discovery greatly increases the market for coal and improves iron production. With the invention of the Newcomen engine (*see* 1712, TECH), this breakthrough marks one of the starting points of the industrial revolution in England.            TECH

c. 1710    The Pennsylvania rifle, invented in England's American colonies, is a substantial improvement on the preferred firearm of the day, the smooth-bore musket.    TECH

1712    English blacksmith Thomas Newcomen invents the steam engine that bears his name, which uses steam to drive a piston to generate power. This device is more efficient than Thomas Savery's steam-driven "miner's friend" of 1698.    TECH

1713    *Ars conjectandi* (*The Art of Conjecturing*), by Swiss mathematician Jakob (Jacques) Bernoulli, published this year posthumously, becomes the first substantial book on the theory of probability. It contains the first full proof of the binomial theorem for positive integral powers as well as the Bernoulli numbers, which will be useful in writing infinite series expansions.    MATH

1713    The first schooner is built, by Scottish-American captain Andrew Robinson in Massachusetts. Schooners will become important to the growing American fishing industry.    TECH

1714    German physicist Daniel Gabriel Fahrenheit invents the mercury thermometer and the Fahrenheit scale.    TECH

1715    The total solar eclipse of April 22, visible in Britain and parts of Europe, is the first such eclipse to be so widely anticipated as to draw large numbers of astronomers as observers. English astronomer Edmund Halley prepares maps charting the predicted path of totality.    ASTRO

1715    English mathematician Brook Taylor publishes the Taylor series, along with other components of the calculus, in his *Methodus incrementorum*.    MATH

1716    Boston preacher Cotton Mather writes that his slaves practiced smallpox inoculation in Africa by applying serum from the pustule of an infected person into an incision made on a healthy individual. Greek physician Emmanuel Timoni draws attention to primitive smallpox vaccination techniques in such places as China and central Europe, where smallpox scabs are used to infect healthy people and produce immunity. *See also* 1796 and 1967, MED.    MED

1718    English astronomer Edmund Halley discovers the "proper motion" (independent movement) of stars, contradicting the ancient theory that the stars are fixed.    ASTRO

1718    Potatoes, originally an Andean crop, are introduced to Boston, their first appearance in England's American colonies.    BIO

1718    French-born English mathematician Abraham de Moivre publishes his *Doctrine of Chances*, an important work on probability.    MATH

1723    M. A. Capeller publishes *Prodomus crystallographiae*, the earliest known treatise on crystallography.    PHYS

| 1724 | The possibility of cross-fertilizing corn is established. | BIO |

| 1724 | German physicist Daniel Gabriel Fahrenheit describes the supercooling of water. | PHYS |

| 1725 | John Flamsteed's three-volume *Historia coelestis Britannica*, published posthumously, catalogs some 3,000 stars. | ASTRO |

| 1725 | Plaster of Paris is used to cast metal printing plates, reducing cost and improving the efficiency of printing by letterpress. | TECH |

| 1727 | In *Vegetable Staticks* English botanist Stephen Hales reports on his work with plant fluid flows and plant respiration. This book also describes experiments determining the generation or depletion of air by various substances. These discoveries later reinforce the claim of Hales to be one of the founders of modern plant physiology. | BIO |

| c. 1727 | Swiss mathematician Leonhard Euler introduces the letter *e* to represent the base of the system of natural logarithms. This notation will become standard. | MATH |

| 1727 | Sir Isaac Newton, the first scientist to be buried in Westminster Abbey, dies in London on March 20. | MISC |

| 1727 | A basic property of photography is discovered by German chemist J. H. Schulze, who determines that light, not heat, activates the chemicals involved in the process of deriving silver salts from silver nitrate. | TECH |

| 1728 | English astronomer James Bradley discovers the aberration of starlight, the apparent displacement of a star's position due to the orbital motion of Earth. This finding gives the most definite proof to date that Earth moves in space. | ASTRO |

| 1728 | Danish navigator Vitus J. Bering, working for Russian tsar Peter the Great, discovers the passage that will come to be called the Bering Strait, proving that North America is not connected by land to Asia. | EARTH |

| 1728 | British-educated physician John Hunter (*d.* 1793), considered the founder of experimental and surgical pathology, is born in Scotland. He will introduce a flexible tube passed into the stomach for artificial feeding, be the first to study teeth scientifically, and argue that aneurysms from arterial disease should be tied off. Hunter will also posit that blood is alive and that the human embryo is, in each stage, a completed form of a lower order of the species. | MED |

| 1729 | French scientist Louis Bourget distinguishes between organic and inorganic growth. | BIO |

| 1729 | English scientist Stephen Gray theorizes that electricity is a fluid and discovers that some substances are electrical conductors but others are nonconductors, or insulators. | PHYS |

1729            Dutch physicist Pieter van Musschenbroek is one of the first people in modern
                times to use the word *physics*, a term dating back to Aristotle, to mean natural
                philosophy.                                                                  **PHYS**

c. 1730         English naturalist Henry Baker begins his essays on the microscope, introduc-
                ing it to the lay public. One of his most significant finds will be his observa-
                tion of various crystal shapes.                                              **BIO**

1730            In England, Lord Charles Townshend determines that livestock can be main-
                tained throughout the year, most importantly in the winter, on turnips rather
                than on seasonally grown feed. This discovery allows beef to be available year-
                round.                                                                       **TECH**

1731            American mathematician Thomas Godfrey and English inventor John Hadley
                independently invent the reflecting quadrant, the precursor of the modern
                sextant.                                                                     **ASTRO**

1731            In a posthumous work, Swiss mathematician Jakob (Jacques) Bernoulli intro-
                duces Bernoulli numbers.                                                     **MATH**

1731            French mathematician Alexis-Claude Clairaut identifies what will become
                known as the Clairaut differential equation.                                 **MATH**

1731            English chemist and physicist Henry Cavendish, discoverer of hydrogen, is
                born in Nice (*d.* 1810). His studies on electricity, unpublished until 1879,
                anticipate the work of such later scientists as Charles-Augustin de Coulomb,
                Michael Faraday, and Georg Simon Ohm. He will also perform the first calcu-
                lations of the gravitational constant and of the earth's mass and density.   **PHYS**

1731            American scientist and inventor Benjamin Franklin improves and enlarges the
                postal service by streamlining routes throughout the colonies.               **TECH**

1732–1734       German philosopher Baron Christian von Wolff develops the field of rational
                psychology, a subdivision of empirical psychology that depends more on rea-
                son than experience.                                                         **PSYCH**

1733            Swiss mathematician Leonhard Euler publishes a seminal work of modern
                mathematical analysis.                                                       **MATH**

1733            French physicist Charles-François de Cisternay du Fay distinguishes "vitreous"
                from "resinous" electricity, noting that each attracts the other but repels itself.
                This finding links electricity to magnetism, which exhibits similar patterns of
                attraction and repulsion.                                                    **PHYS**

1733            English weaver John Kay invents the flying shuttle, which vastly improves on
                the hand loom and simplifies the industrialization of textile production.    **TECH**

1734            French entomologist René de Réaumur begins publishing the first of a six-
                volume work on insect history that will be completed in 1742.                **BIO**

1734        Anglo-Irish empirical philosopher George Berkeley, who becomes bishop of
            Cloyne this year, publishes *The Analyst*, an attack on the intellectual basis of
            the calculus, the theory of fluxions. It spurs Scottish mathematician Colin
            Maclaurin to respond with a defense entitled *Treatise of Fluxions* (1742).    MATH

1735        Swedish naturalist Carolus Linnaeus presents his first system of plant classifi-
            cation, *Systema naturae*, sorting flora according to the number of their stamens
            and pistils. This will lead to further botanical classification, including
            Linnaeus's use of binomial nomenclature to record plant genera and species.
            Linnaeus's method of classification will still be in use more than two and a
            half centuries later.                                                          BIO

1735        Swedish chemist Georg Brandt discovers the element cobalt.                      CHEM

1735        Spanish scientist Antonio de Ulloa discovers the element platinum.             CHEM

1735        Geographic expeditions to equatorial and polar regions confirm Isaac Newton's
            1687 theory that the earth is an oblate spheroid.                              EARTH

1736        French chemist Henri-Louis Duhamel du Monceau is the first to distinguish
            potassium salts from sodium salts.                                            CHEM

1736        French surgeon Jean-Louis Petit distinguishes between cerebral compression
            and concussion, paving the way for different—and more successful—treat-
            ments of injury.                                                               MED

1736        New York City's Bellevue Hospital has its beginnings as a room in the public
            workhouse for those who are physically sick or mentally ill. It will open offi-
            cially as a hospital in 1812.                                                  MED

1736        Swiss mathematician Leonhard Euler writes *Mechanics*, the first book to be
            devoted to that subject. With analytical methods it develops Newton's dynam-
            ics of the mass point.                                                         PHYS

1736        The chronometer is introduced by British inventor John Harrison. Used in
            conjunction with a quadrant, it can aid navigators in pinpointing their longi-
            tudinal position.                                                              TECH

1737        *Biblia naturae*, by Dutch naturalist Jan Swammerdam, is published posthu-
            mously. It introduces discoveries and conclusions drawn from
            Swammerdam's microscopic experiments with dissected insects.                   BIO

1738        Planned excavation begins at Pompeii and Herculaneum in Italy. *See* 79 and
            1592, ARCH.                                                                    ARCH

1738        Dutch diplomat Benoit de Maillet suggests an "ultra-Neptunian" theory, that
            the earth's surface was shaped by the action of a universal ocean.            EARTH

1738    Swiss physicist Daniel Bernoulli publishes what will be called the Bernoulli theorem, stating that at any point in a pipe of flowing fluid the sum of the pressure energy, kinetic energy, and potential energy is constant. Following Boyle (*see* 1662, PHYS), Bernoulli makes his explanation in terms of the movement of atoms comprised by the fluid.                                    PHYS

1739    Swiss naturalist Abraham Trembley discovers the hydra, giving rise to speculation by some scientists that the freshwater polyp may represent a link between the animal and vegetable kingdoms.                                    BIO

1739    French explorers Pierre and Paul Mallet become the first Europeans to see the Rocky Mountains.                                    EARTH

1739    Physicist George Martine shows that the amount of heat in an object is not proportional to its volume.                                    PHYS

1739    Scottish philosopher and diplomat David Hume publishes *A Treatise on Human Nature*, claiming that complex ideas are formed from simple ones based on three laws of association: resemblance, continuity, and cause-and-effect relationships.                                    PSYCH

1739    Glass is manufactured in what will become the state of New Jersey in a factory set up by German-American businessman Caspar Wistar.                                    TECH

1740    Swiss entomologist Charles Bonnet discovers that female aphids can reproduce without fertilization.                                    BIO

1740    Antonio Moro publishes a significant study of marine fossils.                                    PALEO

1740    Sheffield steel, a superstrong cast steel produced in airless crucibles, is introduced by Englishman Benjamin Huntsman.                                    TECH

1741    Indigo is harvested in South Carolina, beginning a dyestuffs industry in that region. *See also* 1870, CHEM.                                    BIO

1741    Danish navigator Vitus J. Bering, working for the Russians, is the first European to reach Alaska, on his second voyage. Shipwrecked on what will become known as Bering Island, he dies there on December 8. Russia claims and explores the Aleutian Islands and Alaska until selling them to the United States in 1867.                                    EARTH

1741    Irish physician Fielding Ould draws attention to the benefits of cutting the female perineum during delivery to prevent the baby's head and shoulders from tearing the woman's pelvic floor. This procedure will become known as episiotomy.                                    MED

1742    Using the hydra (*see* 1739, BIO), Swiss naturalist Abraham Trembley makes the first permanent tissue graft.                                    BIO

1742    German-Russian mathematician Christian Goldbach frames "Goldbach's conjecture"—that any even number greater than 2 can be expressed as the sum of two prime numbers. Two and a half centuries later, the conjecture will have been neither proven nor disproven.                                                                    MATH

1742    Swedish astronomer Anders Celsius invents what will become known as the Celsius scale of temperature, which will eventually supersede the Fahrenheit scale (*see* 1714, TECH) except in the United States. The freezing point of water is represented on the Celsius scale as zero degrees and on the Fahrenheit scale as 32 degrees above zero.                                                                    TECH

1742    Coal is mined in Virginia for the first time in the region.                    TECH

1742    American scientist and inventor Benjamin Franklin invents what will become known as the Franklin stove, which heats a room by circulating preheated air. Also known as a Pennsylvania fireplace, this lean-to-shaped stove contains an air box and is positioned inside a fireplace.                                        TECH

1743    French mathematician Alexis-Claude Clairaut explains how to compute gravitational force at a given latitude.                                                            ASTRO

1743    French physicist Jean Le Rond d'Alembert formulates what will become known as d'Alembert's principle: In a closed system of moving bodies, actions and reactions are in equilibrium.                                                                    PHYS

1744    Swiss mathematician Leonhard Euler discovers transcendental numbers, numbers that can never be a solution to a polynomial algebraic equation (as distinguished from algebraic numbers).                                                            MATH

1744    Leonhard Euler publishes the first exposition of the calculus of variations, including Euler's equations.                                                            MATH

1744    Russian physicist Mikhail V. Lomonosov proposes correctly that heat is a form of motion.                                                                            PHYS

1744    French physicist Pierre-Louis de Maupertuis formulates the principle of least action: Nature operates in such a way that action—the product of force, distance, and time—is at a minimum.                                                            PHYS

1745    Dutch physicist Pieter van Musschenbroek of the University of Leyden and German physicist Ewald Georg von Kleist independently invent what will become known as the Leyden jar, the first practical device for storing static electricity.                                                                            PHYS

1746    Swiss mathematician Leonhard Euler uses the wave theory of light to develop a mathematics of the refraction of light.                                                PHYS

1746    American philosopher and theologian Jonathan Edwards writes on psychological questions in relation to religion. His *Treatise Concerning Religious Affections* will be considered one of the first books of psychology written by an American. Edwards believes that there is no free will—that all human choices are made by God.         **PSYCH**

1747      French mathematician Alexis-Claude Clairaut publishes the first approximate resolution of the three-body problem, examining how three celestial bodies interact.         **ASTRO**

1747      Scottish physician James Lind shows, in the first controlled dietary study, that citrus fruits cure scurvy.         **MED**

1747      American scientist and inventor Benjamin Franklin argues against du Fay's theory of 1733 that there are two electrical fluids, suggesting instead that there is only one, of which an excess could be called positive, a deficiency negative.   **PHYS**

1747      Theorizing about vibrating strings, French physicist Jean Le Rond d'Alembert publishes the general solution of the partial differential-wave equation in two dimensions.         **PHYS**

1748      English astronomer James Bradley discovers the nutation (wobbling) of Earth's axis, an irregular periodic oscillation of Earth's poles caused by perturbation from the sun and the moon.         **ASTRO**

1748      Swiss mathematician Leonhard Euler publishes *Introductio in analysin infinitorum*, which establishes the strictly analytic treatment of trigonometric functions. This book contains the Euler identities, an algebraic theory of elimination, an exposition on infinite series, and a chapter on the Zeta function. **MATH**

1748      Welsh Catholic priest Father John T. Needham claims to have proven spontaneous generation by cooking meat, cooling it, then reheating it. He claims to have identifed *animalcules* in the broth, which appear spontaneously. The erroneous belief that living organisms can be generated directly from lifeless matter will persist for another century. *See* 1864, **MED**.         **MED**

1748      French physicist Jean-Antoine Nollet discovers osmosis, the passage of a solvent such as water through a semipermeable membrane separating two solutions that have different concentrations. He also identifies osmotic pressure, the pressure required to stop the flow from a pure solvent into a solution. In osmosis the solvent tends to flow from the weaker to the stronger solution until the two are equal in concentration or osmotic pressure is applied. *See also* 1877, **BIO**.         **PHYS**

1748      French philosopher Julien Offroy de La Mettrie pioneers French materialism. In his book *L'Homme-machine* (*Man as Machine*), he argues that the body and soul are mortal and that life and thought are nothing more than the nervous system's mechanical action.         **PSYCH**

1749      In the first volume of his book *Natural History*, French naturalist Georges-Louis de Buffon disagrees with earlier classifications by Andrea Cesalpino and

Carolus Linnaeus, claiming that nature's life chain has small gradations from one type to another and that the discontinuous categories are artificially constructed. He also suggests that organic species are descended from a few primordial types in a process of evolution through degeneration from perfect to less perfect forms.                                                                    BIO

1749    Swedish naturalist Carolus Linnaeus describes how nature limits competition between species by allotting each its own geographical location and placement in the food chain. He claims that reproductive rates and predators maintain a species' proper numerical proportions.                                             BIO

1749    In *Natural History*, French naturalist Georges-Louis de Buffon speculates that Earth formed 75,000 years ago through collision of the sun with a comet. Though incorrect, these speculations open the door to further study of the age of the planet.
                                                                                EARTH

1749    French scientist Pierre Bouguer notes what will become known as the "Bouguer correction," observing that the lessening of the pull of gravity with altitude is partially compensated for by the gravitational attraction of the intervening rock.                                                                    EARTH

1749    Scottish astronomer Alexander Wilson is the first to use kites for exploring the properties of the atmosphere, attaching thermometers to them to try to measure temperatures at various heights.                                               EARTH

1749    British physician and biologist David Hartley describes positive afterimages for both auditory and visual stimuli, such as the glow from a flame after it goes out and the sound of a bell after it has stopped ringing. This theory of vibrations and his hypothesis that the mind and body always work together make Hartley the originator of physiological psychology.                          PSYCH

1750    Swiss mathematician Gabriel Cramer publishes what will become known as Cramer's rule (actually discovered by Scottish mathematician Colin Maclaurin as early as 1729) for solving simultaneous equations by determinants.      MATH

1751    Swedish mineralogist Baron Axel F. Cronstedt discovers the element nickel.   CHEM

1751    Paris physician Jean-Étienne Guettard discovers a region of extinct volcanoes in Auvergne, France.                                                          EARTH

1751    The first public hospital in Britain's American colonies is organized in Philadelphia.                                                              MED

1751–1752  French scientist Denis Diderot and French physicist Jean Le Rond d'Alembert publish the first volume of the *Encyclopédie*. The first modern encyclopedia, it takes a rational approach to "the sciences, the arts, and customs."      MISC

1751    Flying a kite attached to a metal key in a thunderstorm, American scientist and inventor Benjamin Franklin proves that lightning is electricity.             PHYS

1752            French physician Jacques Daviel originates modern lens extraction as a cure for
               cataracts.                                                                        MED

1752            French physicist Jean Le Rond d'Alembert formulates certain principles of
               hydrodynamics.                                                                    PHYS

1752            Building on his studies of lightning and electricity, American scientist
               Benjamin Franklin invents the lightning rod.                                      TECH

# Components of Air

Since the time of the ancient Greeks, air was considered to be a discrete element,
impossible to break down further. Not until the 18th century did it become clear that
air was made up of several different gases. In experiments beginning in 1754, Scottish
chemist Joseph Black discovered that a lump of lime (calcium oxide) would gradually
turn to limestone (calcium carbonate) if left in the open air. He had already established
that this reaction required the addition of the gas we call carbon dioxide. Clearly, carbon dioxide was part of air, but experimentation showed that it could not support
combustion or animal life. What was it living things breathed?

Black asked his student Daniel Rutherford to look into this question. In 1772,
Rutherford burned a candle in a closed container until the flame went out. When
Rutherford used chemicals to draw out the carbon dioxide in the container, he found
that a large quantity of another gas was left over. It, too, failed to support life or combustion. Eventually this gas became known as nitrogen, because it was also found in
niter (potassium nitrate).

The breathable component of air was discovered by English chemist Joseph Priestley
in 1774. He found that mercury, heated in air, produced a red powder called mercuric
oxide. When it was heated in a sealed container, it broke down into mercury and some
sort of gas. Priestley found that this gas not only supported the processes of respiration
and combustion but enhanced them.

It was French chemist Antoine-Laurent Lavoisier who determined the relative proportions of the known atmospheric gases. In experiments in the 1770s he found that
after mercury was heated to produce mercuric oxide, about four-fifths of the air in the
closed container still remained. He concluded that this gas, which did not support life,
was Rutherford's gas nitrogen. Priestley's gas, which Lavoisier called oxygen (Greek for
"acid producer"), represented the other fifth of the air. Black's gas, "fixed air" or carbon
dioxide, was present in only marginal quantities.

Lavoisier's figures were roughly correct, though other components of air have since
been found and more exact estimates made. It is now known that nitrogen and oxygen
make up about 99 percent (nitrogen 78.08 percent, oxygen 20.95 percent) of the gases
in Earth's atmosphere. The rest is mostly argon (0.93 percent) and carbon dioxide
(0.03 percent). Neon, helium, methane, krypton, hydrogen, xenon, and ozone are also
present in minute traces.

| 1752 | Publicly funded globe streetlights are installed in Philadelphia, the first such lighting system in Britain's American colonies. TECH |
|---|---|

1753 · Chemist C. G. Junine demonstrates that the element bismuth is different from lead. CHEM

1753 · The Conestoga wagon, a transport vehicle for persons and cargo, is popularized by the Pennsylvania Dutch. It is named for the town where it was developed. TECH

1754 · French scientist Denis Diderot revives the theory of Empedocles of Acragas that, in the past, various animal organs thrived independently and random combinations of these organs were eventually joined to create modern animals. BIO

1754 · Swiss naturalist Charles Bonnet details the nutritional value of plants. BIO

1754 · Scottish chemist Joseph Black heats limestone (calcium carbonate), producing carbon dioxide and lime (calcium oxide). On finding that the process can be reversed by combining calcium oxide with carbon dioxide or simply leaving calcium oxide in the open air, he discovers that carbon dioxide must be a component of air. This experiment is the first to apply quantitative analysis to chemical reactions. CHEM

1754 · Scottish physician William Smellie pioneers midwifery by men. MED

1754 · Swiss-French philosopher Jean-Jacques Rousseau publishes his *Discourse on the Inequalities of Men*. In this and *The Social Contract* (1762), Rousseau expounds his influential views about the "noble" natural condition of humans, the corrupting influence of civilization, and the formation of the social contract to correct inequalities. SOC

1755 · German philosopher Immanuel Kant speculates about the existence of distant collections of stars or "island universes." ASTRO

1755 · A disastrous earthquake befalls Lisbon on November 1, sparking increased interest in geological phenomena. EARTH

1755 · Swiss mathematician Leonhard Euler publishes *Institutiones calculi differentialis*, an influential textbook on differential calculus, followed in 1768–1774 by *Institutiones calculi integralis*. MATH

1756 · English physicist John Canton observes magnetic storms in the earth's magnetic field. EARTH

1756 · In Berlin, physician and metallurgist Johann Gottlob Lehman publishes *Versuch einer Geschichte von Flotzgeburgen*, an important geologic account. He recognizes stratified rocks (*Flotzeburge*) as being of sedimentary origin, distinct from igneous (veined) rock. EARTH

1757    The modern sextant, used to chart navigation, is developed by British seaman John Campbell.                                                                    TECH

1758    As predicted by English astronomer Edmund Halley in 1705, the comet of 1682 returns on December 25, the first such return ever predicted. The object is named Halley's comet in his honor.                                            ASTRO

1758    French botanist Henri du Monceau describes tree structure and physiology. BIO

1758    German chemist Andreas Sigismund Marggraf invents a technique called the flame test for identifying different substances by noting the colors of their flames when heated.                                                           CHEM

1758    Swedish chemist Axel F. Cronstedt begins to classify minerals by their chemical structure as well as appearance when he distinguishes four classes of minerals: bitumens, earths, metals, and salts.                                    CHEM

1760s   French physicist Jean Le Rond d'Alembert formulates the limit concept in calculus.                                                                          MATH

1760s   English researcher Arthur Young uses questionnaires to survey the population. Questionnaires and case studies, which are also developed in this century, will become basic tools of social science research.                       SOC

1760    In London, the Kew Botanical Gardens opens.                               BIO

1760    Scottish chemist Joseph Black shows that mercury has a greater heat capacity than water in that a quantity of mercury heats faster than an equal one of water does. This experiment marks the beginning of the scientific study of heat. CHEM

1760    German physicist Johann H. Lambert introduces the term *albedo* for the varying rates of reflection of the planets.                                         PHYS

1760    Between now and 1762, Dutch anatomist Pieter Camper publishes *Demonstrationum anatomico-pathologicarum*, a two-volume work comparing anatomy in different human races.                                                         SOC

1761    Russian scientist, scholar, and writer Mikhail V. Lomonosov discovers the atmosphere of Venus while observing that planet's transit across the sun. ASTRO

1761    Joseph-Nicolas Delisle of Paris organizes astronomers around the world to observe the transit of Venus across the sun. Jeremiah Horrocks and Edmund Halley had already promoted the idea that simultaneous observations of this rare event from different places on Earth would allow scientists to determine the distance to Venus and the sun. Observations are made this year and at the transit of 1769. The next transits (which always come in pairs eight years apart) will not occur until 1874 and 1882.                               ASTRO

1761        French evolutionist Jean-Baptiste Robinet publishes the first volume of a five-volume work in which he claims that the Creator made organic beings on a scale and that they all have the internal energy to move upward toward the top, where humans are. Robinet also posits that all matter contains life and even inorganic matter can evolve into a living organism.                    BIO

1761        In France the first veterinary school is founded.                    BIO

1761        German physician George Christian Fuschel publishes a lengthy article dealing with the stratigraphy of the area of Thuringa. Notably, it introduces the sense of formation as the primary unit of the study of rock strata. Fuschel correlates strata by means of index fossils.                    EARTH

1761        German mathematician Johann H. Lambert proves that pi is an irrational number.                    MATH

1761        Italian physician Giovanni Morgagni becomes the father of morbid anatomy after linking certain postmortem findings with disease symptoms experienced by the dying.                    MED

1762        English astronomer James Bradley completes a catalog of 60,000 stars.    ASTRO

1762        Scottish chemist Joseph Black discovers latent heat, the quantity of heat absorbed or released when a substance changes its physical phase at constant temperature (e.g., from solid to liquid or liquid to gas).                    CHEM

1763        German botanist J. G. Kölreuter publishes the results of his experiments on plants by animals carrying pollen.                    BIO

c. 1764     Scottish philosopher Thomas Reid becomes the first member of the so-called Scottish School of psychology, objecting to empiricism and associationism. Reid disagrees with the principles of association, which say that all human knowledge is experience. He proposes, in part, the theory that will become known as faculty psychology, that the mind is an organized unity with the ability to act on needs such as self-preservation, desire, self-esteem, pity, and gratitude.                    PSYCH

1764        The first system for measuring and naming sizes of type is developed by French engraver Pierre Simon Fournier and outlined in his work *Manuel typographique*.                    TECH

1765        Geologist Nicolas Desmarest, the inspector-general of manufactures in France, discovers prismatic basalt to be of volcanic origin.                    EARTH

1765        Swiss scientist Horace Bénédict de Saussure invents the electrometer, a device for measuring voltage differences without drawing an appreciable amount of current from the source.                    TECH

1765        The steam engine is refined by Scottish engineer James Watt. With its separate
            chamber for condensing the steam, this engine supersedes the Newcomen
            engine of 1712 in its efficiency. It will be patented in 1769. *See also* 1698, TECH.
                                                                                    TECH

1766        German astronomer Johann Daniel Titius or Tiety of Wittenberg discovers that
            the distances of the known planets from the sun are proportional to the terms
            of the series 0, 3, 6, 12 . . . . . Johann E. Bode, director of the Berlin
            Observatory, will publish this finding in 1772, hence its common name,
            Bode's law. This "law" will later be found to apply to Uranus (discovered in
            1781) but not Neptune (1846) or Pluto (1930). It is therefore not a universal
            law but an observation about certain planets of the solar system.          ASTRO

1766        English chemist Henry Cavendish discovers an inflammable gas he calls fire
            air, produced by reaction between acid and certain metals. It will later be iden-
            tified as the element hydrogen. *See* 1784, CHEM.                          CHEM

1766        The as-yet-unidentified element nitrogen (*see* 1772, CHEM), generated by an
            electrical charge, is used to enrich soil in experiments conducted by chemist
            Henry Cavendish.                                                           TECH

1767        Nevil Maskelyne, Astronomer Royal, begins annual publication of the *British
            Nautical Almanac*, describing the position of celestial bodies at specific times,
            for use in navigation.                                                     ASTRO

1767        British chemist Joseph Priestley publishes *The History and Present State of
            Electricity*, in which he suggests that electrical forces, like gravitational ones,
            increase or decrease in inverse proportion to the square of the distance. This
            book also contains the first detailed account of Benjamin Franklin's kite exper-
            iment. *See* 1751, PHYS.                                                   PHYS

1768        Italian biologist Lazzaro Spallanzani determines that food can be protected
            from microorganisms by being sealed to prevent air penetration.            BIO

1768–1771   During an around-the-world voyage on the H.M.S. *Endeavor*, British naturalist
            Joseph Banks collects 3,607 plant species, some 1,400 of which had not pre-
            viously been identified.                                                   BIO

## Carbonated Water

In addition to discovering ammonia, sulfur dioxide, and the gas now known as oxygen, 18th-century British theologian and chemist Joseph Priestley also invented carbonated water. While living near a brewery in 1768 he appropriated some of its carbon
dioxide and added it to household water. The carbon dioxide infused the water with a
tartness and fizz now associated with refreshments like soda pop (when sugar and fla-
vorings are added), quinine or tonic water (when quinine is added), or club soda or
seltzer (in its unadulterated form).

1768        British chemist Joseph Priestley dissolves carbon dioxide in water to produce carbonated water, later called seltzer or soda water and the basis of all future carbonated soft drinks.                                                        CHEM

1768        On an expedition from 1768 to 1771, British navigator James Cook charts the coasts of New Zealand and the eastern coast of New Holland. He realizes that the latter is large enough to be a continent; it comes to be called Australia.        EARTH

1768        French physicist Antoine Baumé invents the graduated hydrometer, which uses the Baumé scale for specific gravities of liquids.                       TECH

1769        This year's transit of Venus (*see* 1761, ASTRO) is observed from such far-flung places as Tahiti (visited by Captain James Cook), Siberia, and Ireland. Data from the transits of 1761 and 1769 will be used to determine the distance from Earth to the sun—and hence the scale of the solar system—with greater accuracy than ever before.                                            ASTRO

1769        Wine grapes are first planted in California.                          BIO

1769        French chemist Antoine-Laurent Lavoisier develops methods of quantitative chemistry through which he disproves the ancient Greek theory that water boiled long enough will be partially converted into sediment.             CHEM

1769        British surgeon Percivall Pott publishes treatises on tuberculosis of the spine (Pott's disease) and a particular fracture of the leg with the dislocation of the foot outward and backward (Pott's fracture).                         MED

1769        A spinning frame that can produce thread sturdy enough for apparel is invented by British manufacturer Richard Arkwright.                         TECH

1769        Scottish engineer James Watt patents the steam engine he developed in 1765; it will play a major role in the industrial revolution.           TECH

c. 1770s    Danish biologist Otto F. Müller devises a dredge to collect samples of living organisms from the ocean's floor.                              EARTH

1770        Swiss entomologist Charles Bonnet maintains his stand that the female of a species contains miniature forms of all future generations in her body, stating further that catastrophic evolution occurs occasionally, at which time these miniatures within the females evolve upward. Bonnet will later predict that this type of evolution will allow inorganic matter to live, animals to reason, and humans to become angels.                                      BIO

1770        British chemist Joseph Priestley collects and studies water-soluble gases, including ammonia, sulfur dioxide, and hydrogen chloride.               CHEM

1770        Swedish pharmacist Carl Wilhelm Scheele discovers tartaric acid.       CHEM

c. 1770    Italian physiologist Lazzaro Spallanzani experiments with artificial insemination in dogs and proves that sperm is necessary for fertilization of the ovum.   MED

1770       British inventor James Hargreaves patents the spinning jenny, which helps automate textile manufacturing.   TECH

1771       *Opera botanica*, by Swiss naturalist Konrad von Gesner, is published in Nuremburg, some 200 years after his death in 1565.   BIO

1771       British chemist Joseph Priestley discovers that plants produce a substance that supports combustion and animal breathing, a substance he will identify as oxygen three years later.   CHEM

1771       Fossils of ancient humans and extinct cave bears are found together in Germany.   PALEO

1771       British philologist Sir William Jones discovers relationships among Latin, Greek, and Sanskrit that will lead to reconstruction of Indo-European and the development of modern comparative philology.   SOC

1772       British chemist Daniel Rutherford discovers the element nitrogen.   CHEM

1772       French chemist Antoine-Laurent Lavoisier disproves Georg Ernst Stahl's 1700 theory that a loss of a combustible substance called phlogiston causes burning, discovering instead that burning or rusting objects combine with some substance in the air, later shown to be oxygen.   CHEM

1772       By burning a diamond and producing carbon dioxide, Lavoisier discovers that diamonds consist of carbon and are related to coal.   CHEM

1772       Austrian physician Franz Anton Mesmer claims that mental power (magnetism) exerts extraordinary influence on the human body. By 1775, he will be calling this healing power animal magnetism and claiming it has medicinal value. Mesmer will use this popular form of hypnotism to cure patients until the French government investigates his theory, finds against it, and causes mesmerizing to fall into disrepute.   PSYCH

1772       Rubber is named by British chemist Joseph Priestley for its ability to erase penciled errors through rubbing.   TECH

1773       Through microscopic and chemical experiments, French scientist Hilaire-Marin Rouelle discovers potassium and sodium in human and animal blood.   BIO

1773       In Virginia, the Williamsburg Eastern Lunatic Asylum, the first official American asylum for the mentally ill, is founded.   PSYCH

1774       British chemist Joseph Priestley discovers the element oxygen. Swedish chemist Carl Wilhelm Scheele had independently discovered it in 1772, but he failed to publish before Priestley and forfeited the credit for the discovery.

CHEM

1774    The same year that he is scooped by Priestley, Scheele discovers the elements chlorine, manganese, and barium—but fails to get undisputed credit for any of them. Swedish mineralogist Johan Gottlieb Gahn completes his discovery of manganese this year and gets credit for it. Barium (will be rediscovered in 1808) and chlorine (1810) will usually be credited to British chemist Humphry Davy, who will be the first to identify them as elements rather than compounds.                                                                      CHEM

1774    The Shakers, a religious sect from England, settle in the American colonies. Among their accomplishments in animal husbandry will be the breeding of the Poland China hog, a hybrid of an American domestic hog and the Big China hog. Versatile and hardy, it will become a staple of American livestock.

TECH

1775    Danish entomologist Johann Christian Fabricius classifies insects by their mouth structure.                                                                          BIO

# Diamonds and Carbon

One of the more expensive ways of producing carbon dioxide is to burn a diamond. This is precisely what French chemist Antoine-Laurent Lavoisier did in 1772. First he and some fellow chemists pooled their funds to buy a diamond; then, they heated it to a high enough temperature to burn it. The result was a container of very costly carbon dioxide, demonstrating a link between diamonds and carbon. By the end of the 18th century, British chemist Smithson Tennant and French chemist Guyton de Morveau had shown that diamonds are a form of pure carbon that can be converted to graphite. Graphite and diamond are allotropes of carbon—two forms of the same element that differ only in their crystal structures. A diamond's atoms bond to one another in a tightly packed tetrahedral arrangement, making it very dense and very hard. But the more stable form of the element is the lowly graphite, in which the atoms are arranged in loosely stacked but durable layers. Over millions of years, in fact, diamonds tend to turn into graphite..

If diamonds can be turned into graphite, can graphite be turned into diamonds? Only by applying temperatures and pressures great enough to break up the graphite and pack the carbon atoms into the dense tetrahedral shape. Chemists tried to do this all through the 19th and early 20th centuries but could not achieve the necessary conditions. Then, in 1955, scientists at General Electric succeeded in reaching pressures of 100,000 atmospheres and temperatures of 2,500° C. Using chromium as a catalyst, they succeeded in turning graphite into diamonds. In 1962, at still higher temperatures and pressures, they did the same thing without the aid of a catalyst. They had shown that Lavoisier's conversion process of 200 years earlier was reversible, though considerably easier in one direction than the other.

1775        British chemist Joseph Priestley identifies hydrochloric and sulfuric acid. CHEM

1775        British surgeon Percivall Pott gives the first clear example of occupation-relat-
            ed cancer, in chimney sweeps who develop cancer of the scrotum from pro-
            longed exposure to soot and ashes.                                        MED

1775        Italian physicist Alessandro Volta invents a device that can both generate and
            store static electricity. *See also* 1800, TECH.                          PHYS

1775        The first patent for a flush toilet is issued to British inventor Alexander
            Cummings, though such toilets will not become common until the 19th cen-
            tury.                                                                     TECH

1776        Swedish chemists Carl Wilhelm Scheele and Torbern Olof Bergman indepen-
            dently discover uric acid.                                               CHEM

1776        American physician and statesman Benjamin Rush signs the Declaration of
            Independence. He will become surgeon general of the Continental Army, on
            the staff at Pennsylvania Hospital (where he will found the first free dispen-
            sary in the United States), and be treasurer of the U.S. Mint from 1797 to 1813.
            In 1812 he will publish the first U.S. treatise on psychiatry.            MED

1776        Chemist Matthew Dobson proves that the sweetness in diabetic blood and
            urine is due to sugar and suggests that diabetes is not a kidney problem but
            rather a malfunction of metabolism and digestion.                         MED

1776        British social philosopher Adam Smith publishes *An Inquiry into the Nature and
            Causes of the Wealth of Nations*, the founding work of classical economics. In it
            Smith argues for a laissez-faire approach to the market in which individuals
            pursuing their own interests improve the condition of society as a whole.  SOC

1776        German anatomist and natural historian Johann Friedrich Blumenbach pub-
            lishes *On the Natural Varieties of Mankind*, which divides the human race into
            American Indian, Caucasian, Ethiopian, Malayan, and Mongolian branches.
            He also postulates that Caucasians were the original human race which then
            "degenerated" into the other divisions under different environmental
            demands.                                                                  SOC

1776        The submarine is first used in combat, during the American Revolution. This
            seven-foot vessel, called the *Connecticut Turtle*, was designed in 1773 by David
            Bushnell and made of wood, iron, and pitch. Driven by a hand-cranked pro-
            peller, it attempts unsuccessfully to sink British warships in New York harbor.
                                                                                      TECH

1777        French naturalist Georges-Louis de Buffon introduces what is referred to as the
            needle problem, the first example of a geometric probability.             MATH

1777        French physicist Charles-Augustin de Coulomb invents the torsion balance, a
            weight scale that relies on the force required to twist, or apply torsion to, a
            wire or fiber.                                                           TECH

1778        Swedish chemist Carl Wilhelm Scheele discovers the element molybdenum,
            though he will often lose credit for the discovery to Swedish mineralogist Peter
            Jacob Hjelm.                                                             CHEM

1779        Dutch botanist Jan Ingenhousz proves that green plants give off oxygen in the
            presence of sunlight and that their roots, flowers, and fruits exude carbon diox-
            ide in the absence of light. He also shows that plants get carbon from the
            atmosphere, not from the soil.                                          BIO

1779        Swiss geologist Horace-Bénédict de Saussure coins the term *geology* and cor-
            rectly describes the movement of glaciers.                             EARTH

1780        Italian anatomist Luigi Galvani observes that dissected frog legs twitch when
            electricity is applied and when making contact with two different metals.
            Though he is wrong in concluding that animal tissue is the source of electricity,
            his experiments will spur further research into electricity, including that of
            Alessandro Volta in 1800.                                               BIO

1780        German mineralogist Abraham G. Werner, who will come to be known as the
            father of historic geology, publishes the influential article *Kurze Klassification
            und Beschreibung de verschiedenen Gebirgsatzen*. Although he is a Neptunist (*see*
            1738, EARTH), believing that the earth's lands precipitated out of a universal
            ocean, Werner will be renowned for his careful definitions, classification of
            minerals, and application of chronology to rock formations.             EARTH

c. 1780     British physician David Pitcairn is first to note that rheumatic fever can dam-
            age the heart.                                                          MED

1780        Italian physiologist Lazzaro Spallanzani proves that digestion is a chemical
            reaction to gastric juice.                                              MED

1780        Scottish physician John Brown develops the Brownian (Brunonian) system of
            medicine, which sees all body tissues as excitable and life as the result of stim-
            uli acting on tissues. Disease, Brown believes, is caused by either excessive or
            insufficient stimulation. His preferred treatments for disease are opium and
            alcohol.                                                                MED

1780        In London, British physician James Graham establishes the Temple of Health
            and Hymen for the treatment of fertility. After paying Graham 50 guineas,
            infertile and impotent couples use the elaborate "temple" and "grand celestial
            bed" in hopes of conceiving children of unusual beauty.                 MED

1780        Geologists in central Germany study the Buntsandstein and Muschelkalk fos-
            sil beds, which date respectively from the early and middle Triassic periods. *See*
            1834, PALEO.                                                            PALEO

# The Georgian Planet

Five planets, all of them visible to the naked eye, have been known since ancient times: Mercury, Venus, Mars, Jupiter, and Saturn. By the 18th century, it was understood that Earth itself is a planet orbiting the sun, bringing the tally to six. Not until 1781 was a seventh planet discovered: Uranus.

Ironically, Uranus had been visible all along. On a clear, dark night it can be seen dimly—dimly enough that the ancients never noticed it. It took amateur astronomer William Herschel (1738–1822), using a telescope of his own invention, to discover this object. Born in Germany, Herschel was a musician by trade, working as a chapel organist in Bath, England. However, he had a taste for astronomy and a talent for optics that allowed him to build telescopes with far better resolution than those used by the professionals of the time. On a Tuesday night, March 13, 1781, watching the skies with such a telescope, he noted a "curious either nebulous Star or perhaps a Comet." By its disklike shape and characteristic orbit it was proven to be a planet, the first to be discovered since antiquity.

The name of the new planet was uncertain for some time. Herschel wanted to call it the Georgian Planet after George III, then king of Britain; for his loyalty, Herschel was appointed private astronomer to the king in 1782. Others suggested Planet Herschel, while others pressed for a name drawn from Greco-Roman mythology, in keeping with the names of the other known planets. The latter tradition finally won out and the planet was named for Uranus, the mythological father of Saturn. The precedent was maintained when the eighth and ninth planets were discovered: Neptune in 1846 and Pluto in 1930.

Following is a list of the Greco-Roman mythological associations of the names of the planets. With the exception of Uranus and Pluto, which are the names of Greek gods, and Earth, which originates in Middle English *erthe* from Old English *eorthe*, the names of the planets as we know them are those of Roman gods; their Greek counterparts appear in parentheses:

Mercury (Hermes), the messenger god, is also the god of commerce.

Venus (Aphrodite) the goddess of vegetation, is better known by her Greek associations as the goddess of love, beauty, and fertility. The planet Venus is also known as the evening star or morning star.

Earth (Gaea). In Greek mythology, Gaea is the mother of Uranus, who is also the father of Gaea's children, the Titans, the Cyclopes, and the Hecatoncheires.

Mars (Ares) is the god of war. The planet Mars is also known as the red planet.

Jupiter, also called Jove (Zeus), evolved from a sky deity influencing rain and agriculture to become the supreme god, an apt name for the largest planet of the solar system. He is a son of Saturn.

Saturn (Kronos, or Cronos, a Titan), the god of the harvests, is the son of Uranus and Gaea and the father of Jupiter, Neptune, and Pluto, among other gods.

Uranus is the son of Gaea and the father of her children, including Saturn. Uranus is associated with heaven, the first ruler of the universe.

Neptune (Poseidon), the god of the sea, is a brother of Jupiter and a son of Saturn.

Pluto, also called Hades, god of the underworld, is a son of Saturn and a brother of Jupiter and Neptune.

| | |
|---|---|
| 1780 | An automated flour mill is developed by American merchant Oliver Evans. It operates by a water-powered conveyor belt.     TECH |
| 1781 | French astronomer Charles Messier catalogs 103 nebulae, or patches of luminous cloud that later turn out to be galaxies.     ASTRO |
| 1781 | On March 13, German-English astronomer William Herschel discovers Uranus, the seventh planet from the sun and the first to be discovered solely by telescopic observation.     ASTRO |
| 1781 | French mineralogist René-Just Haüy founds the modern science of crystallography when he discovers that crystals are built on successive additions of a unit cell with a constant geometric shape, which he believes may be determined by its chemical composition.     CHEM |
| 1781 | Czechoslovakian mathematician and priest Bernhard Bolzano makes numerous discoveries that do not become widely known and are later rediscovered by others, including the arithmetization of the calculus and the recognition of pathological functions.     MATH |
| 1781 | A smallpox outbreak among Spanish settlers in Texas spreads north to Canada. More than 130,000 Native Americans die, halving the populations of the Blackfoot, Cree, Arapaho, Shoshoni, and Crow peoples.     MED |
| 1781 | Physicist Johan Carl Wilcke introduces the idea of specific heat, the quantity of heat required to raise the temperature of a given substance by a given amount.     PHYS |
| 1781 | German philosopher Immanuel Kant argues against the empiricist viewpoint that all human beings are born with equal potential and are the product of education and environment. In his *Critique of Pure Reason* Kant takes instead the nativist viewpoint, stressing that inherited characteristics and inborn intuitions frame human experience but are not dependent on it.     PSYCH |
| 1781 | After three months of construction, the first all-iron bridge is put into operation, in Shropshire, England. The 378-ton bridge spans 100 feet. The iron for the bridge was cast by Abraham Darby III, grandson of ironworks master Abraham Darby. *See* 1709, TECH.     TECH |
| 1782 | British astronomer John Goodricke suggests that the star Algol has a dark companion that circles and periodically eclipses it. *See* 1889, ASTRO.     ASTRO |
| 1782 | While investigating a badger intestine, German zoologist Johann Melchior Goeze identifies the hookworm.     BIO |
| 1782 | Scottish engineer James Watt patents a double-acting rotary steam engine, which improves significantly on his steam engine of 1769.     TECH |

1783    Spanish mineralogists Juan José and Fausto d'Elhuyar y de Suvisa discover the element tungsten.                                                                CHEM

1783    Swedish chemist Carl Wilhelm Scheele discovers the compound glycerine.                                                                                      CHEM

1783    In June, Jacques and Joseph Montgolfier of France send up the first flying balloon, filled with smoke.                                               TECH

1783    British ironworks master Henry Cort improves wrought-iron production by developing a system for puddling iron.                             TECH

1783    On November 21, Frenchmen Jean Pilâtre de Rozier and the Marquis François d'Arlandes make the first manned free-balloon flight, reaching a height of about 500 feet and traveling about 5.5 miles during their 20-minute flight.

TECH

1784    German-English astronomer William Herschel observes distorting mists or clouds on Mars.                                                            ASTRO

1784    Austrian mineralogist Franz Joseph Müller discovers the element tellurium.   CHEM

1784    English chemist Henry Cavendish discovers that water is composed of hydrogen and oxygen, from which he gives hydrogen its name (from the Greek for "water former").                                                    CHEM

1784    Swedish chemist Carl Wilhelm Scheele identifies citric acid.                    CHEM

1784    French chemist Gaspard Monge liquefies sulfur dioxide, the first substance normally known as a gas to be liquefied.                            CHEM

1784    American physician John Jeffries flies a balloon over London to collect air samples at various heights.                                                EARTH

## The First Locomotive

The first working steam locomotive did not run on rails and carried no passengers. It was a model 14 inches high and 19¼ inches long, built in England in 1784 by William Murdock, an employee of Scottish inventor James Watt. The terrier-sized steam-driven model ran through the streets at six to eight miles an hour, and, according to legend, scared the village parson half to death. A concerned Watt asked Murdoch not to conduct any more such experiments.

Twenty years later, in 1804, British inventor Richard Trevithick was the first to put a full-sized steam locomotive on rails. The result was far more influential.

| 1784 | French aeronaut Jean-Pierre-François Blanchard invents the parachute and he makes and survives the first jump.                                                    EARTH |
|------|-----------------------------------------------------------------------------------------------------------------------------------------------------------------------------|

1784    American scientist and inventor Benjamin Franklin invents bifocal lenses.    MED

1784    Italian physician Domenico Cotugno discovers cerebrospinal fluid, the water cushion that protects the brain and spinal cord from shock.    MED

1784    British physicist George Atwood determines the acceleration of a free-falling body.    PHYS

1784    Jean-Baptiste Meusnier designs the first powered balloon with a crew cranking three propellers on a single shaft to enable the elliptically shaped balloon to reach speeds of about three miles per hour.    TECH

1785    German-English astronomer William Herschel theorizes that the Milky Way is a flattened system of stars, or a galaxy. His estimates of the galaxy's diameter and thickness will prove overly conservative.    ASTRO

1785    French astronomer and mathematician Pierre Simon de Laplace publishes his *Theory on the Attraction of Spheroids and the Shape of Planets*, which contains what will become known as the Laplace equation. This partial differential equation describes electromagnetic, gravitational, and other potentials.    ASTRO

1785    British physician William Withering describes the correct use of digitalis to treat heart failure.    MED

1785    French physicist Charles-Augustin de Coulomb defines what will become known as Coulomb's law: The force between two stationary electric charges is proportional to the product of the charges and inversely proportional to the square of the distance between them.    PHYS

1785    For the first time, in Nottinghamshire, England, a textile plant is powered by steam.    TECH

1786–1802    German-English astronomer William Herschel will publish three catalogs listing 2,500 nebulae. These will become the basis for the expanded catalogue of 1864 and the *New General Catalogue* (NGC) of 1888.    ASTRO

1786    Frenchmen Michel-Gabriel Paccard and Jacques Balmat become the first to climb to the summit of Mont Blanc, the highest peak in the Alps. This feat wins them a prize and inaugurates the modern sport of mountain climbing.    EARTH

1786    The durable linen-wool fabric called linsey-woolsey is developed and popularized by American tradesmen.    TECH

1786    A nail-making machine is invented and patented by Massachusetts inventor Ezekiel Reed.    TECH

1786    The first grain thresher that works by sandwiching grain between a moving cylinder and a curved metal sheet is invented by Scottish agricultural engineer Andrew Meikle.                                                                    TECH

1786    English gunsmith Henry Nock invents a breech-loading musket.                    TECH

1787    German-English astronomer William Herschel discovers Titania and Oberon, two moons of Uranus.                                                                             ASTRO

1787    French scientists Jacques-Alexandre-César Charles (in 1787) and Joseph-Louis Gay-Lussac (in 1802) rediscover Guillaume Amontons's forgotten 1699 law that at constant pressure, all gases expand by the same amount for a given rise in temperature. This principle will later be called Charles's law or Gay-Lussac's law.                                                                                           CHEM

1787    French chemist Antoine-Laurent Lavoisier and colleagues publish *The Method of Chemical Nomenclature,* a systematic approach to naming chemical substances and processes. This system soon gains universal acceptance among chemists.                                                                                         CHEM

1787    French chemist Claude-Louis Berthollet identifies the composition of ammonia, hydrogen sulfide, and prussic acid.                                                        CHEM

1787    A large fossil bone discovered in New Jersey and reported by Caspar Wistar and Timothy Matlack (though the find goes unpublished and unverified) may be the first dinosaur bone ever collected.                                            PALEO

1787    Danish philologist Rasmus Christian Rask is born (*d.* 1832). In addition to compiling the first usable Anglo-Saxon and Icelandic grammars, he will publish important work on the relationships of the Indo-European languages.           SOC

1787    In two separate events, the first steamboats are demonstrated by American inventors, one on the Potomac River, by James Rumsey, the other on the Delaware River, by John Fitch.                                                           TECH

1787    The processes of grinding grain and sifting flour are automated by a system developed by American inventor Oliver Evans, simplifying the labor and time needed to produce bread.                                                         TECH

1788    In a posthumous publication, Swedish mineralogist Torbern Olof Bergman (*d.* 1784) presents tables of affinities marking the extent to which given chemicals interact, including predictions about reactions as yet unobserved.           CHEM

1788    French mathematician Joseph-Louis Lagrange formulates the function later called Lagrangian that expresses the difference between kinetic and potential energy for every point in an object's path.                                         PHYS

| 1788 | In his *Analytical Mechanics*, Joseph-Louis Lagrange works out general equations through which algebra and calculus, rather than geometry, can be used to solve mechanical problems. **PHYS** |
|---|---|
| 1789 | German-English astronomer William Herschel discovers the Saturnian moons Mimas and Enceladus. **ASTRO** |
| 1789 | Herschel completes the world's then-largest telescope, a reflecting telescope with a 48-inch mirror. **ASTRO** |
| 1789 | French botanist Antoine-Laurent de Jussieu is one of the first to attempt sorting and assigning plants under a "natural" classification system when he classifies them into natural families like grasses, lilies, and palms. **BIO** |
| 1789 | German chemist Martin Heinrich Klaproth discovers the elements uranium and zirconium. **CHEM** |
| 1789 | French chemist Claude-Louis Berthollet shows that, contrary to the accepted belief, not all acids contain oxygen. **CHEM** |
| 1789 | In a textbook on chemistry, Antoine-Laurent Lavoisier states the principle of conservation of mass: In a closed system, the total amount of mass remains the same regardless of physical or chemical changes. This law will eventually be revised by Einstein in his 1905 law of the conservation of mass-energy. **CHEM** |
| 1789 | Corn-based bourbon whiskey is first produced in the Kentucky territory, by a minister named Elijah Craig. **TECH** |
| 1790 | German-English astronomer William Herschel discovers planetary nebulae, shells of gas surrounding certain stars. **ASTRO** |
| 1790 | A French government commission including such scientists as Joseph-Louis Lagrange, Pierre Simon de Laplace, and Antoine-Laurent Lavoisier begins to define the metric system of measurement. **MISC** |
| 1790 | British archaeologist John Frere finds stone tools and the fossil remains of extinct animals at Hoxne, Suffolk, England. **PALEO** |
| 1790 | American engineer and entrepreneur Samuel Slater (1768–1835) opens the first working American cotton mill. With ironmaster David Wilkinson, Slater builds a mill from plans used for similar plants in England. This event marks the beginning of the industrial revolution in the United States. **TECH** |
| 1791 | British minister William Gregor discovers the element titanium. **CHEM** |
| 1791 | German chemist Jeremias Richter defines the principle of stoichiometry, specifying the fixed relative proportions in which chemical substances react. **CHEM** |

1791   From Freiberg, German mineralogist Abraham G. Werner publishes *Neue Theorie von den Enstehung der Gänge,* dealing with the formation of ore deposits, an extension of his theory on the origin of rocks.                           EARTH

1791   French physicist Pierre Prévost shows that cold is the absence of heat and that all bodies radiate heat continuously.                           PHYS

1791   French author Donatien-Alphonse-François de Sade, better known as the Marquis de Sade, publishes the novel *Justine,* in which he describes sexual gratification derived from inflicting pain on another, a practice that comes to be known as *sadism. See also* 1874, PSYCH.                           PSYCH

1791   The French National Assembly recommends making an attempt to standardize measurements for the meter and quadrant. They suggest that the meter represent one 10-millionth part of a quadrant of the surface of the earth and the gram one cubic centimeter of water at 4° C.                           TECH

1792   The Mint of the United States opens, to produce coins based on a decimal system.                           TECH

1792   American Eli Whitney invents the cotton gin, a cylindrical machine that quickly separates cotton fibers from seeds, a task that is slow and laborious by hand. The new invention vastly increases cotton production and increases the South's dependence on slaves to pick the cotton to be processed by the gin.                           TECH

1793   German botanist Christian Sprengel (1750–1816) describes the plant pollination process, emphasizing the influence of winds and insects in cross-pollination.                           BIO

1793   Irish barrister and scientist Richard Kirwan attacks Scottish geologist James Hutton's uniformitarian (*see* 1795, EARTH) and vulcanist ideas. Vulcanists, who also included Scottish philosopher William Rowan Hamilton (1788–1856) and French geologist Nicolas Desmarest (*see* 1765, EARTH), believed that basaltic rocks were the product of volcanic lava flow.                           EARTH

1794   German physicist Ernst Chladni argues that meteorites are extraterrestrial in origin. *See also* 1803, ASTRO.                           ASTRO

1794   Finnish chemist Johan Gadolin discovers yttrium, the first known rare earth element. It will not be completely isolated from other elements until Swedish chemist Carl Gustaf Mosander does so about 1843.                           CHEM

1794   British chemist Elizabeth Fulhame publishes her *Essay on Combustion,* in which she develops a theory of combustion as a process combining oxygenation (combination with oxygen) and reduction (restoration of oxygenated bodies). CHEM

1794   French mathematician Adrien-Marie Legendre publishes his highly influential textbook *Elements of Geometry.*                           MATH

c. 1794    Italian surgeon and anatomist Antonio Scarpa describes certain vital parts of the human anatomy which will be named for him: Scarpa's fascia, Scarpa's fluid, Scarpa's femoral triangle, and Scarpa's membrane and ganglion.    MED

1794    In the first military use of a balloon, Frenchman Jean-Marie Coutelle makes two observation flights over an enemy's camp.    TECH

1794    In Pennsylvania the Lancaster Road, a toll road, opens to join Lancaster and its surrounding areas with the Philadelphia area.    TECH

1795    German poet Johann Wolfgang von Goethe claims two archetype plans exist for living beings. One archetype is for the animal world and one is for the plant world.    BIO

1795    In his *Theory of the Earth* Scottish naturalist James Hutton elaborates the "uniformitarian principle" that geological processes, such as erosion, work at a more or less uniform rate. The principle implies that the earth is much older than previously believed.    EARTH

1795    Between now and 1797, Mungo Park becomes the first European to explore the Niger River.    EARTH

1795    French mathematician Gaspard Monge publishes *Feuilles d'analyse*. In this and his 1802 memoir with Jean-N.-P. Hachette, he systematizes solid analytic geometry and elementary differential geometry in what approximates their present state. These works include the two Monge theorems.    MATH

1795    German mathematician Carl Friedrich Gauss discovers, independently of Leonhard Euler, the law of quadratic reciprocity in number theory.    MATH

c. 1795    German-Austrian physician Franz Joseph Gall begins writing on phrenology, claiming that personality can be judged by physical appearances, especially skull characteristics, and argues that qualities such as honesty and depravity are directly associated with bumps and ridges of the skull over specific brain regions.    PSYCH

1795    Sensory psychophysiologist Ernst Heinrich Weber is born in Wittenberg, Germany (*d.* 1878). He will be the first to study touch and kinesthesis, in elaborate experiments. He will also discover a major psychophysical principle, which Gustav Fechner will identify as Weber's law, that the just-noticeable increment in stimulus intensity is a constant fraction of the intensity already present.    PSYCH

1795    The Springfield flintlock musket is developed and named the first official piece of U.S. weaponry. The musket derives its name from its town of origin, Springfield, Massachusetts, the site of the first American arsenal in 1794.    TECH

1796    French astronomer Pierre Simon de Laplace publishes his "nebular hypothesis" theory that the solar system formed by condensation from a cloud of gas.
ASTRO

1796    German-born Russian chemist J. T. Lowitz isolates pure ethyl alcohol.    CHEM

1796    German mathematician Carl Friedrich Gauss invents a method for constructing a heptadecagon (a polygon with 17 sides of equal lengths) with compass and straightedge, and shows that an equilateral heptagon (a polygon with 7 equal sides) could not be built the same way. His discoveries, which mark the first notable advance in geometry since ancient Greece, show the value in mathematics of proving impossibility.    MATH

1796    British physician Edward Jenner inoculates the arm of eight-year-old James Phipp with pus from a cowpox sore on a milkmaid's arm. The boy develops a similar sore, but does not get sick. Then two months later Jenner inoculates Phipp with smallpox, proving that the mild cowpox infection had protected him from smallpox. By 1823 Jenner's vaccine will be widely used. *See also* 1716 and 1967, MED.    MED

1797    German astronomer Heinrich Wilhelm Olbers devises new methods for calculating the parabolic orbits of comets.    ASTRO

1797    French naturalist Georges Dagobert, Baron Cuvier, publishes his *Tableau élémentaire de l'histoire naturelle des animaux*, the founding work of comparative anatomy.    BIO

1797    Baron Cuvier adopts the term *phylum* (from the Greek for *tribe*) for a taxonomic category more general than the class but more specific than the kingdom. Phylums represent the basic body plans of organisms.    BIO

1797    French chemist Louis-Nicolas Vauquelin discovers the element chromium.    CHEM

1797    Scottish geologist Sir James Hall pioneers work in high-pressure, high-temperature mineralogy (later called experimental petrology).    EARTH

1797    British scientist William Hyde Wollaston links the cause and effect of gout when he discovers the presence of uric acid, the end product of purine metabolism, in gouty joints.    MED

1797    While studying blood pressure, physician Jean Poiseuille formulates laws governing the passage of fluids through narrow tubes.    PHYS

1797    Cigarettes, small cigars in paper wrappers, are produced in Cuba.    TECH

1798    French astronomer Pierre Simon de Laplace proposes the existence of the objects later known as black holes.    ASTRO

1798    French chemist Louis-Nicolas Vauquelin discovers the element beryllium. CHEM

1798    French chemist Louis-Bernard Guyton de Morveau succeeds in liquefying ammonia, until then known only as a gas.                                                CHEM

1798    English chemist Henry Cavendish determines the gravitational constant, the only unknown to date in Newton's law of gravitation, and the mass and density of the earth.                                                                                     PHYS

1798    English physicist Benjamin Thompson demonstrates that heat should be understood as the increased motion of particles when heated, not as a type of fluid (caloric), as previously believed.                                             PHYS

1798    English economist Thomas Robert Malthus publishes his *Essay on the Principle of Population*, in which he argues that population tends to increase in geometric progression but food supply in arithmetic progression, so that population will tend to outstrip food supply until it is reduced by famine, disease, or war.   SOC

1798    American inventor Eli Whitney refines mass production through the development of jigs, metal patterns that allow for consistent duplication of parts. This development will help to form the American system of mass production.   TECH

1798    The printing process of lithography is developed by Bavarian printer Aloys Senefelder. Operating through the incompatibility of oil and water, this technique uses oil-based ink to print images, while paper not meant to accept images is water treated, thus rejecting the ink.                     TECH

1799    A French soldier in Napoleon's army invading Egypt discovers an inscribed black stone near the town of Rosetta. The inscription on this, the Rosetta Stone, is found to date to 197 B.C. and is written in three languages: Greek and two forms of Egyptian hieroglyphics. The stone enables scholars (*see* 1822, ARCH) to learn to read ancient Egyptian texts.                          ARCH

1799    German scientist Alexander von Humboldt observes the Leonid meteor shower, a periodic event that will be found to be associated with a comet.        ASTRO

1799–1825   French astronomer Pierre Simon de Laplace's five-volume work *Celestial Mechanics* is published, summarizing and extending current knowledge. Laplace shows that the solar system is stable, despite periodic perturbations.   ASTRO

1799    French chemist Joseph-Louis Proust articulates what will become known as Proust's law or the law of definite (constant) proportions: The proportions of the elements in a compound are always the same, no matter how the compound is made.                                                                             CHEM

1799    The British Mineralogical Society is founded.                          EARTH

1799    In his dissertation at Helmstedt, German mathematician Carl Friedrich Gauss gives the first rigorous proof of what he calls the "fundamental theorem of algebra."                                                                                   MATH

1799        English geologist William Smith suggests that rock strata can be identified by
            their characteristic fossils. This system of classification will become basic to
            paleontology. *See also* 1815, PALEO.                                    PALEO

1799        In Siberia a mammoth is found preserved in ice.                          PALEO

# The Geologic Time Chart

In 1799, English geologist William Smith noted that different strata, or layers, of rock
had their own characteristic fossils, which turned up in those layers and nowhere else.
He suggested that the various strata of rock, even when bent and interrupted by geologic pressures, could be identified by their typical fossils. This insight, coupled with
the inference that higher strata are more recent and lower strata older, meant that the
layers of the earth's crust offered a guide to the history of life. Paleontologists have
since named the intervals of prehistoric time, with their contemporaneous life forms,
for the rock strata associated with them.

In the following chart, the intervals of geologic time are arranged in descending
order from most recent (top) to earliest (bottom). The columns from left to right indicate increasingly subordinate units of time. Thus, the Pleistocene epoch is one stage of
the Quaternary period, which in turn is a phase of the Cenozoic era. The dates in the
column at the far right indicate (in millions of years) how long ago the interval began.
These intervals are not of equal length. The earliest stage, the Precambrian era, lasted
about 4 billion years, more than 85 percent of the earth's history to date.

| Era | Period | Epoch | Began (millions of years ago) |
|---|---|---|---|
| Cenozoic | Quaternary | Holocene | .01 |
|  |  | Pleistocene | 1.8 |
|  | Tertiary | Pliocene | 7 |
|  |  | Miocene | 23 |
|  |  | Oligocene | 38 |
|  |  | Eocene | 53 |
|  |  | Paleocene | 65 |
| Mesozoic | Cretaceous |  | 136 |
|  | Jurassic |  | 190 |
|  | Triassic |  | 225 |
| Paleozoic | Permian |  | 280 |
|  | Carboniferous |  | 345 |
|  | Devonian |  | 395 |
|  | Silurian |  | 440 |
|  | Ordovician |  | 500 |
|  | Cambrian |  | 570 |
| Precambrian |  |  | 4,600 |

1799        British paleontologist Mary Anning is born (*d.* 1847). She will become a professional fossil collector whose dramatic finds will create a sensation in London geological circles. Among her finds will be remains of an ichthyosaur (1811), a plesiosaur, and a pterodactyl.                    PALEO

1799        French chemist Philippe Lebon pioneers the theory and practice of gas lighting, using flammable gas derived from wood.                    TECH

1799        The first suspension bridge using iron chains for support is built by American engineer James Finley.                    TECH

1800        German scientist Karl Friedrich Burdach introduces the word *biology* for the study of the morphology, physiology, and psychology of humans. In 1882 a broader definition will be proposed by Gottfried Treviranus and Jean-Baptiste de Lamarck to include the study of life in general.                    BIO

1800        French biologist Marie-François-Xavier Bichat publishes a treatise on membranes in which he classifies tissues into 21 types.                    BIO

1800        English chemist William Nicholson constructs his own electric battery and electrolyzes water, breaking it into its components hydrogen and oxygen. CHEM

## Who's in a Name

Italian physicist Alessandro Volta, inventor of the voltaic cell (1800), is one of a select group of scientists whose names have been memorialized in units of measurement. Here are some other members of that group.

| Scientist | Unit (abbreviation) | Physical Quantity |
|---|---|---|
| André-Marie Ampère (1775–1836) | ampere (A) | Electric current |
| A. J. Ångstrom (1814–1874) | angstrom (Å) | Small distance |
| Antoine-Henri Becquerel (1852–1908) | becquerel (Bq) | Radioactivity |
| Charles-Augustin de Coulomb (1736–1806) | coulomb (C) | Electric charge |
| Heinrich Hertz (1857–1894) | hertz (Hz) | Frequency |
| James P. Joule (1818–1889) | joule (J) | Energy |
| Isaac Newton (1642–1727) | newton (N) | Force |
| Georg Simon Ohm (1789–1854) | ohm (Ω) | Electric resistance |
| Blaise Pascal (1623-1662) | pascal (Pa) | Pressure |
| Alessandro Volta (1745–1827) | volt (V) | Electric potential difference |
| James Watt (1736–1819) | watt (W) | Power |

| | |
|---|---|
| 1800 | German physicist Johann Wilhelm Ritter invents electroplating when he passes a current through a copper sulfate solution. He also discovers that water consists of two parts hydrogen to one part oxygen.  CHEM |
| 1800 | British chemist Humphry Davy discovers nitrous oxide and its intoxicating effects. As laughing gas it will become the first chemical anesthetic.  CHEM |
| 1800 | British scientist William Hyde Wollaston develops a method for making platinum malleable.  CHEM |
| 1800 | Robert Fulton, who will gain fame as the constructor of the first successful steamship, is the first to use metal to build a submarine.  EARTH |
| 1800 | German physician Johann A. W. Hedenus is the first to perform a thyroidectomy (thyroid removal) for goiter treatment.  MED |
| 1800 | French and English chemists learn to purify water by chlorination.  MED |
| c. 1800 | In Germany the Keuper fossil beds, dating from the late Triassic period, are described.  PALEO |
| 1800 | In the Connecticut Valley, Pliny Moody finds fossil footprints of Triassic dinosaurs. For many years they will be thought to belong to extinct birds. *See also* 1848 and 1915, PALEO.  PALEO |
| 1800 | Studying the spectrum of sunlight, German-English astronomer William Herschel discovers infrared radiation, the first known form of radiation other than visible light.  PHYS |
| c. 1800 | German physician Ferdinand Autenreith invents the "padded room" for use with mental patients.  PSYCH |
| 1800 | Italian physicist Alessandro Volta invents the voltaic cell, or electric battery, consisting of alternating disks of copper and zinc and of cardboard soaked in a salt solution.  TECH |
| 1801 | On January 1 Italian astronomer Giuseppe Piazzi discovers Ceres, the first known asteroid. German mathematician Carl Friedrich Gauss soon calculates its orbit.  ASTRO |
| 1801 | French naturalist Jean-Baptiste de Lamarck begins to classify invertebrates, constructing such divisions as crustaceans and echinoderms. *See also* 1815, BIO.  BIO |
| 1801 | English chemist Charles Hatchett discovers the element niobium, which he calls columbium.  CHEM |

1801        German mathematician Carl Friedrich Gauss's *Disquisitiones arithmeticae* repre-
            sents the beginning of modern number theory. It contains the theory of quadratic
            congruences, forms, and residues.                                        MATH

1801        English physician and physicist Thomas Young gives the first description of
            visual astigmatism, or eye surface discrepancy.                          MED

1801        Noting the phenomenon of diffraction of light (*see* 1665, PHYS), and interfer-
            ence patterns, English physician and physicist Thomas Young argues that light
            is wavelike rather than particlelike. He speculates that the waves are longitudi-
            nal, like sound, rather than transverse, like water. Seventeen years later he will
            be proven wrong. *See* 1818, PHYS.                                        PHYS

1801        German physicist Johann Wilhelm Ritter discovers ultraviolet radiation.    PHYS

1801        British chemist and physicist William Henry discovers what will become
            known as Henry's law, that the mass of gas dissolved in a liquid at equilibri-
            um and constant temperature is proportional to the pressure of the gas.    PHYS

1801        The usual treatment for the mentally ill in asylums includes bloodletting,
            bathing, and purging.                                                    PSYCH

1801        French physician and humanitarian Philippe Pinel strives to make the treat-
            ment of the mentally ill more humane.                                    PSYCH

1801        Frenchman Joseph-Marie Jacquard invents what will become known as the
            Jacquard loom, which uses punched cards to guide needle motions so as to
            produce patterned textiles. Later in the century (*see* 1834, TECH), Charles
            Babbage will realize that punched cards can control calculator processes,
            preparing the way for the digital computer in the 20th century.           TECH

1802        German archaeologist and philologist Georg F. Grotefend deciphers
            cuneiform, an ancient Mesopotamian script, but his work will fail to attract
            attention. *See also* 1846, ARCH.                                        ARCH

1802        German-English astronomer William Herschel discovers that the two brightest
            stars in Castor are a binary star system.                                ASTRO

1802        German astronomer Heinrich Wilhelm Olbers discovers Pallas, the second
            asteroid to be identified.                                               ASTRO

1802        Swedish chemist Anders G. Ekeberg discovers the element tantalum.         CHEM

c. 1802     Scottish physician Sir Charles Bell diagnoses what will become known as Bell's
            palsy, a one-sided facial weakness or paralysis caused by compression of the
            facial nerve.                                                            MED

1802        French scientist Joseph-Louis Gay-Lussac demonstrates that the volume of a
            fixed mass of gas at constant pressure changes by a constant fraction for each

degree of temperature change. First noted by Guillaume Amontons in 1699, this is known as Charles's law (*see* 1787, CHEM) or Gay-Lussac's law. *See also* 1808, PHYS.                                                                                    PHYS

1802        Italian scientist Gian Domenico Romagnosi discovers that electricity passing through a wire will cause a magnetic needle to orient itself perpendicular to the wire.                                                                                      TECH

1802        In a precursor to photography, British physician Thomas Wedgwood produces the first of what will later be called photograms when he places objects on paper treated with silver nitrate and exposes the paper to sunlight. The paper darkens, leaving the areas covered by objects white, but the white images darken quickly after the objects are removed. *See also* 1839, TECH.        TECH

1802        The first icebox, the progenitor of the refrigerator, is developed by American farmer Thomas Moore. It consists of two wooden boxes, one within the other, with the space between them insulated by charcoal or ashes. A third box, made of tin, sits atop the smaller box.                                                TECH

1803        On April 26 witnesses at Orne, France, observe a bolide, or meteor in the form of an explosive fireball. Meteoritic stones are then found in the area. Physicist Jean-Baptiste Biot describes the event and determines that the meteorites did not originate on Earth. *See also* 1794, ASTRO.                             ASTRO

1803        Swedish chemist Jöns Jakob Berzelius, Swedish mineralogist Wilhelm Hisinger, and German chemist Martin Heinrich Klaproth discover the element cerium.                                                                                  CHEM

1803        British chemist Smithson Tennant discovers the element osmium.        CHEM

1803        British scientist William Hyde Wollaston discovers the elements palladium and rhodium.                                                                               CHEM

1803        In his *Essay on Static Chemistry*, French chemist Claude-Louis Berthollet shows that reaction rates depend on the quantities of the reacting substances as well as their affinities.                                                              CHEM

1803        British meteorologist Luke Howard coins names for types of clouds, including cumulus, cirrus, nimbus, and stratus.                                            EARTH

1803        French mathematician Lazare Carnot publishes *Géométrie de position*, a work establishing him as cofounder (with Gaspard Monge; *see* 1795, MATH) of modern pure geometry.                                                                      MATH

1803        American physician James Conrad Otto describes the first clear account of hemophilia, a blood defect characterized by an inability to clot promptly and difficulty in controlling hemorrhaging.                                           MED

| | |
|---|---|
| 1803 | English chemist John Dalton draws on existing evidence to argue that matter is made up of tiny particles called atoms, a word coined by Democritus in 440 B.C. He establishes the concept of atomic weight—different masses for the atoms of different elements—though his preliminary table of atomic weights contains many errors. *See* 1818, CHEM. **PHYS** |
| 1803 | American engineer Robert Fulton builds his first steam-powered ship. **TECH** |
| 1804 | Swiss botanist Nicolas-Théodore de Saussure stresses the importance of soil nitrogen and carbon dioxide to green plants. He also demonstrates plants' capacity to absorb water. **BIO** |
| 1804 | British chemist Smithson Tennant discovers the element iridium. **CHEM** |
| 1804 | In the first high-altitude research flight, French scientists Jean-Baptiste Biot and Joseph-Louis Gay-Lussac ascend four miles in a balloon to study the atmosphere and the earth's magnetic field. **EARTH** |
| 1804–1806 | Americans Meriwether Lewis and William Clark explore the region known as the Louisiana Purchase, from the Mississippi River to the Rocky Mountains, and push on west to the Pacific Ocean. **EARTH** |
| 1804 | Italian physician Antonio Scarpa describes arteriosclerosis as lesions coating the lining of arteries. **MED** |
| 1804 | French surgeon René Laënnec describes peritonitis, an inflammation of the membranes lining the abdomen. **MED** |
| 1804 | The first steam locomotive to operate on a railroad, constructed by British inventor Richard Trevithick, travels nearly 10 miles and achieves speeds of almost 5 miles per hour. **TECH** |
| 1804 | The first foods to be vacuum packed are packaged by Nicolas Appert at his factory in France. **TECH** |
| 1805 | After 20 years of analyzing the proper motion of stars, German-English astronomer William Herschel shows that the sun, together with its planets, is moving through the galaxy. **ASTRO** |
| 1805 | German scientist Alexander von Humboldt publishes his *Essai sur la géographie des plantes*, about his five-year voyage to the Americas, thus introducing plant geography. **BIO** |
| 1805 | On November 15 the Lewis and Clark expedition (*see* 1804–1806, EARTH) reaches the Pacific Ocean, completing the first overland trip by settlers across what will become the United States of America. **EARTH** |
| 1805 | German mathematician Peter Gustav Lejeune Dirichlet is born (*d.* 1859). He will show the value of applying analysis to number theory and |

introduce Dirichlet's theorem and the Dirichlet series. He will also contribute the Dirichlet principle to the calculus of variations.    MATH

1805     German pharmacist Friedrich Sertürner extracts morphine from opium and uses it to relieve pain.    MED

1805     French scientist Pierre Simon de Laplace proposes his theory of capillary forces.    PHYS

1805     German naturalist Johann Friedrich Blumenbach founds the science of physical anthropology.    SOC

1805     The first covered bridge in America opens, over the Schuylkill River in Pennsylvania.    TECH

1806     Italian botanist Giovanni Amici determines the importance of intercellular space for gas conduction in plants.    BIO

1806     French chemist Louis-Nicolas Vauquelin discovers the compound asparagine, which will turn out to be the first known amino acid.    CHEM

1806     Scotsman Patrick Clark develops a cotton thread that rivals silk thread in strength. With his brother James, Clark opens a thread-making factory. Eventually, cotton thread will become the thread of choice.    TECH

1807     British chemist Humphry Davy discovers the elements sodium and potassium. CHEM

1807     Swedish chemist Jöns Jakob Berzelius distinguishes organic from inorganic compounds, defining organic compounds as those obtained from living or dead organisms or related to such compounds and inorganic compounds as all others.    CHEM

1807     Humphry Davy develops the science of electrochemistry, following the discovery in 1800 by William Nicholson that an electric current can decompose water into oxygen and hydrogen.    CHEM

1807     Swiss geodesist Ferdinand R. Hassler convinces American president Thomas Jefferson and treasury secretary Albert Gallatin to establish a Survey of the Coast.    EARTH

1807     The Geological Society of London is founded on November 13. This scientific organization will maintain a library and study collection, discuss and disseminate geological observations, and adopt a single nomenclature.    EARTH

1807     French mathematician Jean-Baptiste-Joseph Fourier demonstrates what will become known as Fourier's theorem, that any periodic oscillation can be expressed as a series of trigonometric functions later known as a Fourier series. *See also* 1822, PHYS.    MATH

1807    English physician and physicist Thomas Young introduces the physical concept of energy.                                                                                        PHYS

1807    Thomas Young develops the coefficients of elasticity of materials.                PHYS

1807    The 133-foot S.S. *Clermont* begins regular cargo service between New York City and Albany, becoming the first commercially successful steamship. Designed by Robert Fulton, it uses a Boulton and Watt engine and paddlewheels, which will become a distinguishing feature of steamboats for years to come.        TECH

1808    British chemist Humphry Davy discovers the elements barium, calcium, magnesium, and strontium.                                                                            CHEM

1808    Davy and French scientists Joseph-Louis Gay-Lussac and Louis-Jacques Thénard discover the element boron.                                                            CHEM

1808    French scientist Joseph-Louis Gay-Lussac shows that when gases combine chemically, the volume of the reactants and of the product (if gaseous) bear simple relationships to one another, given constant temperature and pressure. This principle is the one usually known as Gay-Lussac's law, though his discovery in 1803 concerning the relationship of volume and temperature change in gases sometimes also goes by the same name.            PHYS

1808    Studying sunlight that has passed through a crystal of Icelandic feldspar (*see* 1669, PHYS), French physicist Étienne-Louis Malus (1775–1812) discovers the polarization of light.                                                            PHYS

1808    English chemist John Dalton publishes his *New System of Chemical Philosophy*, formalizing arguments for atomic theory he advanced as early as 1803.    PHYS

1809    French zoologist Jean-Baptiste de Lamarck introduces the first scientific theory on how evolution occurs, later known as Lamarckism. According to him, offspring can inherit characteristics or traits that their parents have acquired during the parents' lifetimes. This concept will later be discredited by British naturalist Charles Darwin and German biologist August Weismann (*see* 1892, BIO).                                                                              BIO

1809    British naturalist Charles Darwin is born (*d.* 1882). In his 1859 book *The Origin of Species*, he will explain the theory of evolution by natural selection, which will revolutionize the life sciences and humanity's understanding of its own origins. He will acquire much of his evidence during his voyage on the H.M.S. *Beagle* (1831–1836). He will also discuss human evolution (in *The Descent of Man*, 1871) and write on the origin of coral reefs.      BIO

1809    American physician Ephraim McDowell successfully, without anesthesia, removes a 22-pound ovarian tumor. He will be considered the father of ovariotomy.                                                                               MED

| | |
|---|---|
| 1809 | The first ocean voyage by steamboat is completed, by American seaman Moses Rogers traveling from New York City to the Delaware River by way of Cape May, New Jersey. <div align="right">TECH</div> |
| 1810s | The term *morphology* is introduced to describe the study of organic form. It particularly concerns the unity underlying plant and animal diversity. German naturalist and poet Johann Wolfgang von Goethe is among those to popularize the term. <div align="right">BIO</div> |
| c. 1810s | French mathematician Sophie Germain formulates Germain's theorem. A step toward proving Fermat's Last Theorem (*see* 1637, MATH), it shows the impossibility of positive integral solutions to $x^n + y^n = z^n$ if $x$, $y$, and $z$ are prime to each other and to $n$, where $n$ is any prime less than 100. <div align="right">MATH</div> |
| 1810 | British chemist Humphry Davy is the first to identify chlorine as an element. *See also* 1774, CHEM. <div align="right">CHEM</div> |
| 1810 | French mathematician Joseph-Diez Gergonne begins publication of the first private mathematical journal, *Annales de mathématiques pures et appliquées*.MATH |
| 1810 | Between now and 1819, Austrian physician Franz Joseph Gall and his student Johann Spurzheim publish a five-volume work on phrenology, which attempts to develop a perfect knowledge of human nature based on measurements of multiple species' skulls and brains. They theorize that brain functions are localized in the cerebral cortex and that the brain is a bundle of individual organs governing the moral, sexual, and intellectual traits of human behavior. <div align="right">PSYCH</div> |
| 1810 | The differential gear, used to steer and turn carriages, is developed by a German publisher, Rudolph Ackermann. <div align="right">TECH</div> |
| 1810 | British inventor Peter Durand is issued a patent for an early version of a tin-plated receptacle that proves to be the precursor to the can. <div align="right">TECH</div> |
| 1811 | German-English astronomer William Herschel theorizes that stars originate in nebulae. <div align="right">ASTRO</div> |
| 1811 | French chemist Bernard Courtois discovers the element iodine. <div align="right">CHEM</div> |
| 1811 | Massachusetts General Hospital is established in Boston. <div align="right">MED</div> |
| 1811 | Scottish physician Charles Bell discovers spinal nerve functions, which convey impulses to and from the spinal cord. <div align="right">MED</div> |
| 1811 | British paleontologist Mary Anning discovers the first complete ichthyosaur skeleton. <div align="right">PALEO</div> |
| 1811 | Italian physicist Amedeo Avogadro formulates what will become known as Avogadro's hypothesis, or Avogadro's law, that equal volumes of all gases contain equal numbers of particles at the same temperature and pressure. The |

term he uses to describe these particles, which he believes to be combinations of atoms, is *molecules*.                                                                                    PHYS

1811          Scottish physicist Sir David Brewster discovers what will become known as Brewster's law, which says that the extent of the polarization of light reflected from a transparent surface is greatest when the reflected ray is at right angles to the refracted ray.                                                                          PHYS

1811          The S.S. *New Orleans* becomes the first steamship to traverse the Mississippi River Valley, on a 14-day trip from Pittsburgh to New Orleans.                       TECH

1811          New York City planning officials adopt the so-called Commissioners' Plan, requiring streets north of Union Square to be aligned by a grid system, as in older Spanish cities.                                                                       TECH

1812          French botanist Augustin-Pyrame de Candolle is the first to use the word *taxonomy* for classification of species.                                                       BIO

1812          Boiling starch in water and sulfuric acid, German-Russian chemist Gottlieb Sigismund Constantin Kirchhoff discovers glucose. He notes that starch is made up of glucose units and that the sulfuric acid was not consumed in the chemical reaction but made the reaction possible. Swedish chemist Jöns Jakob Berzelius will call the latter phenomenon *catalysis*.                                   CHEM

1812          French scientist and mathematician Pierre Simon de Laplace publishes significant work on analytic probability theory.                                                       MATH

1812          French mathematician Siméon-Denis Poisson discovers what will become known as the Poisson equation in probability theory.                                        MATH

1812          French mathematician Jean-Victor Poncelet formulates the principle of continuity, or the principle of permanence of mathematical relations, although it is not published until 1862–1864.                                                              MATH

1812          New York City's Bellevue Hospital is formally established. *See also* 1736, MED.
                                                                                              MED

1812          French anatomist Baron Cuvier (*see* 1797, BIO) discovers the first known fossil remains of a pterodactyl, a flying reptile. He also publishes his *Inquiry into Fossil Remains*, the first major work of paleontology. Although he describes extinct forms of life in detail and classifies them according to the Linnaean scheme (*see* 1735, BIO), he does not consider them ancestors of living things but products of earlier, separate creations, each ending catastrophically. This view will become known as catastrophism.                                       PALEO

1812          French scientist Pierre Simon de Laplace posits a mechanistic model of the universe—one in which the entire history of the universe could theoretically be derived by knowing the mass, position, and velocity of every particle. This idea will prevail until Werner Heisenberg's uncertainty principle (*see* 1927, PHYS) articulates the limits of human knowledge of the physical universe.  PHYS

1812        English chemist and physicist William Hyde Wollaston invents the camera lucida, a device that projects the image of an object onto a flat surface, such as a piece of paper, where a drawing of it can be traced.          TECH

1813        Swedish chemist Jöns Jakob Berzelius proposes a system of chemical symbols that will be widely adopted: capital letters for atoms—with lowercase letters to distinguish elements such as Ca (Calcium) from Cl (Chlorine)—and, for compounds, combinations of these symbols with appropriate subscript numerals representing the number of atoms in molecules (e.g., $H_2O$ for water).          CHEM

1813        London author Robert Bakewell publishes his influential English geology textbook *Introduction to Geology*. It basically adheres to German mineralogist Abraham G. Werner's ideas (*see* 1791, EARTH) but reveals some of Scottish geologist James Hutton's influence (*see* 1795, EARTH).          EARTH

1813        German physician Johann Peter Frank publishes the first comprehensive treatise on public health. His credo is "to prevent evils through wise ordinances." Frank's suggestions, including food inspection, sanitation, prenatal care, and vital statistics on infectious diseases, are thought to be radical and are not accepted in many areas of Europe.          MED

1813        American physician Thomas Sutton distinguishes between hallucinations caused by organic brain inflammation and delirium tremens, caused by alcohol addiction and withdrawal.          PSYCH

1814        While observing sunlight that has passed through a slit and then a prism, German physicist Joseph von Fraunhofer discovers spectral lines. He maps many of the lines produced by sunlight (later called Fraunhofer lines).          ASTRO

1814        French scientists J. J. Colin and H. G. de Claubry discover the starch-iodine reaction. French scientist F. V. Raspail applies it microscopically, founding the study of cells' chemical components, histochemistry.          BIO

1814        Swedish chemist Jöns Jakob Berzelius theorizes that compounds consist of positively charged and negatively charged components.          CHEM

1814        A gas lighting works is erected in London. In it the gas filters out from a central location to local gas mains.          TECH

1814        The first steam locomotive, developed by George Stephenson, to transport coal through England, inaugurates the mass use of railroad transportation. The design is based on the 1804 one of Richard Trevithick.          TECH

1814        The first commercially sold tinned food is produced, at the Donkin-Hall factory in England. Tinned food had earlier been produced in England but only for the military.          TECH

1815        German astronomer Heinrich Wilhelm Olbers discovers a comet, which he names for himself. He proposes that a comet's tail is a cloud of matter expelled from the nucleus.                    ASTRO

1815        Between now and 1822, French naturalist Jean-Baptiste de Lamarck becomes the first to use the terms *vertebrate* and *invertebrate*. With his seven-volume work, the *Natural History of Invertebrates*, Lamarck founds modern invertebrate biology.                    BIO

1815        French physicist Jean-Baptiste Biot discovers that the plane of polarized light is twisted in different directions by different organic liquids. He speculates that this difference is caused by asymmetry in organic molecules.                    CHEM

1815        French scientist Joseph-Louis Gay-Lussac discovers the poisonous gas cyanogen ($C_2N_2$), noting the extreme stability of the cyano group, the carbon-nitrogen combination, which is the first known organic radical.                    CHEM

1815        British chemist William Prout advances the hypothesis that hydrogen is the fundamental atom out of which all other atoms are made. Prout's hypothesis will appear to be disproven by the subsequent discovery of atomic weights that are not exact multiples of hydrogen's, though the later discovery of thermonuclear fusion will show that Prout was partly right.                    CHEM

1815        British mathematicians Charles Babbage, John Herschel, and George Peacock found the Analytic Society to emphasize the abstract nature of algebra and bring new continental developments in mathematics to Britain.                    MATH

1815        In the final year of the Napoleonic wars (1803–1815), French surgeon Dominique-Jean Larrey introduces "flying ambulances," man-powered litters to pick up the wounded during battle instead of leaving them unattended until the fighting ends.                    MED

1815        English geologist William Smith publishes *The Geological Map of England*, the first book to classify rock strata on the basis of characteristic fossils. *See also* 1799, PALEO.                    PALEO

1815        French physicist Baron Augustin Louis Cauchy develops mathematical formulas for describing turbulence.                    PHYS

1815        Macadam, a moisture-resistant road covering made of bits of stone, is developed by Scottish surveyor John McAdam.                    TECH

1816        French mathematician Sophie Germain develops a mathematical model to calculate the vibration of elastic surfaces.                    MATH

1816        French physician René Laënnec invents the modern stethoscope. He uses the term *rales* for the air sound in bronchii when they contain secretions or are tightened by spasms.                    MED

1816    The celeripede, a two-wheeled forerunner of the modern bicycle, is developed by French physicist Joseph Nicéphore Niepce.    TECH

1816    French physicist Joseph Nicéphore Niepce produces the first photographic negative, on paper. The term *negative* will not be coined until 1840 (*see* 1840, TECH). *See also* 1839, TECH.    TECH

1817    French chemists Pierre-Joseph Pelletier and Joseph-Bienaimé Caventou isolate and name chlorophyll.    BIO

1817    Hereford cattle are imported to America for the first time. The hardy breed will become the most popular cattle in the western United States.    BIO

1817    German chemist Friedrich Strohmeyer discovers the element cadmium.    CHEM

1817    Swedish chemist Johan August Arfwedson discovers the element lithium.    CHEM

1817    British physician James Parkinson describes the shaking palsy that will come to be called Parkinson's disease; it is characterized by a slowly spreading fine muscle tremor.    MED

1817    Between now and 1818, American dentist Anthony Plantson introduces the dental plate.    MED

1817    In *The Principles of Political Economy and Taxation*, British economist David Ricardo advances influential theories on the determination of wages and value.    SOC

1818    French astronomer Jean-Louis Pons discovers the comet with the shortest known period (3.29 years), later called Encke's comet for German astronomer Johann Encke, who will analyze its orbit.    ASTRO

1818    Swedish chemist Jöns Jakob Berzelius discovers the element selenium.    CHEM

1818    Berzelius publishes a table of atomic and molecular weights based on some 2,000 analyses of chemicals performed over 10 years. His tables are considerably more accurate than those of John Dalton. *See* 1803, PHYS.    CHEM

1818    American geologist Benjamin Silliman establishes the *American Journal of Science*, through which he will gain an international reputation.    EARTH

1818    French chemist Louis-Jacques Thénard discovers hydrogen peroxide, a mild antiseptic and germicide.    MED

1818    French chemists Pierre-Joseph Pelletier and Joseph-Bienaimé Caventou isolate strychnine.    MED

1818    Scottish physician John Cheyne and Irish physician William Stokes describe a breathing pattern often associated with impending death: periods of no breathing (apnea), followed by increased respiration.    MED

1818    In the Connecticut Valley, Solomon Ellsworth Jr. and Nathan Smith discover fossil bones of the dinosaur *Anchisaurus*, but the significance of this find is not initially understood.    PALEO

1818    French physicist Augustin-Jean Fresnel argues that light is a transverse, not longitudinal, wave, as Thomas Young (*see* 1801, PHYS) had thought.    PHYS

1819    French chemists Pierre-Louis Dulong and Alexis-Thérèse Petit show that the specific heat of an element, the amount of heat required to raise its temperature by 1° C, is inversely proportional to its atomic weight.    CHEM

1819    German chemist Eilhardt Mitscherlich states what will become known as Mitscherlich's law, or the law of isomorphism, that substances with the same crystal structure have similar chemical formulae.    CHEM

1819    Parisian author J. F. d'Aubuisson de Voisins publishes the first successful French geology textbook, which mainly follows German mineralogist Abraham G. Werner's theories (*see* 1791, EARTH).    EARTH

1819    In New Haven, Connecticut, the American Geological Society is founded.    EARTH

1819    In works published now and in 1832, French mathematican Adrien-Marie Legendre develops several important tools of analysis, including the Legendre functions, which are solutions of the Legendre differential equation.    MATH

1819    French chemists Pierre-Joseph Pelletier and Joseph-Bienaimé Caventou isolate the antimalarial substance quinine. *See also* 1666, MED, and 1944, CHEM.    MED

1819    Rome and Utica, New York, are joined in the opening of the first part of the Erie Canal. The canal was two years in construction, mainly by thousands of immigrants.    TECH

1819    The first transatlantic trip by sail/steamship is completed as the U.S.S. *Savannah* travels from Georgia to Liverpool, England.    TECH

1820s   French biologist Baron Cuvier (*see* 1797, BIO) makes a major contribution to morphology with his arrangement of animals into four groups: vertebrates, articulates, mollusks, and radiates.    BIO

1820s   French mathematician Augustin Louis Cauchy develops the foundation of the calculus as it is now generally known. Using d'Alembert's limit concept (*see* 1760s, MATH) to define the derivative of a function, he takes significant steps toward the arithmetization of analysis.    MATH

1820s          Between now and the 1840s, during the age of reform in the United States, many poorhouses (almshouses) are built to house the poor and mentally ill. They are used as clinical laboratories for medical students.                    MED

c. 1820        The second-century marble statue *Venus de Milo* is discovered on the Greek island of Melos.                    ARCH

1820           British scientists John Herschel and Charles Babbage are among the founding members of the Royal Astronomical Society.                    ASTRO

1820           The Royal Observatory at Cape Town is established to make observations in the Southern Hemisphere comparable with those made at Greenwich.                    ASTRO

1820           French naturalist Henri Braconnot uses hydrolysis to obtain glucose from materials such as linen, bark, and sawdust. He is also the first to isolate, from gelatin, the amino acid glycine.                    CHEM

1820           Twenty-one-year-old American sealer Nathaniel Brown Palmer is the first to explore the Antarctic Peninsula.                    EARTH

1820           French chemists and physicians demonstrate the value of iodine in treating goiter.                    MED

1820           The United States's first pharmacopeia is published.                    MED

1820           Danish physicist Hans Christian Oersted shows that electricity and magnetism are related, noting that a magnetic needle moves when electricity flows through a nearby wire. Experiments by French physicists André-Marie Ampère and Dominique François Arago support Oersted's finding, establishing the concept of electromagnetism.                    PHYS

1820           German physicist Joseph von Fraunhofer invents the diffraction grating, a device composed of fine, closely spaced wires used in producing light spectra.  PHYS

1820           French physicist André-Marie Ampère shows that a wire helix or solenoid acts like a bar magnet when electric current flows through it. He also states the right-hand rule for the action of an electric current on a magnet.                    PHYS

1820           French physicist Dominique François Arago shows that iron is not the only magnetic substance when he magnetizes a copper wire by passing electricity through it.                    PHYS

1820           German physicist Johann Salomo Christoph Schweigger (1779–1857) invents the galvanometer, an instrument for measuring electric current.                    TECH

1821           Austrian chemist Johann Joseph Loschmidt is born (*d.* 1895). He will develop the practice of using single lines for single bonds, double lines for double bonds, and so on. He will also calculate Avogadro's number (*see* 1865, CHEM)

and determine that the structures of aromatic compounds contain a benzene ring.                                                                           CHEM

1821    Anthropologist Gabriel de Mortillet is born in France (*d.* 1898). He will be the first to distinguish the subdivisions of the Paleolithic (Old Stone) Age, which lasted from about 2.5 million to 9,000 years ago. These periods, identified on the basis of their characteristic stone tools, will include the Acheulian and the Mousterian.                                                                         PALEO

1821    British chemist and physicist Michael Faraday is the first to speak about electromagnetism as a field in which particles generate lines of force. In an experiment with magnets and electrified wires he shows that electricity can produce motion.                                                                         PHYS

1821    Russian-German physicist Thomas Johann Seebeck notes what will become known as the Seebeck effect, which states that two different metals will generate electricity if their points of juncture are maintained at different temperatures. This is the first known example of thermoelectricity, or electricity generated by temperature difference.                                       PHYS

1822    French linguist Jean-François Champollion founds modern Egyptology when he deciphers the ancient Egyptian hieroglyphics inscribed on the Rosetta Stone (*see* 1799, ARCH).                                                              ARCH

1822    English antiquarian William Bullock recovers relics and casts of Aztec artifacts from Teotihuacán and other sites in the Valley of Mexico.                       ARCH

1822    After two centuries, the Roman Catholic church lifts its ban on the works of Copernicus, Galileo, and Kepler. *See also* 1835, ASTRO.                         ASTRO

1822    William Daniel Congbeare and William Phillips publish a textbook of stratigraphy that becomes a basic field manual for geologists. Congbeare and Phillips are the first to identify the Carboniferous period.                     EARTH

1822    Mineralogist Alexandre Brongniart and zoologist A. G. Desmarest publish the first extensive study of trilobites. Their work will influence future investigations of Paleozoic stratigraphy.                                         EARTH

1822    French mathematician Jean-Victor Poncelet begins the study of projective geometry, which analyzes the shadows cast by geometric figures, as a field separate from analytical and algebraic geometry.                                           MATH

1822    French chemist Georges Simon Serullas discovers iodoform and its antibacterial action.                                                                    MED

1822    Belgian geologist Jean-Baptiste-Julien d'Halloy identifies the Cretaceous period of earth's history.                                                          PALEO

1822    In his *Analytical Theory of Heat*, French mathematician Jean-Baptiste-Joseph Fourier, who was made a baron in 1808, applies his Fourier theorem (*see* 1807, MATH) to the study of heat flow. This work also introduces the mathematical technique of dimensional analysis, in which an equation or solution is checked by analyzing the consistency of the dimensions in which it is expressed.                                                                                          PHYS

1822    German philologist and folklorist Jakob Grimm formulates what will become known as Grimm's law, a principle of relationships in Indo-European languages dealing with shifts in consonants in the development of English and the other Germanic languages.                                                                              SOC

1822    British engineer George Stephenson builds the first iron railroad bridge in the world.                                                                                                                          TECH

1822    In England, American inventor William Church patents the first typesetting machine, which combines automatic composition with manual activity to set words and line length.                                                                                                          TECH

1822    The first lasting photograph is produced by French physicist Joseph Nicéphore Niepce. The photograph is made permanent through the use of bitumen.   TECH

1823    German astronomer Joseph von Fraunhofer invents the first lensed telescope to be mounted equatorially with a clock drive.                                                            ASTRO

1823    German chemist Johann Wolfgang Döbereiner discovers that powdered platinum is an effective catalyst for hydrogen reactions.                                                CHEM

1823    British chemist and physicist Michael Faraday systematically uses cold and pressure to liquefy gases, beginning with chlorine.                                              CHEM

1823    Exploring Nigeria overland, Scotsman Hugh Clapperton becomes the first European to sight Lake Chad.                                                                              EARTH

1823    British physicist William Sturgeon invents the electromagnet, a magnet powered by an electric current.                                                                              TECH

1823    British inventor Francis Ronalds offers his electric telegraph system to the military but is rejected.                                                                                  TECH

1823    A rubber-treated fabric is developed by Scottish scientist Charles Macintosh. It is a predecessor to the fabric used in the raincoat that will take this inventor's name.                                                                                                              TECH

c. 1824    Swedish chemist Jöns Jakob Berzelius discovers the element silicon.   CHEM

1824    Berzelius and French scientist Joseph-Louis Gay-Lussac identify the first known isomers, chemical compounds with the same molecular formulas but different molecular structures or arrangements of atoms, resulting in different properties. The formulas of the compounds are determined by German chemists Justus von Liebig and Friedrich Wöhler. *See also* 1830, CHEM.      CHEM

1824      Norwegian mathematician Niels Henrik Abel shows that a general algebraic solution for equations of the fifth degree (quintic equations, those involving $x^5$) is impossible. *See also* 1830, MATH.      MATH

1824      English cleric and geologist William Buckland publishes a description of the Cretaceous carnivore *Megalosaurus*, identified from fossils in Stonesfield, England. It is the first dinosaur to be discovered and described, though the term *dinosaur* will not be introduced until the time of Richard Owen (*see* 1841, PALEO).      PALEO

1824      Irish mathematician William Rowan Hamilton states the physical principle that an object or collection of objects always moves in such a way that the action has the least possible value.      PHYS

1824      French physicist Nicolas-Léonard-Sadi Carnot publishes *Reflections on the Motive Power of Fire*, the first scientific analysis of steam engine efficiency. Carnot will be considered the founder of thermodynamics, the study of the laws governing the conversion of energy from one form to another.      PHYS

1824      The Hartford (Connecticut) Retreat is founded for the humane treatment of mental illness. By the end of the century it will shift from being an innovative curative institution to a custodial one.      PSYCH

## An Iguanodon Proper

Dinosaurs have been part of popular culture since life-sized models of Mesozoic animals were first exhibited at the Crystal Palace in Sydenham, England, in 1854. Few have taken dinosaur mania closer to heart than the British town fathers who embedded a dinosaur in their civic coat of arms.

The dinosaur was the *Iguanodon*, a large, bipedal herbivore of the Cretaceous period (136–65 million years ago). The second dinosaur to be discovered (after the *Megalosaurus*, in 1824), it was identified in 1825 by English physician and amateur paleontologist Gideon Mantell on the basis of fossil teeth discovered in Tilgate Forest by his wife, Mary Ann. Then in 1834 Mantell identified a partial skeleton of the creature, which W. H. Bensted had dug out of his quarry in Maidstone, Kent. Mantell acquired the skeleton for his collection, but the town of Maidstone continued to feel a sense of ownership. In 1949 the town fathers persuaded the Royal College of Arms to let them show the dinosaur as a part of their civic shield. In the words of the official citation, the coat of arms now reads, "On the dexter side an *Iguanodon* proper Collared Gules suspended therefrom by a chain Or a scroll of Parchment."

1824        Portland cement, a resilient water-resistant compound made from a heated mixture of chalk and clay, is patented by British bricklayer Joseph Aspdin. TECH

1824        The first round barn is built, by the Shakers in Hancock, New York. Its design, with dividers that form feeding stations, will make it popular with farmers of dairy livestock.                                                                                TECH

1825        Danish physicist Hans Christian Oersted discovers the element aluminum.   CHEM

1825        British chemist and physicist Michael Faraday isolates benzene from whale oil.
                                                                                                       CHEM

1825        The full length of the 363-mile Erie Canal opens in New York State, providing a link between the Hudson River and the Great Lakes and sparking other canal projects.                                                                                            EARTH

1825        French mathematician Augustin Louis Cauchy develops complex function theory and introduces Cauchy's integral theorem with residues.                    MATH

c. 1825     French physician François Broussais (1772–1838) believes that the body's vital functions depend on inflammation or irritation. He claims that nature has no healing power and that it is necessary to stop disease by taking active measures—notably bloodletting and leeching. He prescribes applying up to 100 leeches a day to control a patient's inflammation and disease.           MED

1825        French physician Jean-Baptiste Bouillaud is the first to describe aphasia, an inability to speak or difficulty in speaking, and link it to trauma in the brain's anterior lobe.                                                                                  MED

1825        English physician and amateur paleontologist Gideon Mantell claims that fossil teeth discovered by his wife, Mary Ann, in Tilgate Forest, England, belong to an extinct reptile, which he calls *Iguanodon*, the second dinosaur to be discovered. *See also* 1824, PALEO.                                                 PALEO

1825        On a 27-mile line the first steam train passenger service is offered, by the Stockton & Darlington Railway in England. The first American passenger line, the Baltimore & Ohio, will begin to be built in Maryland in 1828.              TECH

1826        German astronomer Heinrich Wilhelm Olbers poses what will become known as Olbers's paradox: If the stars are distributed uniformly through infinite, unchanging space, why is the sky dark at night? Everywhere one looks, there should be a star. The Big Bang theory in the 20th century will explain this paradox by showing that the universe is lumpy, finite, and changing.           ASTRO

1826        German-Russian embryologist Karl Ernst von Baer discovers the eggs of mammals and states that "every animal which springs from the coition of male and female is developed from an ovum, and none from a simple formative fluid." His later two-volume *History of the Development of Animals* is one of the founding works of modern embryology. His studies in comparative embryology will offer ways of demonstrating the likeness of various animal forms.                    BIO

1826        French chemist Antoine-Jérôme Balard discovers bromine.                    CHEM

1826        French chemist Jean-Baptiste-André Dumas invents a technique for measuring the vapor densities of substances that are not gases at normal temperatures.                    CHEM

1826        Scottish explorer Gordon Laing becomes the first European to visit Timbuktu in Africa.                    EARTH

1826        Russian mathematician Nikolai Ivanovich Lobachevsky publishes the first non-Euclidean geometry, which includes this axiom: "Through a given point, not on a given line, any number of lines can be drawn parallel to a given line." Hungarian mathematician János Bolyai will independently publish the same geometry in 1832.                    MATH

1826        German engineer August Leopold Crelle begins publication of the influential periodical *Journal für die reine und angewandte Mathematik*.                    MATH

1826        French mathematicians Jean-Victor Poncelet and Joseph-Diez Gergonne independently discover the principle of duality in geometry.                    MATH

1826        Norwegian mathematician Niels Henrik Abel and German mathematician Carl Gustav Jacobi begin to develop the theory of elliptic functions. (German mathematician Carl Friedrich Gauss had earlier made some of the same discoveries but had not published them.)                    MATH

1826        Between now and 1837 cholera is pandemic—that is, epidemic at the same time in many different parts of the world.                    MED

1826        The first commercially workable gas stove is developed by British businessman James Sharp, who will begin to mass produce it 12 years later.                    TECH

1827        Between now and 1839, American ornithologist and naturalist John James Audubon (1785–1851) publishes his multivolume ornithology collection, *Birds of America*, containing more than 1,000 paintings of bird life.                    BIO

1827        British chemist and physicist Michael Faraday publishes *Chemical Manipulation*, a manual concerning distillation and other forms of chemical processing.                    CHEM

1827        German mathematicians A. F. Möbius, Julius Plücker, and Karl Wilhelm Feuerbach develop homogeneous coordinates.                    MATH

1827        British physician Richard Bright describes nephritis, a kidney disease which will be named for him.                    MED

1827    Irish physicians William Stokes and Robert Adams discover a form of altered consciousness (later called the Stokes-Adams syndrome) caused by decreased blood flow to the brain as the result of a heart block.    MED

1827    German physicist Georg Simon Ohm defines what will become known as Ohm's law, that the flow of current through a conductor is directly proportional to the potential difference and inversely proportional to the resistance.    PHYS

1827    Scottish botanist Robert Brown discovers the phenomenon that will be called called Brownian motion, the continuous random movement of microscopic solid particles when suspended in a fluid.    PHYS

1827    Irish mathematician and physicist William Hamilton unifies the study of optics and correctly predicts conical refraction in his *Theory of Systems of Rays*.    PHYS

1827    Public transportation begins in New York City with the inauguration by Abraham Bower of a 12-seat horsedrawn bus.    TECH

1828    After classifying four major vegetation periods from the Carboniferous to the Tertiary, French botanist Adolphe Brongniart (1801–1876) sorts the plant kingdom into six classes: agamae, cellular cryptogams, vascular cryptogams, gymnosperms, monocotyledonous angiosperms, and dicotyledonous angiosperms.    BIO

1828    Swedish chemist Jöns Jakob Berzelius discovers the element thorium.    CHEM

1828    German chemist Friedrich Wöhler is the first to synthesize an organic compound from inorganic chemicals in the laboratory when he synthesizes urea, the principal waste product found in urine. This experiment shows that organic compounds do not occur only in living things.    CHEM

1828    English mathematician George Green publishes the first attempt at a mathematical theory of electromagnetism, his *Essay on the Application of Mathematical Analysis to Theories of Electricity and Magnetism*.    PHYS

1828    The *Cherokee Phoenix*, the first Native American newspaper, is published. It is written and printed in the Cherokee alphabet, which was developed in 1824 by Cherokee linguist Sequoia.    TECH

1828    The blast furnace is developed for use in iron production by Scottish inventor James Beaumont Neilson.    TECH

1828    An inexpensive process for making chocolate candy is patented by Dutch chocolatier Conrad van Houten. The solid form of chocolate is made possible by an extraction process that yields the cocoa butter needed to bind the dry ingredients (sugar and cocoa powder) together.    TECH

1829    French geologist Alexandre Brongniart (1770–1847), the father of French botanist Adolpe Brongniart (*see* 1828, BIO), gives the name Jurassic to the period lasting from 190 million to 139 million years ago, following the Triassic. This second period of the Mesozoic era was named for the Jura Mountains of France and Switzerland.                                                                 EARTH

1829        Searching for Noah's ark in eastern Turkey, German Johann Jacob von Parrot scales Mount Ararat.                                                                    EARTH

1829        Scottish physicist William Nicol invents what will become known as the Nicol prism, a device for producing plane-polarized light (*see* 1815, CHEM), that makes possible the technique of polarimetry.                                      PHYS

1829        Scottish scientist Thomas Graham, one of the founders of physical chemistry, formulates what will become known as Graham's law (the law of gaseous diffusion), which says that the rate at which a gas diffuses is inversely proportional to the square root of its density.                                               PHYS

1829        French physicist Gaspard-Gustave de Coriolis coins the term *kinetic energy*.    PHYS

1829        Scottish philosopher James Mill publishes his *Analysis of the Phenomena of the Human Mind*, adding muscle sensation (kinesthesis), disorganized sensation (itching or tickling), and gastrointestinal tract sensations to Aristotle's classic five senses. Mill argues that the sensations from these eight senses are the basic elements of consciousness and ideas.                                         PSYCH

1829        American physicist Joseph Henry improves upon the electromagnet invented by William Sturgeon in 1823. By 1831 Henry's electromagnet can lift a ton of iron.                                                                              TECH

1829        The *Tom Thumb*, developed by American entrepreneur Peter Cooper, becomes the first U.S.-built locomotive.                                                           TECH

1829        Through an accidental discovery by French painter and inventor Louis-Jacques-Mandé Daguerre, silver iodide is found to be light sensitive. *See also* 1839, TECH.                                                                      TECH

1829        The Tremont House, the first modern hotel, with private sleeping quarters, opens in Boston.                                                                          TECH

1830s       British scholar William Whewell coins the word *scientist*.                      MISC

c. 1830     British geologist Charles Lyell dismisses the idea of the comparative stability of the earth when he shows that a given fauna may be older than the land it inhabits. His evidence will be important in the development of biogeography, the study of the distribution of organisms through space and time.            BIO

1830        British optician Joseph Jackson Lister invents the achromatic microscope, which eliminates chromatic aberration, allowing the study of fine detail in cells and bacteria.                                                               BIO

1830        Swedish chemist Nils G. Sefström discovers the element vanadium.          CHEM

1830        Swedish chemist Jöns Jakob Berzelius coins the term *isomerism* to describe
            compounds with identical chemical composition but different molecular
            structures and properties. The phenomenon was first noted by Berzelius and
            Joseph-Louis Gay-Lussac in 1824.                                          CHEM

c. 1830     French geologist Élie de Beaumont develops his influential doctrines on the
            origin of mountain ranges, including the basic ideas of tectonic upheaval and
            stratigraphical methods. In 1852 he will sum up his work in the three-volume
            *Notice sur les systèmes de montagnes*.                                   EARTH

1830        British geologist Charles Lyell begins the publication of his *Principles of Geology*
            (three volumes, 1830–1833), which establishes the principle of uniformitari-
            anism in place of the popular theory of catastrophism. This work lays the
            foundations of modern dynamic geology but also has broad public appeal and
            helps to make geology a popular science.                                  EARTH

1830        American physicist Alexander Dallas Bache sets up the first American magnet-
            ic observatory, in the garden of his home in Philadelphia.                EARTH

1830        French mathematician Évariste Galois invents group theory and shows that no
            equation of any degree higher than the fourth can be solved by algebraic meth-
            ods. His work will not be published until 1846. *See also* 1824, MATH.   MATH

1830        British mathematician Charles Babbage publishes his *Reflections on the Decline
            of Science in England*, a critique of British science that results in the formation
            of the British Association for the Advancement of Science.                MISC

1830        P. C. Schmerling finds stone tools, human skulls, and mammoth bones buried
            together at a site in Belgium.                                            PALEO

1830        French tailor Barthélemy Thimmonier invents the first single-thread sewing
            machine.                                                                  TECH

1830        The T-rail, which will become standard equipment for rail lines, is invented by
            Robert Livingston Stevens, an American.                                   TECH

1830        The world's first platform scale, the Fairbanks scale, is developed by American
            inventor Thaddeus Fairbanks. It operates through a system of levers and slid-
            ing weights as counterbalances.                                           TECH

1831        British botanist Robert Brown discovers the cell nucleus in plants.       BIO

1831        British naturalist Charles Darwin embarks on a five-year voyage aboard the
            H.M.S. *Beagle*. The ship's principal mission is to chart South America's coasts,
            but Darwin will take the opportunity to study the flora and fauna of many
            regions. These researches will form the basis for his theory of evolution by nat-
            ural selection, which will be published in 1859.                          BIO

| | |
|---|---|
| 1831 | American chemist Samuel Guthrie discovers the compound chloroform, which will gain wide use as an anesthetic.     CHEM |
| c. 1831 | English geologist Adam Sedgwick engages in field work studying rock succession and the structure of the mountains in North Wales.     EARTH |
| 1831 | American meteorologist William Redfield publishes a chart of the hurricane of 1821, establishing the rotary motion of tropical storms and identifying their region of origin and paths.     EARTH |
| 1831 | British chemist and physicist Michael Faraday and American physicist Joseph Henry independently discover that a changing magnetic force can generate electricity, a process called electromagnetic induction.     PHYS |
| 1831 | British chemist and physicist Michael Faraday invents the dynamo, or electric generator, capable of producing electricity indefinitely by the turning of a copper wheel across magnetic lines of force.     TECH |
| 1831 | Joseph Henry invents the electric motor, in which electric current is used to turn a wheel.     TECH |
| 1831 | French chemist Charles Sauria invents the first practical phosphorus match, which lights by friction when struck on a rough surface.     TECH |
| 1831 | The new London Bridge, which replaces the 900-year-old original, is opened to traffic across the Thames River.     TECH |
| 1831 | Cast iron that can be molded into a variety of shapes is patented by American inventor Seth Boyden.     TECH |
| 1831 | American farmer Cyrus Hall McCormick develops and demonstrates his horse-drawn reaper. Operating through a series of mechanical knives, claws, and a metal receptacle, it vastly improves the efficiency of farming and will be used widely.     TECH |
| 1832 | British chemist and physicist Michael Faraday studies electrolysis, the production of a chemical reaction by passing electric current through a liquid or solution (an electrolyte), and describes the basic laws of electrolysis.     CHEM |
| 1832 | Irish physician Dominic Corrigan describes a full bounding pulse (later known as Corrigan's pulse) associated with heart disease.     MED |
| 1832 | British physician Thomas Hodgkin describes a disease that enlarges the lymph tissue, spleen, and liver. Eventually known as Hodgkin's disease, it comes to be classified under general lymphomas.     MED |
| 1832 | The first cholera outbreak in the United States is initially considered by some to be the plight of the "sinful," from its concentration among immigrants and in impoverished big cities.     MED |

1832      Scottish phrenologist George Combe continues the earlier work of Franz Joseph Gall and Johann Spurzheim, but he turns phrenology into more of a faddish cult than a science. He will help form more than 45 regional phrenological societies that will flourish in the early 20th century.     PSYCH

1832      The first fully fashioned clipper ship, the *Ann McKim*, is built in Baltimore for local businessman Isaac McKim.     TECH

1833      German astronomer Friedrich Wilhelm Bessel completes a catalog of 50,000 stars begun in 1821.     ASTRO

1833      The first European magnetic observatory is established, at Göttingen, Germany.     ASTRO

1833      French chemist Anselme Payen isolates diastase, a plant enzyme that speeds the conversion of starch into glucose. It is the first known organic catalyst that is not itself an organism.     CHEM

1833      British mathematician Charles Babbage publishes *On the Economy of Machinery and Manufactures*, which presents an early form of operations research.     MATH

1833      American physician William Beaumont studies the reaction of the stomach to food, liquid, and emotions through an unhealed abdominal gunshot wound. The wounded man had lived, but the open gastric fistula left a window into the digestion process.     MED

1833      British geologist Charles Lyell identifies new epochs in the earth's history: the Recent (Holocene), Pliocene, Miocene, and Eocene.     PALEO

---

# The Romantic Mathematician

French mathematical genius Évariste Galois (1811–1832) had a short, tragic life befitting the age of romanticism. The son of a mayor of a small town near Paris, he participated on the republican side in the Revolution of 1830, was arrested for alleged threats against King Louis Philippe, tried without success to get papers published by the Academy, and finally died at age 20 in a duel over a woman. The night before the duel, fearing the worst, he wrote a letter to a friend summarizing his discoveries in the theory of equations and asking that they be submitted for comment to the era's leading mathematicians. "After this," he wrote, "there will be, I hope, some people who will find it to their advantage to decipher all this mess."

As it turned out, Galois's "mess" contained a unifying theory of groups that would prove crucial to modern algebra, geometry, and physics. This work did not reach a wide audience until 1846, when most of Galois's papers were finally published in the *Journal de mathématiques*. Now considered a major figure in 19th-century mathematics, Galois fulfilled the romantic destiny of the hero who burns brightly but briefly.

1833        British physicists Michael Faraday and William Whewell coin a number of terms related to electricity, including *electrode, cathode, anode, ion, cation, anion, electrolyte,* and *electrolysis.*                                                          PHYS

1833        American house construction is revolutionized by the development of the balloon frame house by American builder Augustus Deodat Taylor. Its simple wood frame, made of two-by-fours, is sturdier and easier to build than previous designs.                                                                      TECH

1833        For the first time, rollers rather than millstones are used to grind grain in a practice introduced by a Swiss miller.                                            TECH

1834        Danish scholar Christian Jorgensen Thomsen divides the history of humanity into the Stone Age, the Bronze Age, and the Iron Age.                             ARCH

1834        British mathematician and astronomer John Herschel begins the first thorough survey of the southern stars. He notes that the Magellanic clouds discovered by Ferdinand Magellan in 1519–1521 are composed of many individual stars.                                                                              ASTRO

1834        French chemist Anselme Payen discovers cellulose, the main constituent of the cell walls of higher plants.                                                      CHEM

1834        German mining engineer and geologist Johann H. Charpentier advances the geological understanding of glaciers with his paper *Notice sur la cause probable du transport des blocs erratiques de la Suisse.*                                      EARTH

1834        British paleontologists W. H. Bensted and Gideon Mantell discover a partial skeleton of *Iguanodon* in Maidstone, England (*see also* 1825, PALEO). During this decade five other dinosaur genera are discovered: *Hylaeosaurus, Macrodontophion, Thecodontosaurus, Paleosaurus,* and *Plateosaurus.*                 PALEO

1834        German geologist Friedrich von Alberti introduces the Triassic (threefold) system of dating fossils. The oldest and lowest rocks are the Buntsandstein, with Muschelkalk rocks intermediate and Keuper rocks the highest and youngest. *See also* 1780 and c. 1800, PALEO.                                         PALEO

1834        French physicist Jean-Charles-Athanase Peltier discovers what will become known as the Peltier effect, the change in temperature produced when an electric current passes through a juncture between two different metals or semiconductors.                                                                  PHYS

1834        German physicist Heinrich Lenz formulates what will become known as Lenz's law, that an induced electric current always flows in such a direction as to oppose the change producing it.                                                           PHYS

1834        German scholar J. F. Herbert describes *psychology,* a term originally introduced in 1734 by Christian von Wolff, as a science based on experience, mathematics, metaphysics, and possible experimentation.                                            PSYCH

1834    British mathematician Charles Babbage begins work on an "analytical engine" to perform arithmetic operations on the basis of instructions from punched cards. Later regarded as the forerunner of the modern digital computer, this engine was never completed.                                                            TECH

1834    Hansom cabs, designed by British architect Joseph Aloysius Hansom, are used for passenger trade in London.                                                            TECH

1834    Braille, a system of printing and writing for the blind that uses raised dots, is developed by French teacher Louis Braille, a blind man himself.                       TECH

1834    A compression machine that cools water through a process of heat absorption is introduced by American inventor Jacob Perkins. His invention promotes further study of gas refrigeration.                                                            TECH

1835    This year's edition of the Roman Catholic church's *Index of Prohibited Books* is the first since the 17th century not to include the heliocentric works of Copernicus, Galileo, and Kepler. The decision to lift the ban was actually made in 1822.                                                            ASTRO

1835    British naturalist Charles Darwin visits the Galapagos Islands off Ecuador. This island chain's distinctive life forms, particularly its finches, provide him with evidence for the evolution of species.                                                BIO

1835    French chemist C. S. A. Thilorier creates dry ice when he succeeds in freezing carbon dioxide into a solid form. During his experiments he induces temperatures as low as –110° C, the lowest temperature yet reached on the earth's surface.                                                            CHEM

1835    Swedish chemist Jöns Jakob Berzelius coins the term *catalysis* for the action of agents that aid a chemical reaction but are not affected by it.                       CHEM

1835    French physicist Gaspard-Gustave de Coriolis describes what will become known as the Coriolis force, which is the apparent deflection of a moving body caused by unequal rates of rotation between two rotating systems, such as the earth and the air. The Coriolis force will become important later in understanding winds and ocean currents.                                                            EARTH

1835    Irish physician Robert Graves describes the thyroid condition later called Graves's disease.                                                                            MED

1835    German physician Theodor Schwann discovers pepsin, the chief enzyme in gastric juice.                                                                                    MED

1835    English geologist Adam Sedgwick names the Cambrian period.                     PALEO

1835    Scottish geologist Roderick I. Murchison identifies the Silurian period. PALEO

1835    Zinc-coated galvanized iron is developed in France.                           TECH

| | |
|---|---|
| 1836 | British astronomer Francis Baily notes what will become known as Baily's beads, the bright spots along the moon's edge visible during a total eclipse. |
| | ASTRO |
| 1836 | Observed under a microscope, yeast is found to be a living organism.    BIO |
| 1836 | French chemist Auguste Laurent shows that a chlorine atom can be substituted for a hydrogen atom in a compound without great change in the compound's properties.    CHEM |
| 1836 | French mathematician Joseph Liouville founds the influential *Journal de Mathématiques Pures et Appliquées*.    MATH |
| 1836 | British chemist John Frederic Daniell invents what will go by the name of the Daniell cell, a battery that uses copper and zinc electrodes and is more reliable than the voltaic cell of 1800.    TECH |
| 1836 | American inventor Samuel Colt develops the six-shooter revolver that will bear his name; in its many incarnations over the years it will become a popular firearm.    TECH |
| 1837 | French mathematician Pierre Wantsel proves that duplicating the cube and trisecting the angle are impossible using the ancient Greek rules of employing only a straightedge and compass for geometric constructions. Similarly, squaring the circle is later shown to be impossible.    MATH |
| 1837 | French mathematician Siméon Poisson discovers what will become known as Poisson's law, part of the theory of probability.    MATH |
| 1837 | The *Cambridge Mathematical Journal* is founded in England.    MATH |
| 1837 | Smallpox again ravages Native Americans (*see* 1781, MED), this time carried by whiskey peddlers on the Missouri River. Recourse to the traditional Native American remedy of sweat baths only makes the fever worse. Many Native Americans drown themselves in the river or cut their own throats to escape the pain and suffering of the disease.    MED |
| 1837 | French physiologist Marie-Jean-Pierre Flourens discovers that the lower portion of the brain stem, the medulla oblongata, is the body's respiratory center.    MED |
| 1837 | In England an electric telegraph is patented by scientists Charles Wheatstone and William Fothergill Cooke. In the United States a magnetic telegraph is patented by American inventor Samuel F. B. Morse. A Morse code to translate the English alphabet and Arabic numerals into dots and dashes is developed by Morse and his assistant Alfred Vail.    TECH |
| 1837 | The Pitman shorthand system, which translates phonetic sounds and common phrases into stylized page markings, is developed by Englishman Isaac Pitman.    TECH |

1837       The first steam-powered thresher, more efficient than earlier versions, is patented by American inventors John and Hiram Pitts.                              TECH

1838       German astronomer Friedrich Wilhelm Bessel is the first to correctly calculate the distance of a star. He uses a heliometer, a device for measuring distances in the sky, to measure the parallax of the star 61 Cygni and deduce its distance, about 11 light-years away.                              ASTRO

1838       Dutch chemist Gerardus Johannes Mulder coins the word *protein* (from the Greek for "first") to describe the molecule that is the basic building block of albuminous substances and, as is later found, of all living things.                              CHEM

1838–1839  French archaeologist Jacques Boucher de Perthes presents his theory that unusual stone objects found near Abbeville, France, are tools made by humans sometime before the Great Flood reported in the Bible.                              PALEO

1838       British chemist and physicist Michael Faraday discovers a phosphorescent glow produced by electric discharges in low-pressure gases.                              PHYS

1838       British physician Robert Gerdiner Hill, while serving as a surgeon at the Lincoln Asylum, institutes nonrestraint policies for mental patients, insisting that restraints are "never justifiable and always injurious."                              PSYCH

1838       American psychiatrist J. Esquirol recommends applying leeches to the anus of mentally ill patients suffering from depression, then, after the leeches have formed hemorrhoids, removing the leeches and draining the hemorrhoids. This bloodletting is thought to draw the mental malady away from the patient's brain.                              PSYCH

1838       French philosopher Auguste Comte coins the term *sociology*. He is considered the founder of this discipline as well as of the philosophical school known as positivism, which holds that the only real ("positive") knowledge is that which is gained by observation and experiment.                              SOC

1838       The first transatlantic crossing by ships powered entirely by steam is completed by British ships the S.S. *Sirius* and S.S. *Great Western*, which travel from England to New York.                              TECH

1839       American archaeologists John Lloyd Stephens and Frederick Catherwood begin an expedition in which they will discover Mayan ruins in Copán and other sites in Central America.                              ARCH

1839       British astronomer Thomas Henderson makes a nearly correct estimate of the distance of the star Alpha Centauri at 4.3 light-years away.                              ASTRO

1839       British naturalist Charles Darwin publishes the journal of his 1831–1836 voyage on the H.M.S. *Beagle*. *See also* 1831 and 1835, BIO.                              BIO

1839        Swedish chemist Carl Gustaf Mosander discovers the rare earth element
            lanthanum.                                                                    CHEM

1839        In a joint paper titled "On the Physical Structure of Devonshire" Adam
            Sedgwick and Roderick Murchison name the Devonian period.        PALEO

1839        French painter and inventor Louis-Jacques-Mandé Daguerre introduces the
            first commercially viable photographic process, the daguerreotype, in which a
            direct positive image is made on a silver-coated plate.                    TECH

1839        British physicist William Robert Grove invents the fuel cell, an electric cell that
            runs on hydrogen and oxygen fuel, but the device is never made economical
            enough for general use.                                                    TECH

1839        American inventor Isaac Babbitt invents the group of alloys that will become
            known as Babbitt metals; they will be used in making bearings.        TECH

1839        The first full-fledged bicycle is built in Scotland by blacksmith Kirkpatrick
            MacMillan. *See also* 1861, TECH.                                        TECH

1839        American hardware buyer Charles Goodyear discovers a process to "vulcanize"
            rubber, by which it retains its elasticity in all weather. Previously, rubber
            became sticky in warm weather and brittle in cold.                        TECH

1840s       At Nineveh, Mesopotamia (Iraq), British archaeologist Sir Austen Henry
            Layard begins to excavate the remains of the clay tablet library of King
            Ashurbanipal of Assyria, who reigned from 668 to 627 B.C. Later excavations in

---

## Only the Name Was Goodyear

In 1839, when a combination of rubber, sulfur, and white lead overheated in American
shopkeeper Charles Goodyear's kitchen, a temperature-resistant rubber was created
that would revolutionize transportation, public health, and everyday life. But credit for
the material that lined a raincoat or formed the hose that brought water to a burning
building would not go to Goodyear, who had depleted his savings and could not develop his findings in time to obtain the first patent.

Instead, the spoils for inventing vulcanized rubber went to scientists and entrepre-
neurs who used Goodyear's experimental findings. These included Englishman
Thomas Hancock, who refined Goodyear's work and received the first British patent for
durable rubber; and American surgeon Benjamin Franklin Goodrich, who foresaw the
myriad uses of rubber, beginning with the rubber hose, and founded his own rubber-
manufacturing business, the B. F. Goodrich Co., in 1870.

Even when the Goodyear Tire and Rubber Co. was founded in 1898 by Frank
Sieberling, it was in no way affiliated with the inventor, a regular in debtors' prison who
had died in 1860 nearly one-half million dollars in debt. In the case of the Goodyear Tire
and Rubber Co., only the name was Goodyear.

1852–1854 will uncover parts of the *Epic of Gilgamesh*, among other texts. *See* 1872, ARCH.      ARCH

1840s      British naturalist Charles Darwin earns a reputation as a geologist, especially for his explanation of the formation of coral atolls.      EARTH

1840      German-born Russian astronomer Friedrich Wilhelm von Struve calculates the parallax of the star Vega, 27 light-years away.      ASTRO

1840      Swiss-Russian chemist Germain Henri Hess formulates what will become known as Hess's law, or the law of constant heat summation, which states that the amount of heat developed or absorbed in a chemical reaction is fixed, regardless of the route taken. This discovery marks the beginning of thermochemistry, the study of the interrelationship of heat and chemical reactions.      CHEM

1840      German-Swiss chemist Christian Friedrich Schönbein discovers ozone, later identified as a form of oxygen by Irish physical chemist Thomas Andrews.      CHEM

1840      American physicist Alexander Dallas Bache sets up an elaborate magnetic observatory in a specially designed nonmagnetic building at Girard College.      EARTH

1840      Swiss zoologist and geologist Louis Agassiz advances his theory of ice ages in his *Études sur les glaciers*, using evidence of glaciation in Switzerland and other parts of Europe.      EARTH

1840      American explorer Charles Wilkes concludes that the South Polar land mass is continental in size. Twice as large as Australia, it lies almost completely within the Antarctic Circle.      EARTH

1840      In Philadelphia the Association of American Geologists is formally organized.    EARTH

1840      British physicist James P. Joule formulates the first of what will become known as Joule's laws, describing the heat produced when an electric current flows through resistance for a given time. (Joule's second law states that, for ideal gases, the internal energy of a given mass of gas is a function of temperature alone, not of volume or pressure.)      PHYS

1840      French chemist Alexandre-Edmond Becquerel discovers that light can stimulate chemical reactions that result in an electric discharge.      PHYS

1840      The term *negative*, referring to the reverse image created during the photographic process, is coined by British mathematician and astronomer John Herschel. *See also* 1816, TECH.      TECH

1841      German astronomer Friedrich Wilhelm Bessel discovers that the star Sirius has a dark companion, Sirius B, which will later become the first known white dwarf star.      ASTRO

1841        Swedish chemist Jöns Jakob Berzelius discovers allotropy, the existence of two
            or more different forms of an element, when he converts charcoal into
            graphite.                                                                      CHEM

1841        Scottish missionary David Livingstone reaches Cape Town, then moves north-
            ward, attempting to convert Africans to Christianity, becoming the first
            Westerner to explore the Kalahari Desert.                                      EARTH

1841        English anatomist and paleontologist Richard Owen coins the name
            *Dinosauria* ("terrible lizard") for the group of reptiles that includes
            *Megalosaurus* and *Iguanodon*.                                                PALEO

1841        Richard Owen describes two new genera of dinosaurs, *Cladeidon* and
            *Cetiosaurus*.                                                                 PALEO

1841        The first professional psychiatric association in the world, the Association of
            Medical Officers of Asylums and Hospitals for the Insane, is founded in
            England. In 1971 it will become known as the Royal College of Psychiatrists.   PSYCH

1841        American schoolteacher Dorothea Dix begins campaigning to reform the care
            of the mentally ill and poor in almshouses. Between 1841 and 1880 Dix is
            directly responsible for the creation of 32 new state insane asylums.          PSYCH

1841        Scottish surgeon James Braid witnesses demonstrations of mesmerism (*see*
            1772, PSYCH). Initially skeptical, he eventually proves that this "nervous sleep"
            can be induced by having a patient fixate on an object placed above the line
            of vision. Braid will later realize that suggestion is of primary importance to
            hypnosis. He will be considered the discoverer of hypnosis and credited with
            taking the phenomenon out of a magical, mystical arena and applying it to the
            physiological realm.                                                           PSYCH

1841        British inventor William Henry Fox Talbot patents a new photographic
            process, which involves the making of a paper negative (the calotype) from
            which any number of paper positives can be printed. Talbot's calotype
            method, forerunner of modern photographic processes, will eventually super-
            sede that of French painter and inventor Louis-Jacques-Mandé Daguerre. *See*
            1839, TECH.                                                                    TECH

1841        In Paris, arc lights for streets are demonstrated.                             TECH

1841        The first usable breech-loading (rather than muzzle-loading) rifle, the needle
            gun, is developed by Prussian gunmaker Johann Nikolaus von Dreyse.             TECH

1842        Austrian physicist Christian Johann Doppler discovers what will become
            known as the Doppler effect: an apparent shift in wavelengths toward higher
            frequency as a source of light or sound approaches, and toward lower fre-
            quency as it recedes. *See also* 1848, PHYS.                                   PHYS

1842    American psychologist and philosopher William James is born in Massachusetts (*d*. 1910). Later known as the dean of American psychologists, he will develop the school of functionalism, which seeks to understand the conscious aspects of mental life. James will advocate a pragmatic viewpoint on life and theology and struggle with the issues of mental and emotional freedom. Two of his most important works will be *Principles of Psychology* (1890) and *Varieties of Religious Experience* (1902).    PSYCH

1843    Heinrich Samuel Schwabe discovers the sunspot cycle, a semiregular fluctuation in sunspot activity, later found to occur in periods of 11 years.    ASTRO

1843    England's Rothamsted Experimental Station is established by English agriculturalist John Lawes and chemist Joseph Henry Gilbert for the study of wheat and other grains.    BIO

1843    Swedish chemist Carl Gustaf Mosander discovers the rare earth elements erbium and terbium. About this time he also chemically isolates the rare earth element yttrium, first discovered by Finnish chemist Johan Gadolin in 1794.    CHEM

1843    Irish mathematician William Hamilton invents a self-consistent algebra of hypercomplex numbers, which he calls quaternions.    MATH

1843    British mathematician Arthur Cayley works out the analytic geometry of three or more dimensions, which will be called *n*-dimensional analytic geometry.    MATH

1843    American physician and author Oliver Wendell Holmes reports that childbed fever is contagious, but his peers scoff at the idea.    MED

1843    British physicist James P. Joule formulates the mechanical equivalent of heat, the ratio of a unit of mechanical energy to the equivalent unit of thermal energy. Symbolized as $J$, it has an approximate value of 41,800,000 ergs (4.18 joules) per calorie. This discovery shows a fixed quantity of work results in a fixed amount of heat.    PHYS

1843    German physicist Julius Robert von Mayer discovers that heat and mechanical work have a direct relationship and are different forms of energy.    PHYS

1843    In England, the House of Lords decides that an individual is not responsible for having committed a crime if the person is "laboring under a defect of reason" and has a diseased mind. The M'Naghten rule, as it will be called, will not have an American counterpart until the issuance of the Durham rule in 1954.    PSYCH

1843    French physiologist Marie-Jean-Pierre Flourens refutes phrenologists' claims by proving that the skull's contours do not correspond to those of the brain, thus disproving the basic assumption of phrenology. Despite this evidence, phrenology will long persist as a fad.    PSYCH

1843        British engineer Sir George Cayley designs, on paper, the first practical
            helicopter.                                                            TECH

1843        British inventor Charles Wheatstone popularizes, though he does not invent,
            what will become known as the Wheatstone bridge, an electrical circuit that
            precisely measures the value of a resistance.                          TECH

1843        American farmer Jerome Increase Case develops what will become known as
            the Case threshing machine, more efficient than earlier thresher designs.   TECH

1844        French scientists Jean-Baptiste Boussingault and Jean-Baptiste Dumas, experi-
            menting with plant decomposition, prove that as plants decompose their car-
            bonic acid is reduced to carbon, water to hydrogen, ammonium hydroxide to
            ammonium, and nitric acid to nitrogen.                                 BIO

1844        British writer Robert Chambers anonymously publishes *Vestiges of the Natural
            History of Creation*, a controversial work that promotes the idea of biological
            evolution, influencing Charles Darwin and Alfred Wallace (*see* 1859, BIO).   BIO

1844        Russian chemist Karl K. Klaus discovers the element ruthenium.         CHEM

1844        German mathematician Hermann Grassmann publishes his *Theory of
            Extension*, in which he gives a detailed exposition of *n*-dimensional vector
            space. *See also* 1843, MATH.                                          MATH

1844        Amyl nitrite, a drug later used to dilate blood vessels, is discovered.   MED

1844        The American Psychiatric Association is founded.                       PSYCH

1844        American entrepreneur Henry Wells begins an express delivery service between
            Buffalo and Detroit, beginning a company called Wells & Co., which will even-
            tually develop into Wells, Fargo & Co.                                 TECH

1844        German engineer Gottlob Keller pioneers the wood-pulp paper process, which
            creates an inexpensive paper for periodicals.                          TECH

1844        The first telegraph message, from Washington, D.C., to Baltimore, Maryland,
            is transmitted. Sent on May 24 by Samuel F. B. Morse to his assistant Alfred
            Vail, it reads, "What hath God wrought!"                               TECH

1845        British chemist and physicist Michael Faraday identifies six "permanent
            gases" that cannot be liquefied with the available technology: hydrogen,
            oxygen, nitrogen, carbon monoxide, methane, and nitric oxide.          CHEM

1845        German chemist Christian Friedrich Schönbein discovers the explosive nitro-
            cellulose, or gun cotton.                                              CHEM

1845        British geologist Charles Lyell publishes *Travels in North America*, on his recent
            investigations there.                                                  EARTH

1845    American gynecologist James Marion Sims (1813–1883) is the first to surgically relieve a woman suffering from vesicovaginal fistula, an opening from the bladder into the vagina. Sims will later invent the vaginal speculum.    MED

1845    American dentist Claudius Ash originates the single porcelain tooth, which can be set individually into plates for false teeth. This technique eventually replaces older methods like jamming teeth gathered from the dead into a socket prepared in the recipient's gums and manufacturing dentures from celluloid, cloth, and ivory.    MED

1845    British chemist and physicist Michael Faraday discovers paramagnetism and diamagnetism. Based on his observations of the effects of a magnetic field on light polarization in crystals, he suggests that light is a form of electromagnetism.    PHYS

1845    German psychiatrist Wilhelm Griesinger publishes his textbook *Mental Pathology and Therapeutics*, a turning point in psychiatry that shifts the center of the science from France to Germany.    PSYCH

1845    The hydraulic crane is patented by British inventor William Armstrong.    TECH

1845    The first cable suspension aqueduct bridge, spanning the Allegheny River in Pittsburgh, Pennsylvania, is opened to traffic. The 162-foot bridge was constructed by German-American engineer John Augustus Roebling.    TECH

1846    British archaeologist Henry Creswicke Rawlinson deciphers cuneiform, an ancient Mesopotamian script, from an inscription in three languages carved into a cliff in Persia during the reign of Darius the Great (521–486 B.C.). He works without knowing the studies of German Georg F. Grotefend, whose 1802 translation of cuneiform went unnoticed.    ARCH

1846    Mounds built by Native Americans in the Mississippi River Valley, including the Serpent Mound, built in what would become Ohio, are excavated by E. George Squier and Edward H. Davis.    ARCH

1846    On September 23, German astronomer Johann Gottfried Galle and his assistant Heinrich d'Arrest become the first to sight Neptune, the eighth planet from the sun, corroborating predictions by French mathematician Urbain-Jean-Joseph Leverrier and British mathematician John Couch Adams, who are credited with the discovery.    ASTRO

1846    On October 10, British astronomer William Lassell discovers Triton, a satellite of Neptune.    ASTRO

1846    American explorer and scientist James Dana publishes *Zoophytes*, in which he classifies and describes the physiology and ecology of hundreds of these invertebrate animals. Zoophytes include sea anemones and sponges.    BIO

1846    New York professor Elias Loomis publishes the first synoptic weather map.    EARTH

1846        British naturalist Edward Forbes, a pioneer in biogeography, maintains that
            most of the plants and animals of the British Isles migrated there from the
            European continent over land connections before and after the glacial epoch. EARTH

1846        American dentist William Thomas Morton (1819–1868) introduces ether
            anesthesia. After success with nitrous oxide in tooth removal, Morton collab-
            orates with Massachusetts General Hospital surgeon John Warren to use ether
            during an operation to remove a neck tumor. The surgery is a success, and
            when the patient recovers he says he felt no pain.                        MED

1846        The Smithsonian Institution is founded in Washington, D.C., with a bequest
            from British chemist and mineralogist James Smithson.                     MISC

1846        French chemist Louis Pasteur studies the phenomenon of optical activity, the
            ability of certain substances with asymmetric molecules to rotate the plane of
            plane-polarized light as it passes through a crystal or solution. He discovers
            that molecules of these substances exist in two different forms that are mirror
            images of each other.                                                      PHYS

1846        German physicist Wilhelm Eduard Weber develops a system of fundamental
            units for electricity and a method for deducing the magnetic force on a
            charged particle.                                                          PHYS

1846        German philosopher Theodor Waitz writes a fundamental psychology book,
            *Foundations of Psychology*.                                              PSYCH

1846        Boston machinist Elias Howe patents the first modern sewing machine.
            Powered by a hand-driven wheel, it has a double-thread, lockstitch system.  TECH

1846        The Pennsylvania Railroad, eventually one of the nation's largest, is chartered. TECH

1846        A standard gauge for railroads is adopted in Britain.                     TECH

1846        American manufacturer Richard Hoe patents his rotary, or "lightning,"
            press, which prints more efficiently by being cylinder- rather than flatbed-driven.
                                                                                       TECH

1847        American astronomer Maria Mitchell discovers a comet on October 1. She will
            become the first woman admitted to the American Academy of Arts and
            Sciences (1848) and the American Association for the Advancement of Science
            (1850).                                                                    ASTRO

1847        German botanist Matthias Jakob Schleiden and German zoologist and
            botanist Theodor Schwann publish in *Beiträge zur Phytogenesis* what will
            become the basis for modern cell theory. These scientists argue that cells are
            the fundamental organic units common to all living beings and develop by
            "free formation" out of formless cytoblastemic substances in the same basic
            way. German zoologist Robert Remak and pathologist Rudolf Virchow will fur-
            ther develop the Schleiden-Schwann cell theory.                           BIO

1847        Italian chemist Ascanio Sobrero first produces the explosive nitroglycerin.   CHEM

1847        German chemist Lambert Babo states what will become known as Babo's law, that
            the vapor pressure of a liquid decreases with the addition of a solute, the amount
            of the decrease being proportional to the quantity of solute dissolved.   CHEM

1847        The Association of American Geologists changes its name to the American
            Association for the Advancement of Science, holding regular annual meetings
            from this year forward. Unlike its European counterparts, the AAAS becomes a
            forum for the exchange of scientific information without regard to social sta-
            tus, and it spends little money on formal activities.   EARTH

1847        British mathematician George Boole publishes *The Mathematical Analysis of
            Logic*, the founding work of Boolean algebra or symbolic logic. In it Boole
            develops a set of symbols and algebraic manipulations that express logical
            arguments.   MATH

---

# Thermodynamics

The first and second laws of thermodynamics (the study of the interrelation of energy,
heat, and work) are the bane of wishful thinkers in all walks of life. They can be expressed
mathematically as physical principles, but from their initial formulation in the 19th century they have carried philosophical implications as well.

The first law, credited to German physicist Hermann von Helmholtz in 1847, states
that the total amount of energy in an isolated system does not change. Also known as the
law of the conservation of energy, it implies that if you build a machine to turn a wheel,
cool a room, or cook a hamburger, you must draw off the energy from somewhere else.

The second law, formulated by German physicist Rudolf Clausius in 1850, states that
the disorder of an isolated system (the unavailability of energy to do useful work)
increases with time or, at best, remains constant. This law is also known as the law of
entropy, after a term Clausius coined for a measure of a system's disorder. Another way
of stating it is that whenever work takes place, some energy is converted into waste heat.
This means that having a perpetual motion machine—such as a steamship that would
run eternally by using the heat from its own pipes to heat more water and make more
steam—is in principle impossible. It also means that the total disorder of the universe
will inevitably become greater, not less.

A third law of thermodynamics, the law of absolute zero, was formulated by German
physical chemist Walther H. Nernst in 1906. It states that all bodies at absolute zero
would have the same entropy, though absolute zero, the state at which all molecular
motion stops, can never be completely attained. This law is important to physicists
because it provides an absolute scale of values for measuring entropy.

1847        British mathematician Charles Babbage designs the first ophthalmoscope with
            which to study the retina, but his invention is neglected and Hermann von
            Helmholtz later receives the credit. *See* 1851, MED.                          MED

1847        The American Medical Association is established and holds its first meeting in
            Philadelphia.                                                                 MED

1847        Hungarian physician Ignaz P. Semmelweis discovers that puerperal sepsis
            (childbed fever) occurs when medical students fail to wash their hands before
            delivering babies. After he requires students to wash their hands with soap and
            chlorinated lime, the death rate in that maternity ward drops significantly.
            Semmelweiss and Oliver Wendell Holmes (*see* 1843, MED) will be ridiculed for
            their views on childbed fever but eventually will be proven correct.          MED

1847        German scientist Hermann von Helmholtz states the law of conservation of
            energy, the first law of thermodynamics: In an isolated system, the total
            amount of energy does not change. Fellow German Julius Robert von Mayer
            (in 1842) and British physicist James P. Joule (in 1847) also contribute to the
            development of this law.                                                       PHYS

1847        English-born American scientist John W. Draper discovers that all materials
            begin to glow red at about 525° C (977° F) and change color as the temper-
            ature rises until they finally glow white.                                    PHYS

1847        A process for cutting dressmaker patterns for use with the newly invented
            sewing machine is developed by American tailor Ebenezer Butterick.            TECH

1847        The first adhesive postage stamps are used, in the United States.             TECH

1848        In the industrial regions of England a dark-colored variety of pepper moth
            replaces light-colored ones after industrial air pollution begins blackening trees,
            making the light-colored moths visible to insect-eating birds. This example of
            an adapted organism reproducing in greater numbers than a competing strain
            becomes a classic instance of natural adaptation by a process of differential
            reproduction.                                                                 BIO

1848        French chemist Louis Pasteur notes that one form of tartaric acid bends light
            to the left, the other to the right. This discovery will be important in the devel-
            opment of stereochemistry, the branch of chemistry concerned with molecu-
            lar structure and its impact on chemical properties. *See* 1874, CHEM.        CHEM

1848        New England theologian and geologist Edward B. Hitchcock publishes a study
            of fossil tracks he has collected in the Connecticut Valley. He proposes that the
            tracks are those of extinct birds, but they will turn out to be those of dinosaurs.
            *See also* 1800 and 1915, PALEO.                                              PALEO

1848        French physicist Armand Hippolyte Louis Fizeau notes that the Doppler effect
            in light (*see* 1842, PHYS) is best observed by monitoring the changing positions
            of spectral lines, which will become known as the Doppler-Fizeau effect.      PHYS

| | |
|---|---|
| 1848 | British physicist William Thomson, later Lord Kelvin, theorizes that there is an absolute temperature, absolute zero, below which no further energy loss is possible, and that this temperature should be the starting point of a new absolute scale of temperature. With absolute zero at –273.15° C, this scale is henceforth called the Kelvin scale. **PHYS** |
| 1849 | French astronomer Edouard-Albert Roche formulates what will become known as Roche's limit, specifiying the distance from a planet within which tidal forces will tend to break a satellite apart. The concept is later advanced to explain Saturn's rings, which are within Roche's limit. **ASTRO** |
| 1849 | British physician Thomas Addison describes pernicious anemia. **MED** |
| 1849 | Elizabeth Blackwell, America's first female physician, graduates from Geneva College of Medicine in upstate New York. **MED** |
| 1849 | British physicist William Thomson introduces the term *thermodynamics.* **PHYS** |
| 1849 | A safety pin is patented by American inventor Walter Hunt. **TECH** |
| 1849 | In New York City, American inventor James Bogardus completes a commission to build the first prefabricated homes. His iron-and-glass constructions start a trend that extends to several American cities. **TECH** |
| c. 1850s | British mathematician James Joseph Sylvester begins to coin numerous mathematical terms that will become accepted, such as *syzygy, cogredient, invariant, covariant,* and *contravariant.* **MATH** |
| 1850s | A generation of social theorists who will be known as legalists, including Léon Duguit, Maurice Hauriou, and Georg Jellinek, is born. Near the end of the century they will help establish political science as a separate academic discipline. **SOC** |
| 1850 | The first photographic plates are used to record images of stars and the moon, foreshadowing the later importance of photography as a tool of astronomy. **ASTRO** |
| c. 1850 | Irish engineer Robert Mallet coins the term *seismology* to refer to the study of earthquakes and the earth's mechanical properties. **EARTH** |
| 1850 | British sanitarian Sir Edwin Chadwick establishes that poverty and sickness are linked. In a report to England's Poor Law Commission, Chadwick says eradicating disease is hopeless unless poverty is also eradicated. **MED** |
| 1850 | Epidemiology, the study of the causes, distribution, and control of disease, becomes a branch of medical science as physicians seeking to prevent disease form the London Epidemiological Society. **MED** |
| 1850 | Building on experiments the previous year by French physicist Armand-Hippolyte-Louis Fizeau, his assistant, Jean-Bernard-Léon Foucault, determines the speed of light to within less than 1 percent of its actual value. **PHYS** |

| 1850 | German physicist Rudolf Clausius formulates the second law of thermodynamics, which states that the disorder of a closed system increases with time, or that some energy is always lost as heat in any energy conversion. Clausius later coins the word *entropy* for the ratio of a system's heat content to its absolute temperature, which serves as a measure of its disorder. PHYS |
|------|------|
| 1850 | Using the thermopile he has invented, Italian physicist Macedonio Melloni studies infrared radiation, showing that it consists of waves of the same structure as light but longer. PHYS |
| 1850 | German experimental psychologist Georg Elias Müller is born on July 20 in Grimma, Saxony (*d.* 1934). His 1873 doctoral dissertation will be the first empirical study on attention. An expert on vision and memory, he will codiscover Jost's law, which states that when two associations are of equal strength, a repetition strengthens the older one more than the newer one. Müller will also introduce the memory drum, a device for verbal recall used in learning experiments. PSYCH |
| 1850 | Boston machinist Isaac Singer patents a foot-operated sewing machine that becomes a commercial success. Elias Howe (*see* 1846, TECH) later sues him, successfully, for patent infringement. TECH |
| 1850 | The Illinois Central becomes the first U.S. railroad to receive a government land grant. Eventually it will be the leading north–south line in the United States. TECH |
| 1851 | British astronomer William Lassell discovers Ariel and Umbriel, two satellites of Uranus. ASTRO |
| 1851 | French physicist Jean-Bernard-Léon Foucault constructs a pendulum experiment that demonstrates the rotation of the earth. EARTH |
| 1851 | German mathematician Georg Riemann writes his doctoral thesis on the theory of complex functions, introducing the Cauchy-Riemann equations and laying the basis for the concept of the Riemann surface, an important step in the development of topology. MATH |
| 1851 | German scientist Hermann von Helmholtz invents the ophthalmoscope, an instrument for examining the interior of the eye, independently of Charles Babbage's version (*see* 1847, MED). MED |
| 1851 | Irish-born British physicist George Gabriel Stokes introduces a mathematical formula for the fall of a small body, such as a droplet, through a fluid (liquid or gas), like air. PHYS |
| 1851 | British philosopher and economist John Stuart Mill publishes his *Principles of Political Economy*, which will be a basic economic text for decades. SOC |

1851    The Crystal Palace, assembled from prefabricated elements, is built to house the Great Exhibition of 1851 in London. This huge enclosure of glass and iron is a precursor to the glass and steel buildings of modern architecture.    TECH

1851    The wet collodion process for developing photographs, which will replace the daguerreotype process (see 1839, TECH), is developed by English architect Scott Archer.    TECH

1852    German scientist Hermann von Helmholtz determines the speed at which messages travel along nerves.    BIO

1852    English chemist Edward Frankland formulates the theory of valence, stating that an atom of a given element is able to combine with a fixed number of other kinds of atoms according to certain basic rules. An atom's valence will later be found to be related to its atomic weight.    CHEM

1852    French physicist Jean-Bernard-Léon Foucault invents the gyroscope.    EARTH

1852    French mathematician Michel Chasles publishes Traité de géométrie supérieure, which establishes the use of directed line segments in pure geometry.    MATH

1852    The American Pharmacy Association is founded.    MED

1852    German physician and physiologist Robert Remak shows that tissue growth is due to cell division.    MED

1852    British physicists James P. Joule and William Thomson, Lord Kelvin, demonstrate what will become known as the Joule-Thomson effect, the temperature change (usually a drop) that occurs when a gas expands through a porous plug into a region of lower pressure.    PHYS

1852    British physicist George Stokes gives the name fluorescence to the glow observed when an electric current passes through a partially evacuated tube. This term will later be used for any visible light resulting from a collision of matter and radiation rather than a rise in temperature.    PHYS

1852    An oil distilled from coal tar, developed and sold by Boston drug makers as coal oil, will eventually be known instead by the name given it in 1855 by New York physician Abraham Gesner, kerosene. Among its uses are as a patent medicine and lamp oil.    TECH

1852    On the Michigan Southern Railway, the first train successfully travels from the East Coast to Chicago.    TECH

1852    American mechanic Elisha Graves Otis patents the safety elevator, also known as the safety hoister. This cagelike apparatus, which operates by a combination of rope, grips, and ratchets, will become a key factor in the proliferation of multistory buildings.    TECH

1852        On September 24, French engineer Henri Giffard takes the first dirigible, or airship, for a flight. *See also* 1910, TECH.                    TECH

1853        Italian chemist Stanislao Cannizzaro discovers what will become known as the Cannizzaro reaction, a reaction of aldehydes to yield carboxylic acids and alcohols.                    CHEM

1853        *On Coral Reefs and Islands*, published in New York by American geologist James Dwight Dana, confirms British naturalist Charles Darwin's subsistence theory of the formation of coral atolls and offers new insight on corals and reef-building.                    EARTH

1853        British mathematician William Shanks calculates pi to 707 decimal places, though only the value to 527 places is correct.                    MATH

1853        French physicist Jean-Bernard-Léon Foucault shows that light travels faster in air than in water, a discovery that lends support to the wave theory of light.    PHYS

1853        Scottish physicist William J. M. Rankine develops the concept of potential energy, or the energy stored in a body due to its position or shape.                    PHYS

1853        British engineer George Cayley designs and builds the first successful manned glider.                    TECH

1853        Swedish inventor J. E. Lundstrom patents the safety match.                    TECH

1854        German scientist Hermann von Helmholtz predicts the "heat death" of the universe, which he argues will result when the universe reaches a state of uniform temperature.                    ASTRO

1854        German scientist Hermann von Helmholtz posits that gravitational contraction is the source of the sun's heat. From this hypothesis British physicist William Thomson, Lord Kelvin, calculates that the sun's age is about 25 million years, which conflicts with geology's estimate of a much older age for the earth. The conflict will not resolved until 1938, when thermonuclear fusion is proposed as the principal source of the sun's energy.                    ASTRO

1854        British chemist Alexander W. Williamson explains how a catalyst works. *See also* 1835, CHEM.                    CHEM

1854        While searching for an undersea path for the transatlantic cable, American oceanographer Matthew F. Maury discovers Telegraph Plateau, a shallow section of the Atlantic Ocean. This is the first important physical discovery about the ocean floor.                    EARTH

1854        German mathematician Georg Riemann develops a non-Euclidean geometry that resembles the geometry for the surface of a sphere. In his system all lines intersect and are finite in length, and no two lines are parallel.                    MATH

1854    Between now and the 1870s, English mathematicians Arthur Cayley and James Joseph Sylvester develop the algebra of forms, or *quantics*, representing the beginning of the theory of algebraic invariants.                    MATH

1854    German obstetrician and gynecologist Karl S. F. Credé devises a means of excising the placenta and uterine clots to reduce the risk of hemorrhage following childbirth.                    MED

1854    Spanish vocal instructor Manuel García invents the modern laryngoscope, making otolaryngology—the sciences of the ear, nose, and larynx—possible.    MED

1854    British physician John Snow advises the vestrymen of St. James's parish in London to remove the handle of the Broad Street well pump in Soho. Its removal brings the 1854 London cholera epidemic to an abrupt halt, proving Snow's 1849 theory that cholera is a water-borne disease.                    MED

1854    German bacteriologist Paul Ehrlich is born in Strehlen, Silesia (*d.* 1915). Ehrlich, considered the founder of modern chemotherapy, will become known for his methylene-blue cell-staining method and his classification of white blood cells.                    MED

1854    British philanthropist Florence Nightingale arrives in Scutari, Constantinople, with 38 nurses to administer to the sick and injured soldiers of the Crimean War (1854–1856). She will become known as the founder of modern nursing and hospital sanitary reforms.                    MED

1854    German ophthalmologist Albrecht von Graefe (1828–1870) introduces iridectomy, the surgical removal of a portion of the iris, in glaucoma treatment. Von Graefe will be considered the father of modern eye surgery.                    MED

1854    Life-sized models of dinosaurs are displayed publicly for the first time, at the Crystal Palace, which has been moved from London (*see* 1851, TECH) to Sydenham, England.                    PALEO

1854    English anatomist Richard Owen describes the first dinosaur found in the Southern Hemisphere: *Massospondylus*, discovered in Triassic deposits in the Red Beds of South Africa.                    PALEO

1854    French psychopathologist Jules Baillarger is credited with the first significant description of manic-depressive psychosis, or bipolar disorder (*see* 1962, PSYCH), which he calls *la folie à double forme*. For decades this disorder will be known as "circular insanity" in English.                    PSYCH

1854    American professor Benjamin Silliman effects a fractional distillation of crude petroleum, which creates a clean-burning Pennsylvania "rock-oil." The oil company Silliman forms with a copartner will be one of many such formed over the next few decades.                    TECH

1855            French mathematician and astronomer Urbain Leverrier posits the existence of a planet he designates Vulcan to account for irregularities in Mercury's orbit. These variations will later be accounted for by the general theory of relativity (*see* 1916, ASTRO).                                                                                    ASTRO

1855            Irish astronomer William Parsons observes the spiral structure of galaxies.    ASTRO

1855            American oceanographer Matthew F. Maury publishes *Physical Geography of the Sea*, the first textbook on oceanography.                                                                          EARTH

1855            British physician Thomas Addison describes what will become known as Addison's disease, a syndrome resulting from a deficiency in adrenocortical hormone secretion.                                                                                                          MED

1855            Scottish mathematician James Clerk Maxwell gives mathematical expression to the physical concept of lines of force originated by Michael Faraday in 1821.    PHYS

1855            German physicist Heinrich Daniel Ruhmkorff invents the induction coil that will bear his name.                                                                                                         PHYS

1855            British physicist William Thomson, Lord Kelvin, proposes a theory for transmitting electrical signals through undersea cables, which will be applied in constructing the first transatlantic cable.                                                                       PHYS

c. 1855         German chemist Robert Bunsen begins using the gas burner that will bear his name, the Bunsen burner. It was actually invented by a technician named C. Desaga and earlier was independently invented by Michael Faraday.                              TECH

---

# The Gulf Stream

The first modern oceanographer was American naval officer Matthew F. Maury (1806–1873), who was assigned to the depot of charts and instruments after being disabled in an accident. Maury made the most of his post, developing charts of the Atlantic Ocean's winds and currents to reduce nautical travel time. In 1855 he published the first textbook of oceanography, *Physical Geography of the Sea*.

Maury is best known for his charting of the Gulf Stream. Fifty miles wide at its start and moving at an average speed of four miles per hour, this powerful current annually propels warm water from the Gulf of Mexico to the North Atlantic. In Maury's words, "There is a river in the ocean."

Impressive as it is, the Gulf Stream is actually one part of a larger circular movement of water generated by the Coriolis effect. From the North Atlantic the current continues to flow clockwise in the eastward movement known as the North Atlantic drift. Part of the current, deflected by the European land mass, strays up past Britain to Norway. The rest, the Canaries current, moves south along the Canary Islands, then, as the north equatorial current, is deflected westward back to the Caribbean, where the cycle begins again.

1855          German inventor Johann Heinrich Wilhelm Geissler invents what will become
              known as the Geissler tube, an improved device for producing vacuums.  TECH

1855          Aluminum is produced for commercial use by French chemist Henri-Étienne
              Sainte-Claire Deville.                                                    TECH

1855          Pyroxylin, the first synthetic plastic, made from guncotton (nitrocellulose)
              and camphor, is patented by British chemist Alexander Parkes but is not com-
              mercially successful. *See also* 1869, TECH.                              TECH

1855          A process for condensing milk is patented by an American, Gail Borden.  TECH

1856          British astronomer Norman Robert Pogson quantifies the old stellar magni-
              tude system devised by Hipparchus (*see* 100s B.C., ASTRO), constructing ratios for
              the relative brightnesses of stars.                                       ASTRO

c. 1856       French physiologist Claude Bernard discovers glycogen and its function in the
              liver, and coins the term *internal secretion*.                           BIO

1856          German mathematician Karl Weierstrass becomes professor of mathematics at
              the University of Berlin. Among his many contributions will be the discovery
              of analytic continuation and uniform convergence.                         MATH

c. 1856       English epidemiologist William Budd proves that typhoid is a water-borne
              disease.                                                                   MED

1856          American paleontologist Joseph Leidy becomes the first person in the Western
              Hemisphere to identify dinosaur fossils. The fossils he describes include teeth
              of the *Trachodon* and *Deinodon*, found in Montana.                      PALEO

1856          Johann C. Fuhrott discovers the first known skull of Neanderthal man, in the
              Neander Valley near Düsseldorf, Germany. Experts disagree on whether its
              heavy brow ridge and retreating forehead represent the features of an early
              form of human or of a diseased individual. *See also* 1880s, PALEO.       PALEO

1856          Austrian neurologist Sigmund Freud is born in Moravia (*d.* 1939, London).
              One of the best-known names in early psychiatry, Freud will become recog-
              nized as the founder of psychoanalysis. Freud's theories on the unconscious
              and infantile sexuality will dominate 20th-century psychiatry.            PSYCH

1856          British engineer Henry Bessemer patents the Bessemer converter, which uses
              cold air to convert pig iron to steel, a method that will become known as the
              Bessemer process.                                                         TECH

1856          Mauve, the first chemical dye, is accidentally discovered by British chemistry
              student William Henry Perkin. This dye is coal-tar-based.                 TECH

1856   The American telegraph company Western Union is chartered by businessman Hiram Sibley and financier Ezra Cornell. It proceeds to unite several smaller telegraph companies to become the largest in the United States.   TECH

1857   Scottish physicist James Clerk Maxwell shows that Saturn's rings are not solid but are made up of discrete particles.   ASTRO

1857   German mathematician Georg Riemann makes a conjecture that becomes known as the Riemann hypothesis, stating that the nonreal zeroes of the zeta function are complex numbers whose real part is always equal to one half. The

## Neanderthal Mystery

It is now accepted that the fossils discovered in a cave in the Neander Valley (Neander Thal in Old German) near Düsseldorf, Germany, in 1856 represent an extinct form of early humanity. But that consensus was not reached for decades. Neanderthal (or Neandertal) man first came to light in a storm of controversy.

The fossil remains—a partial skull and some limb bones—were clearly human, but they exhibited such unusual features as a heavy brow ridge, a receding forehead and chin, large teeth, and bowed limb bones. Because reliable methods of dating were not then available, it was anyone's guess how old the bones were. Some suggested that the fossils represented an early race of subhumans, but others believed they were those of a much more recent but diseased individual. A contemporary German anatomist, F. Mayer, proposed a specific date—1814, less than 50 years earlier. He theorized that the bones were those of a Cossack cavalryman who had deserted the Russian army as it forced Napoleon to retreat across the Rhine that year. The bowed legs supposedly came from years of sitting in the saddle, complicated by a childhood case of rickets.

Charles Darwin's *The Origin of Species*, published three years later in 1859, intensified the controversy by supplying theoretical grounds for arguing that humans had evolved from earlier forms of life. English biologist Thomas Henry Huxley and French anthropologist Pierre-Paul Broca both advanced the view that Neanderthal man represented an extinct human form. As late as the 1870s, however, the German pathologist Rudolf Virchow was still claiming that the bones were those of a hapless, relatively recent individual who had been afflicted with rickets, head injuries, and severe arthritis.

It took the discovery of several fairly complete Neanderthal skeletons in Spy, Belgium (1887), and La Chapelle-aux-Saints, France (1908), to make it clear that the Neanderthals were a distinct people who had lived in Europe from about 150,000 years ago to 35,000 years ago. Later discoveries showed that they had also lived in the Middle East. However, controversy has followed the Neanderthals even to the present day. Though it is generally held that they were a subspecies (*Homo sapiens neanderthalensis*) of modern humanity, some scientists argue that they were a distinct species (*Homo neanderthalensis*). And the mechanism of their extinction remains an open question: whether they evolved into modern humans (a view not widely held), interbred with modern humans, or were wiped out by modern humans, either passively by competition or actively by warfare.

conjecture will remain unconfirmed, though a generalized version of it will be proven by Belgian mathematician Pierre Deligne in 1974.  MATH

1857 German physiologist Wilhelm Petters discovers acetone in diabetic urine, implying that the pathology and treatment of diabetes are purely chemical problems.  MED

1857 German physicist Rudolf Clausius proposes a mathematical foundation for the kinetic theory of heat.  PHYS

1857 The first passenger elevator in a commercial building is put in use in the E. G. Haughwort Store in downtown New York City.  TECH

1858 Warren De la Rue of England invents the photoheliograph, a device for photographing the sun.  ASTRO

1858 French physiologist Claude Bernard discovers the vasodilator and vasoconstrictor nerves, which are responsible for constricting and dilating the blood vessels.  BIO

1858 German chemist Friedrich August Kekule von Stradonitz works out the basic structure of organic molecules, showing how they are made up of chains of carbon atoms attached to other atoms.  CHEM

1858 Kekule von Stradonitz and British chemist Archibald Scott Couper, working independently, develop a system of chemical notation in which the valence of each atom is indicated by dashes.  CHEM

1858 American geologist Antonio Snider-Pellegrini is an early advocate of the theory of continental drift, which will be fully developed by German meteorologist and geophysicist Alfred Wegener in the 20th century (*see* 1912, EARTH). Snider-Pellegrini proposes that the Atlantic continents broke up and drifted apart in the distant past.  EARTH

1858 British explorer John Hanning Speke reaches the largest lake in Africa, Lake Victoria Nyanza. In 1862 Speke will prove it to be one of the principal sources of the Nile.  EARTH

1858 *The Geology of Pennsylvania*, the finest geologic map of an American state thus far, appears in print.  EARTH

1858 English mathematician Arthur Cayley publishes an analysis of the theory of transformations, representing the beginning of his study of matrices.  MATH

1858 The United Kingdom passes the Medical Act of 1858, establishing the General Council of Medical Education to regulate and register the medical profession and protect the public from unqualified physicians. *See also* 1878, MED.  MED

1858        British surgeon and anatomist Henry Gray publishes *Gray's Anatomy*, which becomes the standard anatomy textbook for more than a century.          MED

1858        German pathologist Rudolf Virchow publishes what will be his most famous work, *Die Cellularpathologie*, establishing the foundation of cellular pathology, which studies the effect of disease on cells. He determines that there are "no specific cells in disease, only modifications of physiological types." Virchow was the first to describe not only leukemia and the doctrine of embolism (blood vessel obstruction by a clot or foreign substance) but leukocytosis (marked by increased numbers of white blood cells in the blood).          MED

1858        American paleontologist Joseph Leidy describes a partial skeleton of the dinosaur *Hadrosaurus foulkii*, discovered by W. P. Foulke at Haddonfield, New Jersey.          PALEO

1858        German physicist Julius Plücker observes the flow of an electric current through the vacuum of a Geissler tube (*see* 1855, TECH) and notes that the position of the resulting fluorescence shifts when an electromagnetic field is applied.          PHYS

1858        The first usable two-tiered sleeping car is developed, by American cabinetmaker George M. Pullman for the Chicago & Alton Railroad.          TECH

1858        Central Park, designed by Frederick Law Olmsted and landscape designer Calvert Vaux, opens in New York but will not be fully completed for another five years.          TECH

## Gray's Anatomy

British surgeon and anatomist Henry Gray's 1858 book *Gray's Anatomy* was not the only anatomy book on the market in the mid-19th century. Ever since Andreas Vesalius established the modern subject of anatomy with *De humani corporis fabrica* in 1543 there had been numerous texts in the field. But the completeness, clarity, and intelligibility of *Gray's Anatomy* have made it a lasting resource for physicians and medical students.

The Philadelphia publishers Blancard and Lea, purchasers of the American rights to Gray's book in 1859, were right in believing they had their hands on the most understandable anatomy book to date. Colleagues praised Gray for the thoroughness of his dissections, the coherent literary style of the text, and the innovative use of 363 illustrations drawn by Henry Van Dyke Carter.

Despite Gray's other writings on the development of the optic nerve and retina, the human spleen, and the ductless (endocrine) glands, this fellow of the Royal College of Surgeons will probably always be remembered best for the work that begins, "The entire skeleton in the adult consists of 200 distinct bones."

1858    The first mechanically operated refrigerator, which cools through the use of liquid ammonia, is invented by Frenchman Ferdinand P. A. Carré.    TECH

1858    The zinc-lidded Mason jar, which revolutionizes home canning, is devised by American craftsman John Mason.    TECH

1858    On August 16, the first transatlantic cable message is sent, from Queen Victoria to American president James Buchanan. The queen's message reads, "Glory to God in the Highest, peace on earth, goodwill to men." The cable will fail in a few weeks and the first permanent one not be laid until 1866.    TECH

1859    British astronomer Richard Christopher Carrington discovers the differential rotation of the sun's equator and poles.    ASTRO

1859    British naturalist Charles Darwin explains his theory of evolution by natural selection in *The Origin of Species*. He amasses evidence to show that species are modified and new species emerge as individuals better adapted to their environment outreproduce those less well adapted. About the same time, British naturalist Alfred Russel Wallace develops a similar theory. Though controversial, Darwin's theory eventually will be accepted by scientists.    BIO

1859    German chemist Robert Bunsen and German physicist Gustav Kirchhoff invent the spectroscope, a device to chemically analyze elements heated to incandescence. The spectroscope will be used to discover new elements and study the chemical composition of the sun and stars.    CHEM

1859    British physician Alfred Garrod (1819–1907) is the first to clearly distinguish between osteoarthritis (marked by cartilage destruction in joints, and spur formation), and rheumatoid arthritis, which is characterized by joints swelling, deformity, and inflammation.    MED

1859    Scottish mathematician James Clerk Maxwell develops the kinetic theory of gases, which explains the physical properties of gases in terms of the random movement of particles. Existing laws concerning gases, such as Boyle's law (*see* 1662, PHYS), can be deduced from Maxwell's theory.    PHYS

1859    German psychologist Moritz Lazarus and German philologist Heymann Steinthal found the first comparative psychology journal.    PSYCH

1859    American businessmen Orson Fowler and Lorenzo and Samuel Wells establish a firm to sell phrenological equipment and related devices. They write a best-selling book entitled the *Phrenological Self-Instructor*. Phrenology becomes widely accepted as businesses require phrenological exams before hiring, people considering marriage are advised first to consult a phrenologist, and the 1860 presidential candidates are phrenologically analyzed for their leadership abilities.    PSYCH

1859    Scottish physician Lockhart Robinson begins using wet sheets to treat mental illness. In this method a cold, wet sheet is wrapped around a patient's naked body to control agitation. This "wet pack" will be used well into the 20th century, until the advent of antipsychotic drugs in the 1950s.          PSYCH

1859    French anthropologist and surgeon Pierre-Paul Broca founds the Société d'Anthropologie.                                                             SOC

1859    British philosopher and economist John Stuart Mill publishes *On Liberty*, a classic statement of utilitarian political theory, in which he argues that public action should aim to promote the greatest happiness of the greatest number of people.                                                                      SOC

1859    On August 28, American railway conductor Edwin L. Drake drills 69 feet under the ground at Titusville, Pennsylvania, to produce the first petroleum oil well. TECH

---

## The Origin of Species

In 1859, British naturalist Charles Darwin published *On the Origin of Species by Means of Natural Selection* (best known as *The Origin of Species*), the controversial book that introduced the theory of evolution by natural selection as an explanation for the history of life. Meticulously documented and argued, the book was scientifically rigorous but aimed at the general reader. Though sometimes hard to follow through the thicket of supporting examples drawn from plant and animal life, its calm, reasoned prose occasionally approaches quiet grandeur. Nowhere is this more true than in its famous closing paragraph:

It is interesting to contemplate an entangled bank, clothed with many plants of many kinds, with birds singing on the bushes, with various insects flitting about, and with worms crawling through the damp earth, and to reflect that these elaborately constructed forms, so different from each other, and dependent on each other in so complex a manner, have all been produced by laws acting around us. These laws, taken in the largest sense, being Growth with Reproduction; Inheritance which is almost implied by reproduction; Variability from the indirect and direct action of the external conditions of life, and from use and disuse; a Ratio of Increase so high as to lead to a Struggle for Life, and as a consequence to Natural Selection, entailing Divergence of Character and the Extinction of less-improved forms. Thus, from the war of nature, from famine and death, the most exalted object which we are capable of conceiving, namely, the production of the higher animals, directly follows. There is grandeur in this view of life, with its several powers, having been originally breathed into a few forms or into one; and that, whilst this planet has gone cycling on according to the fixed law of gravity, from so simple a beginning endless forms most beautiful and most wonderful have been, and are being, evolved.

1860s    In the early part of the decade, American experimenter L. M. Rutherford devises a telescopic lens system optically designed to take photos—in effect a camera with a telescope as a lens.                                                    ASTRO

1860s    Glass goggles first come into use for underwater exploration.                    TECH

1860     Many notable scientists, including especially the Swiss-American Louis Agassiz, criticize British naturalist Charles Darwin and the evolutionary theory he presented last year in his *Origin of Species*. Agassiz claims, for instance, that each species has been separately—and divinely—created.                    BIO

1860     German chemist Robert Bunsen and German physicist Gustav Kirchhoff discover the elements cesium and rubidium.                    CHEM

1860     By now, French chemist Pierre-Eugène-Marcelin Berthelot has synthesized a number of organic compounds in the laboratory, including some that do not occur in living tissue, disproving the idea that only living tissue can produce organic compounds.                    CHEM

1860     A conference in Karlsruhe, Germany, on the structure of organic molecules is the first international scientific meeting, organized chiefly by German chemist Friedrich August Kekule von Stradonitz. At the conference, Italian chemist Stanislao Cannizzaro urges his fellow colleagues to accept Amedeo Avogadro's 1811 hypothesis that all gases at a given temperature contain the same number of molecules. The acceptance that is eventually given permits the determination of the molecular weight of different gases.                    CHEM

1860     French physician Georges Hayes begins his studies on platelets, the smallest cellular elements of the blood, and is the first to accurately count them. Platelets will prove important in the diagnosis and treatment of many immunological diseases.                    MED

1860     Dutch ophthalmologist Frans Donders introduces the use of glasses to correct visual astigmatism.                    MED

1860     The first nurse-training school is established by Florence Nightingale at London's St. Thomas Hospital.                    MED

1860     German physicist Gustav Kirchhoff hypothesizes that a blackbody, or a body that absorbs all incoming light, will emit all wavelengths of light when heated.PHYS

1860     Scottish physicist James Clerk Maxwell and Austrian physicist Ludwig Eduard Boltzmann independently develop the Maxwell-Boltzmann statistics analyzing the statistical behavior of molecules in a gas.                    PHYS

1860     French physicist Gaston Planté invents a rechargeable storage battery. At this time he also invents the first workable electric storage battery.                    TECH

1860        Samuel Archer King and William Black become the first aerial photographers when they take two photographs of Boston from a balloon.   TECH

1860        Belgian inventor Jean-Joseph Étienne Lenoir patents an early internal combustion engine. A more efficient engine will be created by Nikolaus Otto in 1876.   TECH

1860        The first Winchester repeating rifle is put into production in the United States.   TECH

1860        The Pony Express begins operation transporting mail by relay teams on horseback between Missouri and California.   TECH

1860        Linoleum, a linseed-oil-based hard floor covering, is developed by British inventor Frederick Walton.   TECH

1861        German physicist Gustav Kirchhoff shows that the composition of the sun can be determined by spectral analysis.   ASTRO

1861        German cytologist Max Schultze develops the theory of protoplasm (cell substance), based on his own work and on earlier cell studies by Casper Friedrich Wolff, Félix Dujardin, Jan Purkinje, and Hugo von Mohl. According to Schultze, a cell's most basic components are its nucleus and protoplasm.   BIO

1861        British physicist Sir William Crookes discovers the element thallium.   CHEM

1861        German chemist Friedrich August Kekule von Stradonitz publishes a textbook on organic chemistry that defines its subject as the chemistry of carbon compounds, not living things.   CHEM

1861        Scottish physical chemist Thomas Graham distinguishes colloids from crystalloids. The former are substances such as gelatin and starch that will not dissolve through a membrane, whereas the latter will.   CHEM

1861        American surgeon Gordon Buck introduces the weight-and-pulley apparatus since called Buck's traction to use with orthopedic patients.   MED

1861        American gynecologist James Marion Sims develops a method for the surgical removal of the cervix and describes vaginismus, or painful vaginal spasms. MED

1861        French physician Prosper Ménière describes the syndrome that will bear his name, in which sufferers experience recurrent episodes of vertigo and deafness caused by an inner-ear disorder resulting from an enlarged cochlear duct.   MED

1861        Quarry workers in Solnhofen, Germany, unearth a fossil of *Archaeopteryx*. With both reptilian and avian features, this creature appears to represent an intermediate phase in the evolution of birds, thus lending support to Darwin's theories (*see* 1859, BIO). *Archaeopteryx* flourished during the Jurassic period, about 150 million years ago.   PALEO

1861        French surgeon Pierre-Paul Broca discovers the speech center in the brain,
            which leads to his theory of the localization of function. His work will initiate
            systematic mapping of areas of the brain and their functions, a fundamental
            concept in brain surgery.                                              PSYCH

1861        The velocipede, or bicycle, is developed by Parisian transportation designer
            Pierre Michaux. *See also* 1839, TECH.                                  TECH

1861        The multiround, rapid-firing Gatling gun is developed by American engineer
            Richard Jordan Gatling.                                                 TECH

1861        Steelmaking is streamlined with the development of the open-hearth process,
            invented independently by French engineer Pierre Émile Martin and German-
            born British inventor William Siemens.                                  TECH

1862        German biochemist Ernst Felix Hoppe-Seyler is first to crystallize and name
            the protein called hemoglobin that is responsible for the blood's red color. He
            also establishes the first biochemistry laboratory, at the University of
            Tübingen.                                                               BIO

1862        British naturalist Charles Darwin writes in detail on cross-fertilization, includ-
            ing insects' role in the process, in his *Fertilization of Orchids*.    BIO

1862        German botanist Julius von Sachs establishes that plant starch results from pho-
            tosynthesis with carbon dioxide absorbed from the air.                  BIO

c. 1862     British physicist William Thomson, Lord Kelvin, launches an attack on uniformi-
            tarianism and Darwinian evolution, arguing that, given the present temperature
            of the earth and the presumed rate of its past cooling, the earth is not as old as
            supposed. This argument will seem incontrovertible until the discovery of
            radioactivity in 1898 shows that the earth possesses a source of internal
            heat unknown to Thomson.                                                EARTH

1862        French chemist Louis Pasteur publishes a paper supporting the germ theory of
            disease. His own discoveries will later provide further evidence (*see* 1880, MED).
                                                                                    MED

1863        British astronomer William Huggins argues from stellar spectra that stars are
            composed of the same elements that exist on Earth.                      ASTRO

1863        German mineralogists Ferdinand Reich and Theodor Richter discover the ele-
            ment indium.                                                            CHEM

1863        British chemist John Alexander Reina Newlands makes a list of elements in
            order of their atomic weight, claiming to find a law of octaves governing the
            properties of the elements, a claim later discredited.                  CHEM

1863        German chemist Adolf von Baeyer discovers barbituric acid, the parent com-
            pound of barbiturates.                                                  CHEM

1863        Irish physicist John Tyndall discovers what will become known as the green-house effect: Gases such as carbon dioxide and water vapor in the atmosphere permit visible light to enter but tend to hold in the infrared radiation emitted by the earth's surface. As a result, the earth's surface is warmer than it would be without these gases.                                                          EARTH

1863        British scientist Francis Galton publishes *Meteorographica*, which introduces modern weather-mapping techniques and the concept of anticyclones, or high-pressure areas of the atmosphere.                                                 EARTH

1863        British geophysicist Augustus Love discovers the phenomenon later called the Love wave, a type of earthquake. His subsequent research allows geologists to measure the thickness of the earth's crust.                                        EARTH

1863        American geologist James Dwight Dana writes his *Manual of Geology*, an essential text of American geology for four decades.                                  EARTH

c. 1863     Dutch ophthalmologist Herman Snellen develops the Snellen eye chart to test sharpness of vision.                                                             MED

1863        French chemist Louis Pasteur develops the process of partial heat sterilization later called pasteurization that prevents wine spoilage. It will later be used with milk and beer.                                                            MED

1863        On October 26, Swiss humanitarian Jean-Henri Dunant establishes a relief society in Geneva that will eventually evolve into the International Red Cross.                                                                                MED

1863        The U.S. Congress gives the National Academy of Sciences in Washington, D.C., its charter to advise the government on issues concerning science and technology.                                                                          MISC

1863        English scientist Thomas Henry Huxley publishes *Man's Place in Nature*, the first book to take a scientific approach to human evolution. Using comparative anatomy, Huxley deduces that the chimpanzee and the gorilla are the closest relatives of humans and that all evolved according to the same principles.                                                                                   PALEO

1863        The London Underground, the first subsurface public railway, is opened.  TECH

1863        Granula (not Granola), the first cold breakfast food, made of baked graham, is introduced to the public by American sanitarium operator James Caleb Jackson.                                                                             TECH

1864        The distance of Earth from the sun is calculated as 91 million miles, close to the actual value of 92.96 million miles.                                          ASTRO

1864        British mathematician and astronomer John Herschel publishes *A General Catalogue of Nebulas*, which lists 2,500 nebulas catalogued by himself and an additional 2,500 catalogued by his father, William Herschel. The catalogue will be revised in 1888 by Johann L. E. Dreyer. **ASTRO**

**1864–1867** In his *Principles of Biology* (1864, 1867), British scholar Herbert Spencer coins the term *survival of the fittest* to describe a way of understanding Darwin's theory of natural selection. Spencer will also suggest that the concept implies that the weakest, most "useless" members of society should be allowed to die off. **BIO**

**1864** French chemist Louis Pasteur proves that microorganisms exist in the atmosphere and are not spontaneously generated. *See* 1748, MED. **MED**

**1864** In an effort to cut down the spread of syphilis, England passes the Contagious Diseases Act, requiring compulsory medical exams and registration of all prostitutes in towns and seaports where military personnel are housed. **MED**

**1864** Influential German sociologist Max Weber is born (*d.* 1920). He will develop methodologies for cross-cultural studies and argue that sociology should be value free. *See* 1904, SOC. **SOC**

**1864** Americans George M. Pullman and Benjamin Field patent the design for a sleeping car, the *Pioneer*, with convertible berths. **TECH**

**1865** While daydreaming about a snake coiling in on itself, German chemist Friedrich August Kekule von Stradonitz discovers the ring structure of the benzene molecule. **CHEM**

**1865** Austrian chemist Johann Joseph Loschmidt calculates the number of molecules in one mole of a gas. The number, which has the approximate value $6 \times 10^{23}$, is called Avogadro's number or, later, the Avogadro constant. From this number the mass of molecules and atoms can be calculated. **CHEM**

**1865** Construction of the Union Pacific Railroad begins in the United States. **EARTH**

**1865** On July 14, British artist turned mountaineer Edward Whymper reaches the top of the Matterhorn in the Alps. **EARTH**

**1865** German mathematician August Ferdinand Möbius invents the Möbius strip, a ribbon of paper with only one edge and one side. This invention is one of the founding events of topology, the branch of geometry concerned with the properties of objects that are not changed by continuous deformations such as stretching and twisting. **MATH**

**1865** French physiologist Claude Bernard marks a medical ethics milestone when he concludes that medical experiments can be permissible and even obligatory. He displays no interest in the informed consent issue. **MED**

**1865** After studying French chemist Louis Pasteur's theories on bacteria, English physician Joseph Lister successfully uses carbolic acid as a surgical antiseptic on August 12. This antiseptic will forever change the survival rate and safety of surgery, as gangrene and suppuration (pus formation) are remarkably reduced.

**MED**

1865    Scottish mathematician and physicist James Clerk Maxwell formulates Maxwell's equations, which describe the behavior of electric and magnetic fields and show that a changing field of one type will generate a field of the other type. The value Maxwell calculates for the speed of electromagnetic radiation closely matches the measured speed of light, leading him to propose that light is electromagnetic in nature. These equations represent the first theoretical unification of physical phenomena.                                    **PHYS**

1865    British anthropologist Edward B. Tylor publishes his *Early History of Mankind*. **SOC**

1865    A design and production system to manufacture the interchangeable parts of rifles is developed by American machinists Francis Asbury Pratt and Amos Whitney. The two found their industrial factory, Pratt & Whitney, this year.    **TECH**

## The Electromagnetic Spectrum

Since the publication of James Clerk Maxwell's equations describing electromagnetism in 1865, scientists have come to understand the full range of wavelengths over which electromagnetic radiation extends. Visible light is only one narrow part of this electromagnetic spectrum. The spectrum of radiation, from longest to shortest wavelengths, is as follows:

| Type | Wavelength (Meters) |
|------|---------------------|
| Radio | $10^5 - 10^{-3}$ |
| Infrared | $10^{-3} - 10^{-6}$ |
| Visible light | $10^{-7}$ |
| Ultraviolet | $10^{-7} - 10^{-9}$ |
| X rays | $10^{-9} - 10^{-11}$ |
| Gamma rays | $10^{-11} - 10^{-14}$ |

These varieties of radiation share a common wave structure of crests (high points) and troughs (low points). They differ only in the length of their waves or, to put it another way, their frequency, the number of times per second a wave passes a given point. Wavelength and frequency are inversely proportional; that is, long waves have low frequencies, short waves high frequencies. Frequency is also a reflection of energy level. The highest-frequency (shortest) waves, gamma rays and X rays, are the most energetic and most potentially damaging to delicate systems like living tissue. The lowest-frequency, or longest, waves (infrared and radio) are the weakest waves.

Despite the explanatory power of the wave theory of electromagnetic radiation, it would become clear in the 20th century (see 1900, **PHYS**) that such rays could also be usefully regarded as particles, or quanta.

1865        American inventor Linus Yale patents the Yale lock, a cylinder lock that will
            become popular for both commercial and residential use.                    TECH

1865        A compression ice-making machine is devised by American inventor Thaddeus
            Sobieski Coulincourt Lowe. It will be useful in developing refrigeration
            technology.                                                                 TECH

1866        American astronomer Daniel Kirkwood discovers the gaps later known as
            Kirkwood gaps in the distribution of asteroids. The uneven spacing is caused
            by the influence of Jupiter's gravity.                                      ASTRO

1866        After experimenting with the crossbreeding of sweet peas and making a statis-
            tical analysis of the results, Austrian monk Gregor Mendel concludes that each
            of an organism's inherited traits is determined by two heredity units called
            genes, one from each parent organism. Forgotten until the 20th century,
            Mendel's laws of inheritance will form the basis of classical genetics. It will
            later be understood that Mendel's genes are the material acted upon by natur-
            al selection in Darwin's theory of evolution (*see* 1859, BIO).              BIO

1866        German biologist Ernst Haeckel is first to use the word *ecology* to describe the
            relations and interaction between organisms and their environment.           BIO

## Cholera and God's Will

The U.S. cholera epidemics of 1832, 1849, and 1866 killed tens of thousands of
Americans in several cities. It is now known that cholera is spread through contami-
nated food, contact with afflicted persons, and, particularly, impure drinking water. It
is also recognized that education about disease prevention and a supply of safe drinking water is the best deterrent to the spread of the disease.

But it was not until 1866 that public health agencies like New York City's
Metropolitan Board of Health began to use new sanitation standards and education to
stem the spread of the disease.

It was not until 1883 that the *Vibrio comma* bacterium that causes cholera was discovered.

Before then, most Americans saw cholera as a God-given retribution for moral failure. It was divine punishment for a life not lived in adherence to religious beliefs and
upright behavior. Comments in newspapers like this August 1849 *Scioto Gazette* from
Chillicothe, Ohio, are representative of that belief in divine force as a cause of the
spread of the disease.

> The Almighty conducts the order of Providence by rule—rules that are alike
> at all times and abiding. Who however shall set bounds to the physical and the
> moral—and pretend to tell where one ends and the other begins? Who can say
> whether it is from chemical or moral causes that this community is scourged,
> while that is spared?

| | |
|---|---|
| 1866 | Haeckel popularizes the idea that ontology recapitulates phylogeny, meaning that an embryo's stages of development mirror the evolutionary history of that species. Haeckel calls this the biogenetic law, but the doctrine will later be considered overstated, for embryonic development is far more complex.    BIO |
| 1866 | French geologist Gabriel-Auguste Daubrée proposes the existence of an iron and nickel core at the center of the earth.    EARTH |
| 1866 | F. Zirkel publishes *Lerbuch der Petrography*, a textbook on petrology notable for including an account of H. C. Sorby's technique for examining thin sections of rock with a polarizing microscope.    EARTH |
| 1866 | In Beijing, China, a department of mathematics and astronomy is created at the Tongwenguan, previously a college for interpreters. The move signals growing interest in China in educating modern scientists.    MISC |
| 1866–1867 | English scientist Thomas Henry Huxley names two Triassic dinosaurs from South Africa, *Euskelosaurus* and *Orosaurus*.    PALEO |
| 1866 | German physicist August Kundt designs what will become known as Kundt's tube, a device to measure the speed of sound in different fluids.    TECH |
| 1866 | The first usable dynamo-electrical machine or dynamo, which allows for the mass generation of electricity, is developed by Ernst Werner von Siemens.    TECH |
| 1866 | Dynamite is developed by Swedish engineer Alfred Nobel. The new compound is a mixture of nitroglycerin and diatomaceous earth.    TECH |
| 1866 | American merchant Cyrus West Field lays the first permanent transatlantic telegraph cable. *See also* 1854, EARTH.    TECH |
| 1867 | The International Botanical Congress establishes the rules for botanical nomenclature known as the Paris Code. However, not until 1930 will a botanical code with worldwide acceptance be adopted.    BIO |
| 1867 | Norwegian chemists Cato Maximilian Guldberg and Peter Waage formulate the law of mass action, which states that the rate at which a chemical reaction occurs is proportional to the product of the active masses of the reactants. The active masses are defined by their concentration, the quantity of mass in a given volume.    CHEM |
| 1867 | British gynecologist John Braxton Hicks describes the painless uterine contractions in pregnancy that come to be called false labor.    MED |
| 1867 | British physician Sir Thomas Allbutt introduces the modern clinical thermometer.    MED |
| 1867–1894 | German social philosopher Karl Marx publishes *Das Kapital*, which his colleague Friedrich Engels finishes editing after Marx's death in 1883. The work is |

a systematic exposition of the economic and political philosophy that will become known as Marxism, which places class struggle at the center of history and predicts the overthrow of the capitalist class by the exploited working class. Marx's work strongly influences economics, political science, and sociology, among other fields.                           soc

1867    French engineer Georges Leclanché invents the dry cell, an electric battery employing a chemical paste rather than fluids.                           TECH

1867    The safety boiler, which reduces explosions, is patented by American engineers George Babcock and Stephen Wilcox. In the same year the two found the company Babcock & Wilcox.                           TECH

1867    The first U.S. steam-operated elevated railway opens in New York City.    TECH

1868    British astronomer Sir William Huggins uses the Doppler-Fizeau effect (*see* 1848, PHYS) to calculate that the star Sirius is moving away from Earth at 29 miles per second, a relatively close estimate.                           ASTRO

1868    Italian astronomer Pietro Angelo Secchi completes the first spectroscopic catalog, of about 4,000 stars.                           ASTRO

1868    German botanist Julius von Sachs publishes his textbook *Lehrbuch der Botanik*, which will have a major influence on botany's emergence as a comprehensive discipline.                           BIO

1868    French astronomer Pierre-Jules-César Janssen and British astronomer Joseph Norman Lockyer discover the element helium. *See also* 1895, CHEM.    CHEM

1868    French paleontologist Édouard-Armand-Isidore-Hippolyte Lartet discovers four prehistoric human skeletons in Cro-Magnon, France. First appearing some 40,000 years ago, Cro-Magnon man will become known as one of the earliest anatomically modern human ancestors.                           PALEO

1868    British physician Sir William Gull becomes the first to describe anorexia nervosa, an eating disorder characterized by self-starvation.                           PSYCH

1868    American George Westinghouse invents the Westinghouse air brake, which will be crucial to more efficient train travel. In 1872 he will introduce the automatic air brake.                           TECH

1868    The knuckle coupler, a safer alternative to the old standard link-and-pin system of joining rail cars, is patented by former American military officer Eli Hamilton Janney.                           TECH

1868    Tungsten steel, harder than other types, is developed by English metallurgist Robert Forester Mushet.                           TECH

1868    The typewriter is patented by American inventors Christopher Sholes, Carlos G. Glidden, and Samuel W. Soulé. In 1874, it will be produced in quantity by the Remington Arms Co.                                                  TECH

1869    English naturalist and zoologist Alfred Russel Wallace publishes *The Malay Archipelago: The Land of the Orangutan and the Bird of Paradise*. In it, he points out that Malay animal life is divided into two distinct types. The animal life in the western half is like that in India, that in the eastern half more like Australia's. The line dividing the two regions will henceforth be called Wallace's line. This discovery marks the foundation of biogeography.    BIO

1869    French botanist Jules Raulin proves that zinc is crucial to the growth of the plant *Aspergillus*.                                                            BIO

1869    On March 6, Russian chemist Dmitry Ivanovich Mendeleyev publishes the periodic table of the elements.                                              CHEM

1869    Modern medicine's first demonstration of free skin grafting is performed in Paris in December.                                                           MED

1869    Irish physical chemist Thomas Andrews suggests that every gas has a critical temperature above which it cannot be turned into liquid by an increase in pressure alone.                                                            PHYS

1869    Irish physicist John Tyndall discovers what will become known as the Tyndall effect, a characteristic scattering of light through a medium containing small particles. The effect accounts for the blueness of the sky and the redness of sunsets.                                                               PHYS

1869    British scientist Francis Galton publishes *Hereditary Genius*, concerning the familial tendency to inherit brilliance. He will pioneer the use of statistics in psychological measurement.                                          PSYCH

1869    The 103-mile-long Suez Canal opens for use, uniting the Gulf of Suez on the south with the Mediterranean Sea via the Red Sea. Trade and travel become far less onerous.                                                           TECH

1869    Margarine, an oil-based butter substitute, is produced commercially for the first time by Frenchman Hippolyte Mége-Mouries.                          TECH

1869    The U.S. transcontinental railroad is completed on May 10 at Promontory Point, Utah. There 1,006 miles of line built by the Union Pacific Railway from Omaha, Nebraska, joins 680 miles built by the Central Pacific Railway from Sacramento, California.                                                   TECH

1869    The first commercially successful plastic, celluloid, is developed by American inventor John Wesley Hyatt. Hyatt's substance is an advance on the similar but unsuccessful product pyroxylin (*see* 1855, TECH).                     TECH

1870s   German mathematician Georg Cantor establishes the theory of sets (or *Mengenlehre*, aggregates) as a distinct field of research and develops an arithmetic of transfinite numbers.                                    MATH

1870s   At mid-decade American paleontologist Othniel Charles Marsh makes a number of major contributions to vertebrate paleontology. In addition to clarifying the evolutionary history of the horse, Marsh discovers many dinosaur species and offers evidence of a link between reptiles and birds. *See also* 1877, PALEO.

PALEO

1870s   Austrian physicist Ernst Mach states what will become known as Mach's principle, that a body's inertia can be attributed to the interaction between that body and the rest of the universe and a body in isolation will have an inertia equal to zero.                                    PHYS

1870   German archaeologist Alexander Conze classifies chronological styles of pottery, providing a key to dating of strata.                                    ARCH

1870   English biologist Alfred Wallace publishes *Contributions to the Theory of Natural Selection* and American paleontologist Edward Drinker Cope publishes *Systematic Arrangement of the Extinct Bactrachia, Reptilia, and Aves of North America*. Both publications provide additional support for the theory of evolution.                                    BIO

1870   German chemist Adolf von Baeyer becomes the first person to synthesize the blue dye indigo (*see also* 1193, TECH, and 1741, BIO). In 1905 he will win the Nobel Prize for chemistry for his work concerning organic dyes and hydroaromatic compounds.                                    CHEM

1870   The U.S. Weather Bureau is established, largely due to the efforts of meteorologist Cleveland Abbe.                                    EARTH

c. 1870   German surgeon Friedrich Trendelenburg begins positioning surgery patients on their backs with the pelvis elevated and the head downward at an angle. This Trendelenburg's position will come into common use in abdominal surgery and to combat shock.                                    MED

c. 1870   The passage of pure food laws in western Europe advances the cause of preventive medicine.                                    MED

1870   Drawing on his studies of the Seneca people, American anthropologist Lewis Henry Morgan correlates kinship terminology with cultural traditions of marriage and descent.                                    SOC

1870   English machinists William Hillman and James Kemp Starley patent a light, all-metal bicycle. *See also* 1839, TECH.                                    TECH

1870   A rotary press capable of printing two sides of a page at once is developed by American inventor Richard Hoe.                                    TECH

1870        The first personal mechanical can opener, using a moving wheel to cut metal, is developed by American inventor William Lyman.                                                TECH

1871        German archaeologist Heinrich Schliemann begins excavating a hill called Hissarlik in Turkey and discovers the ruins of ancient Troy, the legendary site of Homer's *Iliad*.                                                                                          ARCH

1871        Russian chemist Dmitry Ivanovich Mendeleyev predicts the properties of three elements missing from his periodic table—eka-boron, eka-aluminum, and eka-silicon. His predictions later prove to be correct.                                     CHEM

1871        German chemist Max Bodenstein develops the concept of a chemical chain reaction, a reaction whose products cause further reactions of the same kind and is therefore self-sustaining.                                                              CHEM

1871        Harvard University supports American physiologist Henry Pickering Bowditch in establishing the first U.S. physiology laboratory. He will become known for childhood-growth investigations that will determine that growth is dependent on good nutrition and that arrested development can be a signal of disease.

MED

1871        British naturalist Charles Darwin publishes *The Descent of Man, and Selection in Relation to Sex*, in which he extends his theory of natural selection to human evolution and invents the concept of sexual selection. He also suggests that humans evolved from apelike ancestors in Africa millions of years ago.     PALEO

1871–1893   British scientist William Crawford Williamson publishes a series of papers describing fossil plants found in coal deposits in northern England. This work proves a major contribution to paleobotany research.                              PALEO

1871        British anthropologist Edward B. Tylor publishes *Primitive Culture*, in which he discusses animistic religion and develops methods for making cultural comparisons.                                                                                                    SOC

1872        From clay tablets discovered at Nineveh (*see* 1840s, ARCH), English archaeologist George Smith deciphers the *Epic of Gilgamesh*, a Sumerian story that may date back to as early as 2500 B.C.                                              ARCH

1872        German Egyptologist Georg Ebers discovers the Ebers papyrus (*see* 1600 B.C., MED) in the ruins of Thebes, Egypt. It becomes important for its documentation of ancient Egyptian medicine.                                                       ARCH

1872        Studying the star Vega, American astronomer Henry Draper becomes the first to photograph a stellar spectrum.                                                                      ASTRO

1872        German botanist Ferdinand J. Cohn founds the science of bacteriology.  BIO

1872        French biology professor Henri de Lacaze-Duthiers establishes a "Laboratory of Experimental Zoology," the Roscoff Biological Station, at Roscoff, Brittany.  BIO

1872            A U.S. government survey of the 40th parallel, led by geologist Clarence King, exposes a mining stock fraud in what will come to be known as "the diamond hoax." EARTH

1872–1876      The *Challenger* expedition, a cooperative effort of the Admiralty and the Royal Society, voyages a total of 68,890 miles, making 492 deep soundings and 133 dredgings. The scope and thoroughness of this research makes the expedition a milestone in the history of oceanography. EARTH

1872            American Ferdinand Hayden has the Yellowstone region—which lies mainly in northwestern Wyoming and extends into Montana and Idaho—declared the first U.S. national park, marking the beginning of the national park movement. EARTH

1872            American geologist John Wesley Powell begins systematically surveying the canyon country of Colorado. EARTH

1872            Contributions are made to the arithmetization of analysis by several mathematicians, including Frenchman H. C. R. Méray and Germans Karl Weierstrass, H. E. Heine, Georg Cantor, and J. W. R. Dedekind. MATH

1872            German mathematician J. W. R. Dedekind introduces the so-called Dedekind cut, with which irrational numbers can be defined. MATH

1872            German mathematician Felix Klein gives an address known as the Erlanger Program, in which he describes geometry as the study of invariant properties under a particular transformation group. MATH

c. 1872         The first U.S. nursing school is founded, at New York City's Bellevue Hospital. MED

1872            Austrian dermatologist Moritz K. Kaposi describes the skin disease later called Kaposi's sarcoma that will become associated with AIDS (acquired immunodeficiency syndrome) in the late 20th century. MED

1872            American physician George Sumner Huntington describes the hereditary neurological disease characterized by grotesque grimaces and dementia that will become known as Huntington's chorea. PSYCH

1872            German aviator Paul Haenlein flies in the first dirigible to be powered by an internal combustion engine, fueled by coal gas from the supporting bag. TECH

1872            Gum is sold in stick form for the first time, by American photographer Thomas Adams, who has flavored his Blackjack gum, made from chicle, with licorice. TECH

1873            British astronomer Richard Anthony Proctor argues that lunar craters were formed by meteoric impact. ASTRO

1873 | The occurrence of mitosis—mitotic nuclear division—in cells is recognized for the first time. BIO

1873 | Italian histologist Camillo Golgi introduces cellular staining with silver salts. The network of membranous vesicles in the cell known as the Golgi apparatus is revealed for the first time with this staining. BIO

1873 | British scientist Walter Bagehot publishes *Physics and Politics*, applying the theory of natural selection to the rise and fall of human customs and institutions. BIO

1873 | The International Meteorological Organization is founded. EARTH

1873 | German mineralogist Ferdinand Zirkel and German geologist Karl Rosenbusch simultaneously but independently publish similar treatises on microscopical petrography, thus laying the foundations of the field. EARTH

1873 | French mathematician Charles Hermite identifies the first proven transcendental number: *e*, equal to 2.71828... Transcendental numbers are those that are not algebraic, meaning they cannot serve as solutions to any conceivable polynomial equation. MATH

1873 | Dutch physicist Johannes Diderik van der Waals modifies existing gas laws, observing that only perfect gases (those with zero-volume molecules and no attraction between molecules, which exist only in theory) obey the unmodified laws. Most real gases under most conditions approximately follow the existing laws, but van der Waals introduces the more accurate van der Waals equation of state that is especially useful at extremes of temperature and pressure. PHYS

1873 | The magnetic effect of electric convection is observed by American physicist Henry Augustus Rowland. TECH

1873 | The world's first cable-car streetcar begins operation, in San Francisco. TECH

1873 | In France the perfume industry is simplified with discoveries of a process to extract floral essences. TECH

1873 | Barbed wire, or the devil's rope, is demonstrated for fencing off farmers' crops and ranchers' stock by American Henry Rose. Machines to produce barbed

## The Real Henry Higgins

The insufferable Henry Higgins in George Bernard Shaw's 1912 play *Pygmalion*, the basis for the 1956 musical *My Fair Lady*, was based on an actual professor of phonetics and Old English, Henry Sweet (1845–1912). This British scholar invented his own system of phonetics and published such influential works as *History of English Sounds* (1874) and *Handbook of Phonetics* (1877).

wire are subsequently patented by Americans Jacob Haish and Joseph Farwell Glidden.                                                                                    TECH

1874    American photographer William Henry Jackson explores the ancient Native American cliff dwellings of Mesa Verde, Colorado. The area will become a national park in 1906.                                                                    ARCH

1874    German zoologist Anton Dohrn founds the Naples, Italy, Zoological Station, the model for many biological stations in the future.                          BIO

1874    Dutch chemist Jacobus van't Hoff (and, independently, French chemist Joseph-Achille Le Bel) advances the tetrahedral carbon atom as a three-dimensional representation of organic molecules. This model proves useful in explaining optical activity, the ability of some substances to rotate the plane of polarized light (see 1815, CHEM, and 1846, PHYS), and creates interest in stereochemistry, the study of how the structural arrangement of atoms in molecules affects a substance's chemical properties.                                        CHEM

1874    German mathematician Georg Cantor investigates transfinite numbers. MATH

1874    British physician and feminist Sophia Louisa Jex-Blake founds the London School of Medicine for Women, to ensure the availability of medical education for women.                                                                          MED

1874    German physicist Ferdinand Braun notes that in certain crystals an electric current will flow in one direction but not in the opposite one.                    PHYS

1874    Austrian novelist Leopold von Sacher-Masoch writes Die Messalinen Weins, in which he describes sexual gratification derived from subjection to pain. His name is the origin of the term masochism. See also 1791, PSYCH.              PSYCH

1874    American psychologist Edward Lee Thorndike is born on August 31 in Williamsburg, Massachusetts (d. 1949). He will be the first psychologist to study animal behavior in laboratory experiments. At Columbia University he will teach that psychology should study behavior, not conscious experience or mental conditions. He will apply work with animals to human education, leading to his pioneering work in the field of mental measurement.       PSYCH

1874    The steel-arched Eads Bridge at St. Louis, Missouri, becomes the first bridge to span the Mississippi River.                                                      TECH

1874    Chilled ridged rollers are used, along with millstones, to produce flour more efficiently at the C. C. Washburn mill in Wisconsin.                              TECH

1875    In California, Luther Burbank establishes a plant nursery for the study and cultivation of grasses, grains, fruits, and vegetables.                            BIO

1875    German embryologist and anatomist Oscar Hertwig demonstrates that sperm enters the ovum, at which point fertilization takes place.                          BIO

1875            French chemist Paul-Emile Lecoq de Boisbaudran discovers the element galli-
               um, which corresponds to Mendeleyev's eka-aluminum and thus supports the
               validity of the periodic table. *See* 1869 and 1871, CHEM.                      CHEM

1875            V. V. Dokudaev begins field studies of Russian soils from which he evolves the
               basic principles of the new science of pedology, or soil science.               EARTH

1875            Russian mathematician Sonya (Sofya) Vasilyevna Kovalevskaya furthers the work
               of Augustin Louis Cauchy in solving partial differential equations; the result is
               called the Cauchy-Kovalevskaya theorem.                                         MATH

1875            German pathologist Rudolf Virchow becomes the first to describe congenital
               spina bifida, a defect in the spinal canal walls that allows the spine to protrude
               in the form of a tumor.                                                          MED

1875            Scottish physicist John Kerr discovers what will become known as the Kerr
               effect, or the ability of some substances, when placed in an electric field, to dif-
               ferentially refract light waves whose vibrations occur in two directions.       PHYS

1875            British physicist William Crookes invents the radiometer, a set of pivoted
               vanes whose spontaneous motion in sunlight in a partial vacuum lends sup-
               port to the kinetic theory of gases.                                             PHYS

1875            Swiss psychiatrist Carl Gustav Jung is born in Zurich (*d.* 1961). He will be best
               known for his theories on the collective unconscious, the interpretation of
               dreams, and the division of personalities into introvert and extrovert. He will
               form his own school of analytical psychology after breaking with his teacher
               and colleague Sigmund Freud.                                                     PSYCH

1875            French neurologist Guillaume Duchenne (*b.* 1806) dies. Duchenne, a pioneer
               of neurology, was the founder of electrotherapy and described numerous neu-
               rological disorders. The spinal cord and brain stem degeneration which devel-
               ops in untreated syphilis, Duchenne's disease, was named for him.               PSYCH

1875            British philosopher and economist John Stuart Mill proposes post-hoc analy-
               sis: the examination of different childhood experiences to discover the effect
               they have on moral character development. In the 20th century, developmen-
               tal psychologists will use this procedure in longitudinal studies of children.  PSYCH

c. 1875         American sociologist William Graham Sumner, a follower of British scholar
               Herbert Spencer, teaches the first American course in sociology, at Yale
               University.                                                                      SOC

1875            British engineer Thomas Moy demonstrates the "Aerial Steamer," at the Crystal
               Palace in London. This steam-powered pilotless monoplane lifts itself six inch-
               es off the ground while attached to a tether on a circular track.                TECH

1875            British physicist William Crookes invents an improved vacuum tube that will
               be called the Crookes tube.                                                      TECH

1875        Blasting gelatin is developed by Swedish engineer Alfred Nobel.        TECH

1875        The telephone is invented by Scottish-American teacher Alexander Graham
            Bell. It comes about as a result of Bell's efforts to develop a system of electric
            speech for the deaf students he teaches.        TECH

1875        A wax-stencil duplicating process that is developed now will be patented next
            year by American inventor Thomas Alva Edison. These machines will be
            improved by both Edison and the machine's authorized licensee, American
            Albert Blake Dick, who will eventually develop the A. B. Dick Diaphragm
            Mimeograph.        TECH

1875        Synthetic vanillin, which in its natural form is an ingredient of vanilla beans, is
            patented by British scientist Ferdinand Tiemann.        TECH

1875        The first milk chocolate for public consumption, made by mixing chocolate
            and sweetened condensed milk, is pioneered in Vevey, Switzerland, by staff
            members of the Nestlé and Daniel Peter chocolate factories.        TECH

1876–1878   German archaeologist Heinrich Schliemann excavates ancient Mycenae, the
            legendary city of the Greek leader Agamemnon.        ARCH

1876        English naturalist Alfred Wallace systematizes the science of biogeography. In
            his *Geographical Distribution of Animals*, Wallace discusses extinct fauna and
            their geographical movement or relocation over the years.        BIO

1876        British naturalist Charles Darwin publishes *The Effects of Cross and Self
            Fertilization in the Vegetable Kingdom*, in which he claims that cross-fertilization
            increases the strength and vigor of offspring.        BIO

1876        American physicist Josiah Willard Gibbs begins publishing a two-year series of
            papers outlining the laws of chemical thermodynamics. In discussing the basic
            forces that operate in chemical reactions he defines the concepts of free ener-
            gy, chemical potential, equilibrium between phases of matter, degrees of free-
            dom, and the equation called the phase rule.        CHEM

c. 1876     Between now and 1880 Henrik Mohn and Cato Maximilian Guldberg devel-
            op the field of dynamical meteorology, with the results known as the Mohn-
            Guldberg equations.        EARTH

1876        British explorer Henry M. Stanley circumnavigates Africa's Lake Victoria and
            charts the course of the Congo River.        EARTH

1876        German bacteriologist Robert Koch isolates the anthrax bacillus, the cause of
            an infectious disease that attacks cattle, sheep, horses, and goats and is passed
            to humans by these animals.        MED

1876        American librarian Melvil Dewey invents the Dewey decimal system for classi-
            fying and organizing books in libraries.        MISC

1876        American paleontologists Edward Drinker Cope and Charles H. Sternberg explore the Judith River formation in Montana, discovering fossils of Cretaceous dinosaurs.                    PALEO

1876        Near his home in Trenton Falls, New York, geologist Charles Doolittle Walcott discovers the first U.S. trilobite with its appendages preserved.                    PALEO

1876        German physicist Eugen Goldstein discovers cathode rays, streams of fluorescence flowing from the cathode, or negatively charged electrode, in an evacuated tube.                    PHYS

1876        Scottish psychologist Alexander Bain founds the periodical *Mind*, the first general psychological journal. At this time psychology begins to develop as a discipline apart from philosophy and physiology.                    PSYCH

1876–1896   British scholar Herbert Spencer applies evolution concepts to human sociology. In his three-volume *Principles of Sociology*, Spencer writes about the process of individual differentiation from a group and its effect on individual freedom.                    SOC

1876        German inventor Nikolaus August Otto constructs the first practical internal-combustion engine, in which the piston makes four strokes (a four-cycle engine).                    TECH

1876        The Corliss 160-horsepower engine or dynamo, named for American designer George Henry Corliss, is used to run exhibits at Machinery Hall in the U.S. Centennial Exposition. It is immortalized as a 19th-century icon in the essay "The Virgin and the Dynamo" by American author Henry Adams.                    TECH

1876        American shopowner Henry Sherwin, cofounder of the Sherwin-Williams Co., develops a process for grinding and dispersing pigments into linseed oil and popularizes ready-to-use paint.                    TECH

1876        On an improved, patented version of his telephone, Scottish-American inventor Alexander Graham Bell utters to his assistant, Thomas Watson, the first complete sentences transmitted over the telephone: "Mr. Watson, come here. I want you." The call takes place at Bell's residence in Boston, Massachusetts. The first telephone will be sold in May 1877.                    TECH

1876        The Howe floater, which distributes melted solder onto can tops and bottoms, mechanizes the soldering process and speeds it 100-fold.                    TECH

1877        American astronomer Asaph Hall discovers Phobos and Deimos, the two satellites of Mars.                    ASTRO

1877        Italian astronomer Giovanni Schiaparelli observes lines on Mars that he calls *canali*, or channels. American astronomer Percival Lowell will misinterpret these to be canals, considering them evidence of waterworks constructed by intelligent beings.                    ASTRO

1877        German botanist Wilhelm Pfeffer explains osmosis (*see* 1748, PHYS) by relating
            it to movement of molecules across a membrane. As a result of his studies,
            Pfeffer is able to calculate the molecular weight of proteins involved in osmo-
            sis.                                                                    BIO

1877        German biochemist Ernst Felix Hoppe-Seyler coins the term *biochemistry* for
            the study of the chemistry of living organisms, which will emerge as a full-
            fledged discipline early in the 20th century.                          BIO

1877        French physicist Louis-Paul Cailletet (and, independently, Swiss chemist
            Raoul-Pierre Pictet) liquefies oxygen, nitrogen, and carbon monoxide.  CHEM

1877        British anesthesiologist Joseph Clover invents equipment that will come to be
            called Clover's apparatus to more safely administer anesthesia.        MED

1877        American paleontologists and lifelong rivals Othniel Charles Marsh and
            Edward Drinker Cope make the first of several separate collections of late
            Jurassic dinosaur skeletons in the Morrison formation in Colorado and at
            Como Bluff, Wyoming. The dinosaurs they discover in the next few years will
            include *Allosaurus*, *Apatosaurus*, (*Brontosaurus*), *Diplodocus*, *Camarasaurus*, and
            *Stegosaurus*.                                                         PALEO

1877        Russian physiologist Ivan Pavlov begins his studies on conditioned response
            in relation to dogs' digestion. He will be awarded the 1904 Nobel Prize for
            physiology or medicine.                                                PSYCH

1877        American anthropologist Lewis Henry Morgan publishes his *Ancient Society*,
            which proposes that human societies go through stages of evolution from sav-
            agery to barbarism to civilization.                                    SOC

1877        Swiss physician A. E. Fick invents the first practical contact lenses, but as they
            are made of glass and cover the entire eyeball, they are not widely used.  TECH

1877        The first telephone switchboard is put into operation, in Boston.      TECH

1877        The first phonograph is tested by its inventor, American Thomas Alva Edison,
            who sings and hears played back "Mary Had a Little Lamb." His phonograph
            operates by recording sounds on indented metal cylinders.              TECH

1878        German archaeologist Heinrich Schliemann discovers ancient Ithaca, the leg-
            endary home of Greek hero Odysseus.                                    ARCH

1878        Swiss chemists Jacques Louis Soret and Marc Delafontaine discover the ele-
            ment holmium; Swedish chemist Per Teodor Cleve will discover it indepen-
            dently next year (*see* 1879, CHEM).                                   CHEM

1878        The United Kingdom passes the Dentists Act, requiring dentists to be regis-
            tered. Like the earlier Medical Act (*see* 1858, MED), it is an attempt to protect the
            public from unqualified practitioners.                                 MED

1878        A large group of well-preserved *Iguanodon* skeletons is discovered in a coal
            mine at Bernissart, Belgium. *See* 1882, PALEO.                        **PALEO**

1878        German neurologist Wilhelm Erb discovers a disease marked by muscular weak-
            ness and fatigability, later called myasthenia gravis.                **PSYCH**

1878        The Japanese term *shinrigaku,* later formally adopted as the Japanese word for
            *psychology,* is first used in Japanese literature.                   **PSYCH**

1878        John Broadus Watson, the founder of behavioral psychology, is born on
            January 9 in Greenville, South Carolina (*d.* 1958). This area of work restricts
            the field to the study of objectively observable behavior, which it then explains
            in stimulus-response terms. His writings will include *Animal Education* (1903),
            *Behavior* (1914), and *Psychological Care of the Infant and Child* (1928).    **PSYCH**

## A Better Light Bulb

By the 1870s, the streets of Paris and other cities were lit by electric arc lamps that
worked by forcing a continuous discharge of current across the air space between two
conductors. These lights were harsh, flickering, and hazardous. A better electric light
was needed, and inventors set out to make one.

Scientists had noted that a wire or filament heated by electricity would sometimes
glow. Unfortunately, filaments never survived long enough to be practical as a light
source, because oxygen in the air quickly oxidized and destroyed them. By 1875,
British physicist William Crookes had invented a method for removing much of the air
from a tube. It was found that a wire filament placed in a vacuum tube glowed longer,
but it still disintegrated too soon to make it useful in a lamp.

American inventor Thomas Alva Edison came on the case. He tried hundreds of different materials until, on October 21, 1879, he found his filament: an ordinary
scorched cotton thread. A bulb with this carbon filament burned for 40 hours. Edison
quickly registered a patent and on New Year's Eve 1879 lit up the main street of Menlo
Park, New Jersey. Three years later, his Pearl Street Power Station provided electric
power to 203 Manhattan customers enjoying the light of more than 3,000 electric
lamps.

Only one factor threatened to detract from Edison's success. A rival British inventor,
Joseph Swan, had also discovered the carbon filament and patented his own lamp in
1878. However, Swan was slower in setting up a system to distribute public electricity,
so Edison received the fame. The two men initially sued each other for patent viola-
tions but eventually settled their claims and became partners.

The electric bulb continued to improve. In 1910 the heat-resisting element tungsten
replaced cotton as the filament of choice. In 1913 the inert gas nitrogen replaced the
vacuum as the preferred environment for filaments. Today's incandescent bulbs burn
for about 2,000 hours.

1878        American inventor Thomas Alva Edison discovers that, by subdividing or step-
            ping down electrical current, electric power can be supplied for mechanical use
            in the home.                                                              TECH

1878        British chemist Joseph Swan demonstrates the first carbon filament incandes-
            cent bulb, in London.                                                     TECH

1878        At the Remington Arms Co., in Connecticut, a shift key is developed for the
            Remington typewriter, for the first time allowing the production of upper- and
            lowercase letters.                                                        TECH

1878        In Cincinnati, Ohio, the first floating soap is invented by a Procter & Gamble
            factory worker who accidentally injects air into the mixture. The product is
            named Ivory soap by company owner Harley Procter, who had come across
            the word *ivory* in a psalm in church.                                    TECH

1879        Paintings done by Cro-Magnon man some 14,000 years ago are discovered on
            cave walls in Altamira, Spain. At the time, they are the earliest known evidence
            of artistic ability in prehistoric people.                                ARCH

1879        English astronomer George H. Darwin suggests a theory of lunar formation
            (later discredited), that the moon formed from material ejected from the
            earth's crust during an early period of rapid rotation.                    ASTRO

1879        French chemist Paul-Émile Lecoq de Boisbaudran discovers the element
            samarium.                                                                 CHEM

1879        Swedish chemist Lars Fredrik Nilson discovers the element scandium,
            Mendeleyev's predicted eka-boron (*see* 1871, CHEM). Swedish chemist Per
            Teodor Cleve discovers the elements thulium and holmium (*see* 1878, CHEM).
                                                                                      CHEM

1879        American chemists Ira Remsen and Constantine Fahlberg synthesize
            orthobenzoyl sulfimide, or saccharin. This sweet organic compound will
            become the first commercial sugar substitute.                             CHEM

1879        Russian chemist Vladimir Markovnikov shows that carbon rings with four
            atoms, rather than the more commonly known six, can be formed.            CHEM

1879        American astronomer E. C. Pickering invents the meridian photometer and
            lays the foundations of exact stellar photometry.                         EARTH

1879        Clarence King heads the newly organized U.S. Geological Survey. After two
            years, he will resign and John Wesley Powell will take his place.          EARTH

1879        French mathematician Jules-Henri Poincaré writes a doctoral thesis that leads
            him to discover the properties of automorphic functions, one of his many con-
            tributions to the theory of differential equations and other areas of mathe-
            matics (including topology; *see* 1895, MATH).                             MATH

1879      German physician Albert Neisser identifies the bacteria that cause gonorrhea and meningococcal meningitis, inflammation of the brain or spinal cord. MED

1879      French dermatologists coin the term *biopsy* for the process of taking skin for microscopic examination.      MED

1879      Austrian physicist Josef Stefan formulates what will become known as Stefan's law (derived theoretically by Ludwig Eduard Boltzmann; *see* 1884, PHYS), which states that the total energy radiated per unit surface area of a blackbody in unit time is proportional to its absolute temperature raised to the fourth power. PHYS

1879      American physicist Edwin H. Hall discovers what will become known as the Hall effect, the production of an electromotive force within a conductor or semiconductor carrying current in the presence of a strong transverse magnetic field.      PHYS

1879      American physicist Albert Einstein is born in Germany (*d.* 1955). With his theories of special relativity (1905) and general relativity (1916), he will do away with the concept of absolute space and time that had dominated physics since Newton in the 17th century. He will develop new understandings of gravity and matter-energy, contribute to quantum theory and atomic physics, and attempt (unsuccessfully) to develop a unified field theory. He will immigrate to the United States in 1933.      PHYS

1879      German physiologist and psychologist Wilhelm Wundt founds the world's first psychological research laboratory, at the University of Leipzig.      PSYCH

1879      American inventor Thomas Alva Edison demonstrates an incandescent vacuum light bulb that burns for nearly 50 hours and is still not extinguished. TECH

1879      English-American photographer Miles Ainscoe Seed introduces the portable Seed Dry-Plate for taking photographs, which is more stable and less dependent on chemicals than the wet plate.      TECH

1879      Milk bottles are developed to replace household pitchers. The first bottles are used by a Brooklyn dairy. Within two decades they are commonplace nationally.      TECH

1880s      American mathematician Josiah W. Gibbs develops a system of vector analysis. MATH

1880s      Two skeletons of Neanderthal man (*see* 1856, PALEO) are found in a cave at Spy, Belgium, buried with stone tools and bones of extinct mammals. This discovery, excavated according to scientific methods, helps convince scientists that Neanderthal man represents an earlier variety of human being.      PALEO

c. 1880s      American psychologist William James and Danish scientist Karl Lange independently develop the theory that emotion is a consequence of physiological stimulus. For example, the sight of an oncoming train produces a physical response followed by an emotion, fear. This theory is the reverse of the then-popular view of emotion.                                                                    PSYCH

1880    Archaeologist M. Kalokairinos discovers the walls of the labyrinth at the palace of King Minos at Knossos, Crete.                                          ARCH

1880    Russian chemist Friedrich K. Beilstein begins publication of his classic *Handbook of Organic Chemistry*.                                               CHEM

1880    American schoolteacher and autodidact William Ferrel designs an analog machine to predict tidal maxima and minima.                                     EARTH

1880    In Ecuador, British mountaineer Edward Whymper climbs the Andean peak Chimborazo—twice.                                                              EARTH

1880    British geologist John Milne becomes a leading figure in modern seismology by inventing, then progressively improving, the seismograph, a device to measure the strength of earthquakes.                                                   EARTH

1880    French chemist Louis Pasteur discovers streptococcus, staphylococcus, and pneumococcus bacteria and develops a vaccine for anthrax and chicken cholera.                                                                          MED

1880    English ophthalmologist Warren Tay describes the cherry-red spots at the backs of the eyes that are characteristic of the fatal inherited metabolic disorder Tay-Sachs disease, which causes physical and mental retardation.     MED

1880    Paleontologist Othniel Charles Marsh learns to protect fossils by encasing them in a jacket of burlap and plaster of Paris. This method will become standard practice for fossil hunters.                                            PALEO

1880    British physicist William Crookes shows that cathode rays are not electromagnetic waves like light but streams of electrically charged particles.     PHYS

1880    French physicist Émile Hilaire Amagat begins experiments in inducing and studying high-pressure conditions, reaching a pressure equal to 3,000 atmospheres.                                                                         PHYS

1880    French physicist Pierre Curie discovers piezoelectricity, arising from the interaction of pressure and electric potential in certain crystals, which can be manipulated to produce ultrasonic vibrations. This phenomenon will prove important to the development of microphones and other electronic sound instruments.                                                                     PHYS

1880    Austrian physician Josef Breuer begins treating the hysterical patient he calls Anna O. Her case study will be essential in the development of psychoanalysis. *See also* 1895, PSYCH.                                                  PSYCH

1880    In Britain, parcel post service begins. It will begin in the United States in 1913. TECH

1881      German Egyptologist Heinrich Karl Brugsch and French Egyptologist Gaston Maspero discover mummies of over 30 members of Egyptian royalty, including Ramses the Great, in a cave at Dier el-Bahri near Luxor, Egypt.    ARCH

1881      American photographer and self-taught astronomer Edward Emerson Barnard makes the first photographic discovery of a comet.    ASTRO

1881      British naturalist Charles Darwin writes about how earthworms shape the landscape by creating an upper layer of soil through loosening soil and grinding down rock particles. His point is once again to show the great effects that occur through the accumulation of small changes over time.    BIO

1881      British mathematician John Venn invents what will become known as the Venn diagram, in which intersecting circles are used to represent logical statements.    MATH

1881      German-born American physicist Albert A. Michelson invents the interferometer, which he uses to attempt to measure the speed of Earth's absolute motion through the luminiferous ether, a light-carrying substance then thought to fill all of space. The experiment fails to detect any motion through the ether. Michelson then refines his procedure and repeats the experiment, definitively, with Edward W. Morley (*see* 1887, PHYS).    PHYS

1881      British physicist J. J. Thomson discovers electromagnetic mass when he deduces that the mass of an object changes with the addition of an electric charge.    PHYS

1881      German psychologist Georg Elias Müller founds a psychophysic laboratory in Göttingen, Germany, a rival of Wilhelm Wundt's at Leipzig (*see* 1879, PSYCH).    PSYCH

1881      British anthropologist Alfred Reginald Radcliffe-Brown is born (*d.* 1955). His structural-functional analyses of the interdependent parts of social systems will help build social anthropology as a science. *See* 1952, SOC.    SOC

1881      The Savoy Theatre in London becomes the first public building in Britain to be electrically lighted.    TECH

1881      Roll film for photographs is invented by American inventor David Henderson Houston.    TECH

1881      The refrigerator car for transporting fresh meat is developed by American meat merchant Gustavus Swift. This car speeds transport time to the East Coast, increasing the national availability of meat and lowering prices.    TECH

1882      American scientist Henry Augustus Rowland invents a machine to make precision diffraction gratings for spectroscopic use. In 1886 he will use it to create an improved map of the solar spectrum incorporating some 14,000 lines.    ASTRO

1882        German botanist Eduard Adolf Strasburger observes changes in plant cells dur-
            ing division. He proceeds to sort protoplasm into its nucleoplasm (inside the
            nucleus) and cytoplasm (outside the nucleus).                                BIO

1882        American geologist Clarence Edward Dutton shows that a relatively short geo-
            logical time is enough for a river to cut through rock, creating a valley. His
            work strengthens geologists' acceptance that landforms are principally a result
            of surface denudation, the exposure of rock strata by erosion.               EARTH

1882        German mathematician Ferdinand von Lindemann determines that pi is a
            transcendental number, like *e* (*see* 1873, MATH), and that the ancient Greek
            problem of squaring the circle with straightedge and compass in a finite num-
            ber of steps is impossible.                                                  MATH

1882        German bacteriologist Robert Koch, who will be awarded the 1905 Nobel
            Prize for physiology or medicine, discovers mycobacterius tuberculosis, the
            bacteria causing tuberculosis in mammals. He also establishes what will
            become known as Koch's postulates or laws, the criteria to prove that a given
            disease is caused by a specific microorganism.                               MED

1882        French paleontologist Louis Dollo publishes the first of a series of papers on the
            *Iguanodon* skeletons of Bernissart, Belgium (*see* 1878, PALEO). He becomes the
            founder of ethological paleontology, the study of the relationships of ancient
            animals to their environment.                                                PALEO

1882        Argentine paleontologist Florentino Ameghino identifies fossils of late
            Cretaceous dinosaurs from Neuquén, Argentina, in Patagonia.                  PALEO

1882        German-American physicist Albert A. Michelson attains a more accurate figure
            than Foucault did in 1849 for the speed of light: 186,320 miles per second, very
            close to the currently accepted value.                                       PHYS

1882        American psychologist G. Stanley Hall (1844–1924) founds the first formal
            psychological laboratory in the United States, at Johns Hopkins University in
            Baltimore.                                                                   PSYCH

1882        German-English psychoanalyst Melanie Klein is born on March 30 (*d.* 1960).
            A pioneer of child psychoanalysis who will focus on early infantile stages, she
            will contribute to the understanding and psychoanalytic treatment of psy-
            chosis.                                                                      PSYCH

1883        Russian-born French bacteriologist Ilya Ilich Mechnikov demonstrates that
            certain cells called leukocytes move to damaged areas in the body, where they
            ingest bacteria.                                                             BIO

1883        British anthropologist Francis Galton coins the term *eugenics* for the study of
            improving human qualities by selective, careful breeding.                    BIO

1883        The volcano Krakatau (Krakatoa) between Sumatra and Java in the Dutch East Indies erupts on August 27, destroying most of its island, killing nearly 40,000 people, and sending ocean waves as far away as Cape Horn.                    EARTH

1883        On a trip to Egypt, German bacteriologist Robert Koch discovers the cholera bacillus.                    MED

1883        German chemist Ludwig Knorr compiles the compound antipyrine, which reduces fever and pain. It is the first significant entirely artificial drug and its invention will begin the synthetic drug industry.                    MED

1883        American feminist activist, birth control advocate, and nurse Margaret Sanger is born (d. 1960). She will educate working-class women on issues concerning reproduction.                    MED

1883        The *Journal of the American Medical Association* (JAMA) is established as a weekly magazine.                    MED

1883        American inventor Thomas Alva Edison discovers what will become known as the Edison effect, the flow of electric current from a hot carbon filament to a cold metal wire across a gap inside a light bulb. Useless as it is to Edison, this discovery nevertheless will become important in the development of electronics.                    PHYS

1883        German-American anthropologist Franz Boas begins his field work with observations of Eskimos. He will become the most influential voice in American anthropology, training or inspiring such scholars as Ruth Benedict, Alfred Kroeber, Margaret Mead, and Edward Sapir. He and his followers will stress the importance of culture and language in shaping human behavior. Boas's strict methodology and painstaking research will build the credibility of cultural anthropology.    SOC

1883        Croatian electrical engineer Nikola Tesla constructs an induction motor that can make use of alternating current, which can be more easily obtained from generators than can direct current.                    TECH

1883        French chemist Louis-Marie-Hilaire Bernigaud de Chardonnet invents rayon, the first synthetic fiber.                    TECH

1883        The 1,595.5-foot Brooklyn Bridge, also known as the Great East River Bridge, opens to pedestrian traffic. Featuring a steel web truss for stabilization, it was designed by German-American John Augustus Roebling. Following Roebling's death in 1869, the bridge was completed by his son, Washington Augustus Roebling.                    TECH

1883        The Maxim/Vickers gun, the first completely automatic machine gun, is developed by English engineer Hiram Maxim.                    TECH

1884        At an international meeting in Washington, D.C., the prime meridian is established as running through Greenwich, England.                    ASTRO

1884      Danish bacteriologist Hans Christian J. Gram stains bacteria with dye and finds that when it is washed with iodine and alcohol the dye stain is removed from some bacteria but not from others. The bacteria retaining the dye he calls Gram-positive, the bacteria free of the dye Gram-negative. This identification will become crucial once antibacterial agents are developed.    BIO

1884      Swedish chemist Svante Arrhenius advances the theory of ionic dissociation, which describes the breakdown of electrolytes, when in solution, into positively and negatively charged particles called ions.    CHEM

1884      German chemist Emil Fischer analyzes the structure of sugars and studies the compounds called purines. In 1902 he will win the Nobel Prize for chemistry.    CHEM

1884      German chemist Otto Wallach, who will be awarded the 1910 Nobel Prize for chemistry, begins to isolate terpenes from essential oils in a process that will become invaluable to the perfume industry.    CHEM

1884      Russian mathematician Sonya (Sofya) Vasilyevna Kovalevskaya demonstrates the possibility of expressing certain kinds of Abelian integrals in terms of simpler elliptic integrals.    MATH

1884      German mathematician Gottlob Frege provides an influential definition of cardinal numbers in his *Die Grundlagen der Arithmetik* (*The Foundations of Arithmetic*).    MATH

1884      German obstetrician and gynecologist Karl S. F. Credé introduces the use of 1 percent silver nitrate solution in the eyes of newborn infants to prevent blindness from gonorrheal infection.    MED

1884      German bacteriologist Arthur Nicolaier discovers the tetanus (lockjaw) bacillus. MED

1884      The first American tuberculosis sanitarium is founded, in New York's Adirondack Mountains, by physician Edward Trudeau, whose studies on tuberculosis make him an authority on this lung disease.    MED

1884      German bacteriologist Friedrich Loeffler discovers the diphtheria bacillus. MED

1884      Canadian paleontologist Joseph Burr Tyrrell discovers the partial skull of a Cretaceous carnivorous dinosaur, *Albertosaurus sarcophagus*, at the Red Deer River in Alberta, Canada.    PALEO

1884      Austrian physicist Ludwig Eduard Boltzmann, the founder of statistical mechanics, shows how Stefan's law (*see* 1879, PHYS) can be derived from thermodynamic rules. As a result, this principle will sometimes be known as the Stefan-Boltzmann law.    PHYS

1884      American psychologist William James founds the U.S. branch of the Society for Psychical Research.    PSYCH

1884 In Vienna, Austrian psychologist and psychotherapist Otto Rosenfeld (later Rank) is born (*d.* 1939). A specialist in the psychology of myth and dreams, Rank will become Sigmund Freud's assistant, and each will profit from the other's intuitions and ideas. Many in the field will later say that Freud groomed Rank to take over as the leader in psychoanalysis, but Rank will break from Freud in 1923. Rank's early work *Myth of the Birth of a Hero* (1909) will become a psychoanalytic classic. PSYCH

1884 British inventor and physicist Hertha Marks (later Marks Ayrton) invents an instrument for dividing a line into equal parts, since used by architects, artists, and engineers. TECH

1884 The compound steam turbine is developed by British engineer Charles Algernon Parsons. TECH

1884 German-American mechanic Ottmar Mergenthaler patents the Linotype type-setting machine. By its mechanizing of parts of the typesetting process, it becomes an extremely popular choice for printing daily periodicals. TECH

1885 German archaeologist Heinrich Schliemann excavates Tiryns. ARCH

1885 A supernova appears in the Andromeda galaxy (M31), the only extragalactic supernova visible to the naked eye until 1987. ASTRO

1885 Austrian chemist Carl Auer discovers the elements neodymium and praseodymium. CHEM

1885 German biochemist Albrecht Kossel, who will win the Nobel Prize for physiology or medicine in 1910, studies the molecular structure of nucleic acids, isolating two purines, three pyrimidines, and a sugar molecule. CHEM

1885 Dutch chemist Jacobus van't Hoff proposes what will become known as the van't Hoff factor, which appears in equations for osmotic pressure and other colligative properties. CHEM

c. 1885 Austrian geologist Eduard Suess begins publication of *The Face of the Earth* (1885–1909), a five-volume work based on the premise that the earth contracted while cooling, and that this contraction can explain the earth's geological features. EARTH

1885 *Snow Cover: Its Effect on Climate and Weather,* by A. I. Voeykov, considered the first important work on the subject, is published in St. Petersburg. EARTH

1885 German bacteriologist Paul Ehrlich expounds his side-chain theory, which states that there is a special affinity between certain drugs and specific cells. This theory lays the foundation for Ehrlich's development of arsenic as a treatment for syphilis. MED

1885        Swiss physicist Johann Jakob Balmer develops an equation to interrelate the
            wavelengths of the hydrogen spectrum. This principle yields the so-called
            Balmer series of lines in the visible hydrogen spectrum.            PHYS

1885        German psychologist Herman Ebbinghaus (1850–1909) publishes the first
            experimental research on memory. In it he asserts that learning and practice
            are linked in that if the time spent learning is doubled, then the amount
            learned will be doubled also. Ebbinghaus is later considered the father of
            memory research.                                                    PSYCH

1885        Between now and 1905, Ebbinghaus works on developing a sentence comple-
            tion test (the first successful test of higher mental abilities) and the nonsense
            syllable, the latter of which will revolutionize the study of learning and
            association.                                                        PSYCH

1885        British anthropologist Francis Galton notes the uniqueness of each individ-
            ual's fingerprints and works out a system for classifying them. His work will
            become important in law enforcement.                                SOC

1885        American electrical engineer William Stanley invents the transformer, to shift
            the voltage and amperage of alternating current and direct current.   TECH

1885        The first skyscraper, built for the Home Insurance Co., goes up in Chicago.
            Designed by American architect William LeBaron Jenney, it is a 10-story steel-
            framed marble building.                                             TECH

1885        A workable transmitter for dispersing mass electrical power is refined by
            American inventor and manufacturer George Westinghouse and electrical
            engineer William Stanley. It employs alternating current.            TECH

1885        German engineer Carl Friedrich Benz develops the first working motor car
            powered by gasoline.                                                TECH

1885        The safety bicycle, with two wheels of equal size, is developed by English
            inventor J. K. Stanley. Previously (see 1861, TECH), bicycles were built with large
            front wheels.                                                       TECH

1886        French chemist Paul-Émile Lecoq de Boisbaudran discovers the element dys-
            prosium. Boisbaudran and Swiss chemist Jean-Charles de Marignac discover
            the element gadolinium. French chemist Henri Moissan, who will win the
            Nobel Prize for chemistry in 1906, discovers the element fluorine. German
            chemist Clemens Alexander Winkler discovers the element germanium,
            Mendeleyev's predicted eka-silicon (see 1871, CHEM).                 CHEM

1886        French physical chemist François-Marie Raoult discovers what will become
            known as Raoult's law, that the partial pressure of solvent vapors in equilibri-
            um with a solution is proportional to the ratio of the number of solvent mol-
            ecules to solute (dissolved) molecules. This law provides a new way of deter-
            mining the molecular weight of solutes.                             CHEM

| | |
|---|---|
| 1886 | S. Ziesel develops what will become known as the Ziesel reaction to determine how many methoxy groups are in an organic compound. CHEM |
| 1886 | Two Englishmen build the first electric submarine, the *Nautilus*, which can cruise for 130 kilometers between recharges. EARTH |
| 1886 | German physicist Eugen Goldstein discovers channel rays or canal rays, streams of positive ions produced by boring holes in the cathode of an evacuated tube. PHYS |
| 1886 | American chemist Charles Martin Hall and French metallurgist Paul-Louis-Toussaint Héroult independently invent a cheap way of isolating aluminum. This Hall-Héroult process will make the once-precious metal a common structural material. TECH |
| 1886 | A halftone photo-engraving process, using multi-sized dots on the page, is devised by American printer Frederic Eugene Ives. TECH |
| 1887 | The Lick refracting telescope, on Mount Hamilton near San Francisco, becomes the world's first mountaintop telescope. Its lens is 36 inches (91 centimeters) across. ASTRO |
| 1887 | British scientist William Abney invents a way to photograph infrared radiation, a technique he uses to observe the solar spectrum. ASTRO |
| 1887 | French physiologist Raphael Dubois is the first to show the chemical nature of bioluminescence, the technique of producing light by living organisms such as glowworms, some bacteria, certain fungi, and various deep-sea creatures. BIO |
| 1887 | English paleontologist H. G. Seeley divides dinosaurs into two orders, based primarily on the structure of their pelvic girdle: the *Ornithischia* (bird hipped) and the *Saurischia* (lizard hipped). This classification will become standard, as will his grouping of dinosaurs into one category with crocodiles, birds, and the extinct reptiles known as thecodonts. (The reptiles in this group will later be known as archosaurs.) PALEO |
| 1887 | German-American physicist Albert A. Michelson, who will win the 1907 Nobel Prize for physics, and American chemist Edward W. Morley perform a definitive version of an earlier Michelson experiment (*see* 1881, PHYS) demonstrating that Earth has no discernible motion through the luminiferous ether postulated by scientists. These puzzling results will not be completely explained until Albert Einstein does so in 1905. PHYS |
| 1887 | German physicist Heinrich Hertz discovers the photoelectric effect, later defined as caused by the liberation of electrons from matter exposed to electromagnetic radiation. PHYS |
| 1887 | Austrian physicist Ernst Mach studies what happens when solid objects move rapidly through the air, particularly above the speed of sound. The standard |

measurement for the speed of a body relative to the speed of sound will be called a Mach number, with Mach 1 being a speed equal to that of sound.   PHYS

1887   The *New York World* publishes a series of articles chronicling the treatment and abuse of the mentally ill in the New York City Lunatic Asylum on Blackwell's (later Roosevelt) Island. American journalist Nellie Bly fakes insanity, is admitted under a pseudonym, and writes of her ordeal in articles for that periodical that will be published collectively in *Ten Days in a Mad House* (1887).   PSYCH

1887   French sociologist Émile Durkheim teaches the first course in sociology, at the University of Bordeaux. He develops scientific methods for sociology and emphasizes the importance of collective beliefs and values for social cohesion.   SOC

1887   The first Daimler motorized vehicle appears on the market. *See also* 1901, TECH.   TECH

1887   The first multicolumn adding machine, the Comptometer, is manufactured by the Felt & Tarrant Manufacturing Co. Its inventor is the business's cofounder Dorr Eugene Felt.   TECH

1887   The first motorized phonograph, playing recordings imprinted on wax cylinders, is invented by American inventor Thomas Alva Edison.   TECH

1888   British astronomer Joseph Lockyer outlines the evolutionary cycle of a star from birth to death.   ASTRO

1888   Danish-born British astronomer Johann L. E. Dreyer publishes *A New General Catalogue of Nebulas and Clusters of Stars*, known simply as *NGC*, which revises John Herschel's 1864 catalogue and now includes nearly 8,000 items. An *NGC* number becomes a standard designation for celestial objects.   ASTRO

## Agoraphobia

The most common phobia (persistent, irrational fear) for which people seek psychiatric treatment was also the first to be diagnosed: agoraphobia. The term was coined in the late 1800s by the German neurologist Alexander Karl Otto Westphal. From the Greek for "fear of the marketplace," it signifies an abnormal fear of open or public places.

Westphal observed that this phobia keeps its sufferers away from such places as streets, stores, tunnels, public transportation, churches, theaters, and the post office at noon. Most agoraphobics were, and continue to be, women. Onset is usually between 18 and 35 years of age and often begins with an unexpected, spontaneous panic attack that cannot be easily controlled, creating a generalized fear in future and similar situations.

1888        The Marine Biological Laboratory at Woods Hole, Massachusetts, holds its first
            classes.                                                                    BIO

1888        German anatomist Heinrich Wilhelm Gottfried von Waldeyer-Hartz suggests
            the name *chromosomes* for the chromatin threads that appear during cell
            division.                                                                   BIO

1888        French chemist Henri-Louis Le Châtelier states what will become known as Le
            Châtelier's principle, that in a system in equilibrium any change in one of the
            factors tends to shift the equilibrium so as to minimize the original change.
                                                                                        CHEM

---

# Victorian Botanists

In the 19th century, most fields of scientific study were closed to women. One partial
exception was botany, which, according to a 1980 article in the *British Journal for the
History of Science*, was exempt because it reinforced "both of the contemporary alternative ideals of femininity"—the "unintense intellectualism" of the upper class and the
"sentimentalized womanhood" of the middle class.

Given these restraints and others imposed by academic and professional institutions
like the Linnean Society, which would not admit women, female botanists demonstrated their talents primarily as illustrators, researchers, and writers of popular works
on botany. During the Victorian era hundreds of such works were written, many limited by the prevailing science of the era but some, though now relatively forgotten, still
vivid. Among the more diligent female nature writers of the 19th and early 20th centuries are:

- Elizabeth Knight Britton (1858–1934). American botanist and early conserva-
  tionist active in the founding of the Wild Flower Preservation Society of America
  and unofficial curator of the moss collection at Columbia University. She was the
  author of 346 scientific papers and is said to have provided the idea for establishing the New York Botanical Garden, which opened in 1891.

- Mary Agnes Meara Chase (1869–1963). American botanist. After working as a
  meat inspector in the Chicago stockyards (1901–1903), she became a botanical
  artist and botanist specializing in agrostology, the study of grasses, for the U.S.
  Department of Agriculture (1903–1939).

- Alice Eastwood (1859–1953). American botanist. Researched flowering plants of
  the Rocky Mountains. Author of *A Popular Flora of Denver, Colorado* (1893), she was
  also the curator of botany for the California Academy of Sciences (1892–1949),
  and she was noted during her lifetime in *American Men of Science*.

- Ethel Sargant (1863–1918). British botanist. Specialist in the cytology and morphology of plants. Her work in plant embryology led to scientific papers on the
  life cycle of monocotyledons.

1888    Chemist Friedrich Wilhelm Ostwald discovers that catalysts affect reactions' speed, but not their equilibrium.    CHEM

1888    The American Mathematical Society is established.    MATH

1888    In the Judith River beds of Montana, John Bell Hatcher and Othniel Charles Marsh discover the first known skull of *Triceratops*, a three-horned Cretaceous dinosaur.    PALEO

1888    German physicist Heinrich Hertz verifies James Clerk Maxwell's equations (*see* 1865, PHYS), using an oscillating current to produce radio waves that are electromagnetic in nature and behave as light does. Hertz's experiments confirm Maxwell's theory that light is an electromagnetic phenomenon and represent the first known detection of radio waves.    PHYS

1888    Electric trolley cars, designed by American engineer Frank Julian Sprague, run successfully in Virginia.    TECH

1888    American inventor George Eastman markets the first low-cost, easy-to-use camera, the Kodak. This $25 device holds a roll of stripping paper for up to 100 photographs.    TECH

1888    A new shorthand system, called Light Line Phonography, is developed by American inventor John Robert Gregg.    TECH

1888    Construction is completed on the Washington Monument, the tallest masonry building in the world, rising 553 feet.    TECH

1889    German astronomer Hermann Karl Vogel, analyzing the Doppler shift of the star Algol, demonstrates that it is an eclipsing binary star.    ASTRO

1889    Working with Italian histologist Camillo Golgi, Spanish histologist Santiago Ramón y Cajal improves Golgi's cell stain and works out the cellular structure of the brain and spinal cord, further establishing neuron theory as a field on its own. The two histologists will share the 1906 Nobel Prize for physiology or medicine.    BIO

1889    Swedish chemist Svante Arrhenius analyzes the concepts of activation energy (i.e., the minimum energy required for chemical reactions to occur) and chain reactions.    CHEM

1889    German mountaineers Hans Meyer and Ludwig Purtscheller succeed in climbing Mount Kilimanjaro, at 5,895 meters the tallest peak in Africa.    EARTH

1889    American geologist William Morris Davis introduces a new method of landscape analysis using a concept of the cycle of erosion by which rivers contribute to the shaping of landscapes. Davis systematizes this field of study as geomorphology, the study of the development of landforms.    EARTH

1889        Between Uganda and Zaire, British explorer Henry Stanley discovers the
            Ruwenzori mountain range, flanked by Lakes Albert and Edward.        EARTH

1889        American geologist Clarence Edward Dutton introduces the term *isostasy* for
            the equilibrium that tends to exist in the earth's crust between the forces that
            elevate land masses and those that depress them. Dutton argues that the con-
            tinents are made from lighter rock than the ocean floor.              EARTH

1889        Russian mathematician Sonya (Sofya) Vasilyevna Kovalevskaya publishes a sig-
            nificant paper on the rotation of asymmetric bodies about a fixed point.  MATH

1889        Italian mathematician Giuseppe Peano publishes *A Logical Exposition of the
            Principles of Geometry*, which applies symbolic logic to the task of building up the
            fundamental axioms of mathematics.                                    MATH

1889        At the Sorbonne, French psychologist Alfred Binet founds the first French psy-
            chological laboratory.                                               PSYCH

1889        British chemists Frederick Augustus Abel and James Dewar invent the smoke-
            less explosive called cordite.                                        TECH

1889        The first electric sewing machines are introduced to the market by the I. M.
            Singer Co.                                                            TECH

1889        The coin-operated telephone, patented by American inventor William Gray, is
            first used in a Connecticut bank.                                     TECH

c. 1890s    In the United States, geology is put to widespread practical use, particularly in
            assessing the lands of the newly acquired western territories and developing the
            mining industry. Universities such as Harvard, Columbia, and Yale develop
            research groups in economic geology whose work will rival that of the U.S.
            Geological Survey.                                                    EARTH

1890s       By now, Americans Othniel Charles Marsh and Edward Drinker Cope have dis-
            covered 136 species of dinosaurs, although some of these will later be disal-
            lowed. Bitter personal enemies, these paleontologists have spent their careers
            trying to outdo each other.                                           PALEO

1890        German astronomer Hermann Karl Vogel discovers spectroscopic binary
            stars.                                                                ASTRO

1890        American astronomers Edward C. Pickering and Williamina Paton Fleming
            introduce the alphabetical Harvard Classification System for classifying stars
            according to their spectral characteristics. *See also* 1901, ASTRO.    ASTRO

c. 1890     Biometrics, the branch of science jointly studying biology and mathematics,
            begins when statistical methods to analyze human and biological evolution
            are developed.                                                        BIO

1890       Swiss cytologist Richard Altmann reports the discovery of bioblasts within cells. In 1898 German biologist C. Benda will name these organisms *mitochondria*, mistakenly believing them threads of cartilage.    **BIO**

1890       Acetylene chemistry is discovered.    **CHEM**

c. 1890     During the next two decades the theoretical foundations are established for measuring geological time by analyzing the products of radioactive decay.    **EARTH**

---

# Ye Olde Medical Dictionary

When Victorian novelists consign their characters to the mountains to cure consumption or to their boudoirs to treat la grippe, they refer to medical conditions now known by more modern names. Here are a few medical terms common in their time but now remembered only through literature and long-lived relatives.

Catarrh. Marked inflammation of the mucus membranes of humans or animals.

Chilblains. Inflammation and swelling of the body, particularly feet and hands, through exposure to cold.

Consumption. Process of the diminution or wasting away of the human body, usually through tuberculosis.

Croup. An intermittent laryngitis, marked by a raspy cough and labored breathing, usually affecting infants.

Dropsy. Edema, the excess accumulation of liquid in human tissue or a bodily cavity.

Grippe. A contagious, feverish viral infection similar to influenza.

Halitosis. Bad breath.

Lumbago. Severe muscular inflammation in the lumbar region.

Melancholia. An emotional state marked by depression, perceived medical problems, and, at times, hallucinations.

Neurasthenia. An emotional or psychological state marked by feelings of fatigue, lethargy, and low self-worth.

St. Vitus' dance. Colloquial name for chorea or Sydenham's chorea. A group of diseases of humans and dogs marked by uncontrollable body and facial movements. The name is derived from St. Vitus, a Christian martyr tortured by Roman emperor Diocletian by immersion in a vat of molten lead and pitch.

Trench mouth. Also known as Vincent's angina or Vincent's infection. A contagious disease marked by ulceration of the mucus membrane of the mouth or respiratory system. The disease was discovered to be generated in the presence of the fusiform bacillus and spirillum, two germs active in hospital gangrene, by French bacteriologist Jean-Hyacinthe Vincent (1862–1950).

Vapors. A nervous condition marked by depression and hysteria.

1890        At Kedung Brebus, Java, in the Dutch East Indies, Dutch paleontologist Marie
            Eugène Dubois discovers fossils of Java man, now considered a variety of *Homo
            erectus*, a prehistoric ancestor of humans. *Homo erectus* evolved in Africa about
            1.8 million years ago.                                                    PALEO

1890        American psychologist William James publishes his classic text *The Principles of
            Psychology*.                                                              PSYCH

1890        The U.S. Congress passes legislation dividing each state into districts, mandat-
            ing a state hospital for each one. *Hospital*, not *asylum*, is from now on the pre-
            ferred term for such institutions.                                        PSYCH

1890        Scottish anthropologist James George Frazer publishes *The Golden Bough*, a mon-
            umental comparative study of folklore and religion.                       SOC

1890        British economist Alfred Marshall, founder of the neoclassical school of eco-
            nomics, publishes the landmark work *Principles of Economics*.            SOC

1890        French physicist Édouard-Eugène Branly invents a radiowave detector that
            improves on Hertz's of 1888.                                              TECH

1890        An electromechanical punch-card system for recording data is refined by
            American engineer Herman Hollerith. Eventually patented, this Hollerith sys-
            tem is first used to record U.S. census findings. Hollerith's company will ulti-
            mately develop into International Business Machines (IBM).                 TECH

1891        German astronomer Maximilian Wolf makes the first photographic discovery
            of an asteroid. He will go on to discover some 500 other asteroids.       ASTRO

1891        German zoologist Karl Gottfried Semper introduces the idea of a food chain,
            claiming that an ecosystem maintains its energy flow by passing along materi-
            als through the process of eating and being eaten.                        BIO

1891        American inventor Edward Goodrich Acheson produces silicon carbide, or car-
            borundum, a useful abrasive.                                              CHEM

1891        German aeronautical engineer Otto Lilienthal successfully launches himself
            on the first flight of a glider, of his own design.                       EARTH

1891        Hungarian physicist Roland Eötvös detects no significant difference between
            inertial and gravitational mass.                                          PHYS

1891        Irish physicist George Johnstone Stoney suggests the name *electron* for the as-
            yet-undiscovered fundamental unit of electricity. *See* 1897, PHYS.       PHYS

1891        American inventors Thomas Alva Edison and W. K. L. Dickson patent the
            Kinetograph camera and Kinetoscope viewer, the world's first motion picture
            system.                                                                   TECH

1891          The Carpenter Electric Heating Manufacturing Co. in the United States offers
             the first commercially available electric ovens.                    TECH

1892          Photographer and self-taught American astronomer Edward Emerson Barnard
             discovers Amalthea, a fifth satellite of Jupiter and the first to be found since
             the work of Galileo in 1610.                                        ASTRO

1892          Edward Emerson Barnard presents evidence that novae are exploding stars.
                                                                                ASTRO

1892          French physician and physicist Jaques-Arsène d'Arsonval applies an electro-
             magnetic field to himself and finds that it produces warmth without mus-
             cle contraction. Thus begins the study of the interaction of electromagnetic
             energies with biological systems, or bioelectromagnetics.           BIO

1892          British geneticist and biologist J. B. S. Haldane is born in Oxford (d. 1964). He
             will formulate a mathematical approach to the study of natural selection that
             leads the way in establishing the rates of genetic change in populations. His
             writings will include *The Inequality of Man* (1932), *New Paths in Genetics*
             (1941), and *The Biochemistry of Genetics* (1954). Haldane will predict human
             cloning and acclaim its potential uses.                             BIO

1892          German biologist August Weismann introduces the germ plasm theory of
             heredity, that a special hereditary substance embodies the only organic conti-
             nuity between generations. He also claims that in germ cell formation, the
             process of meiosis, the nuclei undergo a halving of the chromosomes.
             Weismann repudiates Lamarckism (*see* 1809, BIO) which argues that organisms
             inherit physical characteristics acquired by their parents during the latter's life-
             time.                                                               BIO

1892          Russian physicist Pyotr Nikolayevich Lebedev shows that light exerts
             pressure.                                                           PHYS

1892          Irish physicist George FitzGerald derives an equation (modified by Hendrik
             Lorentz in 1895) that accounts for the negative result of the 1887 Michelson-
             Morley experiment by predicting the contraction of a moving body in the
             direction of its motion. *See also* 1905, PHYS.                     PHYS

1892          German physicist Philipp Eduard Lenard studies open-air cathode rays.   PHYS

1892          Czechoslovakian psychiatrist Arnold Pick (1851–1924) is the first to describe
             a presenile disorder marked by a deterioration in functional intelligence. This
             degenerative malody, Pick's disease, is characterized by mood swings, lack of
             social restraint, fatigue, confusion, and memory difficulties.      PSYCH

1892          British chemist James Dewar invents the Dewar flask, a double-walled con-
             tainer with a vacuum between the walls to slow temperature changes in its
             contents. It will achieve commercial use as a thermos bottle.       TECH

| | |
|---|---|
| 1892 | The tin-plated bottle cap with a cork seal for beverage bottles is patented by Baltimore shop foreman William Painter. TECH |
| 1892 | The cultured pearl, grown by implanting a small foreign object as an irritant in an oyster shell, is grown successfully by Japanese businessman Kokichi Mikimoto. TECH |
| 1893 | English astronomer Edward Walter Maunder discovers what will become known as the Maunder minimum, a 70-year period from 1645 to 1715 when there were virtually no reports of sunspot activity. Not until the 1970s will Maunder's findings be corroborated. ASTRO |
| 1893 | German physicist Wilhelm Wien shows that the peak wavelength at which an object radiates electromagnetic energy is inversely proportional to the object's absolute temperature. Hotter bodies radiate at shorter, more energetic, wavelengths. PHYS |
| 1893 | The journal *Physical Review* begins publication. PHYS |
| 1893 | French sociologist Émile Durkheim, one of the founders of sociology, publishes *The Division of Labor*. SOC |
| 1893 | The first American department of sociology is founded, at the University of Chicago. Chicago will be the center of American sociological studies in the period between the two world wars. SOC |
| 1893 | German-American electrical engineer Charles Proteus Steinmetz develops the mathematics of alternating current (AC) circuitry, thus making AC equipment more useful and ensuring the demise of direct current. TECH |
| 1893 | The first Ford motor vehicle, or gasoline buggy, is test driven by its inventor, American machinist Henry Ford. TECH |
| 1893 | American inventor Whitcomb L. Judson introduces the clasp locker, forerunner to the zipper (*see* 1923, TECH). TECH |
| 1894 | British physicist J. W. Strutt and British chemist William Ramsay discover the element argon. CHEM |
| 1894 | British neurologist John Jackson publishes his paper "The Factors of Insanity," in which he defines the difference between positive and negative symptoms and describes their relationship to the nervous system. PSYCH |
| 1894 | British physicist Oliver Joseph Lodge invents the device called the coherer, for detecting radio waves from a distance of half a mile. He is the first to send and receive radio-wave messages in the Morse code used in telegraphy. TECH |

1894        Oil is accidentally discovered in Texas for the first time during the boring of a well (for water) in Corsicana. The Texaco oil company will be founded in 1902.                                                                       TECH

1894        The Hershey chocolate bar, created from sugar, milk, cocoa beans, and chocolate liqueur, is marketed. It sells for five cents, a price that will not rise for 74 years. (The size of the bar, however, will have decreased by about 34 percent by 1968.)                                                                               TECH

1894        The diesel engine, mechanically simpler than a gasoline engine, is developed by German engineer Rudolf Diesel.                                                         TECH

1894        Pneumatic tires for automobiles are introduced in France by André and Édouard Michelin. In America pneumatic tires are manufactured by the Hartford Rubber Works.                                                                      TECH

1895        In his work *Astronomical Constants*, Canadian-American astronomer Simon Newcomb calculates such constants as nutation, precession, and solar parallax.                                                                                  ASTRO

1895        In Germany the first germfree isolator, a sterilized glass chamber, is built.  BIO

1895        British chemist William Ramsay is the first to discover helium on earth rather than in the solar spectrum. *See* 1868, CHEM.                                              CHEM

1895        German chemist Carl Paul Gottfried von Linde develops a system for liquefying gases in quantities that are commercially useful.                                        CHEM

1895        Coal-tar chemistry undergoes new developments.                                  CHEM

1895        French mathematician Jules-Henri Poincaré publishes his *Analysis situ*, the first systematic treatment of topology.                                                     MATH

c. 1895     German mathematician Georg Cantor proposes his continuum hypothesis, that there are no cardinal numbers between aleph-null (the smallest transfinite cardinal number) and the cardinal number of the points on a line. American mathematician Paul J. Cohen will revisit this hypothesis in 1963.

                                                                                              MATH

1895        German physicist Wilhelm Röntgen discovers X rays.                              PHYS

1895        French physicist Jean Perrin performs experiments proving that cathode rays are streams, not waves, of negatively charged particles.                                  PHYS

1895        Dutch physicist Hendrik Lorentz independently derives George FitzGerald's equation (*see* 1892, PHYS) and proposes also that mass increases with velocity, becoming infinite at the speed of light. The resulting formulation becomes known as the Lorentz-FitzGerald contraction.                                        PHYS

1895        French physicist Pierre Curie discovers that each ferromagnetic metal, such as iron or nickel, has a specific point during heating at which it loses its magnetism, an interval that will become known as the Curie point or Curie temperature.                                                                                          PHYS

1895        Dutch physicist Hendrik Lorentz discovers what will become known as the Lorentz force, a perpendicular force exerted by electric and magnetic fields on a moving, charged particle. He will share the 1902 Nobel Prize for physics with Dutch physicist Pieter Zeeman (*see* 1896, PHYS).                                        PHYS

1895        American psychologist Gardner Murphy is born (*d.* 1979). He will create the biosocial theory of personality, which emphasizes that personality observation is possible only when a patient is interacting in a social environment. Murphy will also underscore the importance of developmental and environmental factors in determining personality.                                                            PSYCH

1895        Austrian physician Josef Breuer and psychiatrist Sigmund Freud publish *Studies on Hysteria*, the first book on psychoanalysis. This work contains Breuer's case history of Anna O. (*see* 1880, PSYCH), often considered the founding case of psychoanalysis.                                                                                    PSYCH

1895–1896   Sigmund Freud identifies the phenomenon of projection, a defense mechanism in which a person projects feelings or wishes onto another person, group, or thing.                                                                                      PSYCH

1895        French inventors Louis and Auguste Lumière patent and demonstrate the Cinématographe, a motion picture system that, unlike Edison's Kinetoscope (*see* 1891, TECH), allows film to be projected on a screen rather than viewed peep-show style.                                                                                       TECH

1895        Italian electrical engineer Guglielmo Marconi and Russian physicist Aleksandr Stepanovich Popov invent the radio antenna, a device for strengthening radio signals.                                                                                        TECH

1896        American biologist Edmund Beecher Wilson publishes *The Cell in Development and Inheritance*. In this landmark work Wilson claims that higher life forms, whether animal or plant, consist of the structural units known as cells.       BIO

1896        French mathematicians Jacques Hadamard and C. J. de la Vallée-Poussin independently prove the prime number theorem.                                          MATH

1896        French physicist Antoine-Henri Becquerel discovers radiation from potassium uranyl sulfate, a uranium compound. *See also* 1898, PHYS.                    PHYS

1896        Dutch physicist Pieter Zeeman discovers what will become known as the Zeeman effect, the splitting of lines in a spectrum when the spectrum's source is exposed to a magnetic field. He will share the 1902 Nobel Prize for physics with Dutch physicist Hendrik Lorentz (*see* 1895, PHYS).                                      PHYS

1896      American physicist Wallace Clement Ware Sabine founds the science of architectural acoustics when he develops mathematical equations to aid in designing lecture and concert halls.      PHYS

1896      Scottish physicist C. T. R. Wilson succeeds in producing an artificial cloud of water droplets. He now determines that dust and electrically charged ions in the atmosphere encourage the formation of clouds. In 1927 he will share the Nobel Prize for physics with American physicist Arthur H. Compton (*see* 1923, PHYS).      PHYS

1896      American psychologist David Wechsler is born (*d.* 1981). He will devise the widely used individual intelligence scales that bear his name: the Wechsler Bellevue, the Wechsler Adult Intelligence, the Wechsler Intelligence Scale for Children, and the Wechsler Preschool and Primary Scale of Intelligence tests. He will be the first to combine verbal and nonverbal tests into a composite scale and to introduce the concept of nonintellective factors of intelligence.      PSYCH

1896      American inventor Edward Goodrich Acheson patents the Acheson process for manufacturing graphite by heating a mixture of coke and clay.      TECH

1896      The first motorized car to be put on the market in the United States is the Haynes-Duryea, made by the Duryea Motor Wagon Co. of Massachusetts. Also this year another American motorcar will be introduced, the Stanley Steamer, invented by brothers Francis Edgar and Freelan O. Stanley.      TECH

1897      With the support of American astronomer George Ellery Hale, the world's largest refracting telescope begins operation at the Yerkes Observatory in Williams Bay, Wisconsin. The lens of this telescope, which will remain the world's largest refracting telescope for at least a century, is 40 inches (102 centimeters) in diameter. *See also* 1917 and 1948, ASTRO.      ASTRO

1897      German scientist Eduard Buchner (1860–1917) finds the chemical zymase to be the cause of alcohol's fermentation, putting to rest the vitalist notion that a 'vital substance' found only in living cells was responsible for fermentation. In 1907 he will win the Nobel Prize for chemistry.      BIO

1897      French chemist Paul Sabatier, who will share the 1912 Nobel Prize for chemistry with French chemist Victor Grignard (*see* 1901, CHEM), shows that nickel can be used as a catalyst.      CHEM

1897      Henry F. Osborn and the American Museum of Natural History collect dinosaur skeletons in the first of several expeditions to the Morrison beds of Wyoming.      PALEO

1897      Polish-born French physicist and chemist Marie Curie publishes her first paper on the magnetism of tempered steel.      PHYS

1897      British physicist J. J. Thomson calculates the mass of the electron (named by Stoney; *see* 1891, PHYS), the fundamental particle that carries the electric charge responsible for electricity. Thomson is considered the discoverer of the electron, the first known subatomic particle. *See also* 1899, PHYS.                    PHYS

1897    British physicist Ernest Rutherford distinguishes two types of uranium radiation, the massive and positively charged alpha rays and the lighter and negatively charged beta rays.                    PHYS

1897    German psychiatrist Sigbert Ganser is first to describe a rare psychotic syndrome that seems to occur as a response to severe stress. Ganser's syndrome is characterized by incorrect, absurd, and silly responses to questions requiring only a factual answer, hallucinations, delusions, amnesia, and clouded consciousness. Some war veterans will be diagnosed as having Ganser's syndrome before post-traumatic stress disorder is documented in the late 20th century.                    PSYCH

1897    With his work *Suicide: A Study in Sociology*, French sociologist Émile Durkheim becomes one of the first to present a rigorous statistical study.                    SOC

1897    German physicist Ferdinand Braun invents the oscilloscope, a modification of the cathode-ray tube that is a predecessor to the television set. In 1909 he will share the Nobel Prize for physics with Italian electrical engineer Guglielmo Marconi (*see* 1901, TECH).                    TECH

1897    American engineer Simon Lake builds the *Argonaut*, which sails from Norfolk, Virginia, to New York under its own power, the first submarine to succeed in the open sea.                    TECH

---

## Nobel's Second Thoughts

By the age of 33 in 1866, Swedish engineer Alfred Nobel had secured a place in technological history by combining nitroglycerin and diatomaceous earth to produce dynamite. Over the next 30 years of his life Nobel would see his new substance, a safer alternative to the then-common blasting powder, used to tunnel through rock for construction and to detonate explosives in the many wars of the late 19th century.

Dismayed at the violence generated by his invention and believing that "inherited wealth is a misfortune which merely serves to dull a man's faculties," Nobel willed shortly before his death in 1896 that much of his fortune be used to establish annual prizes for science, literature, and peace. In 1901 the first Nobel Prizes were awarded: for physics, to Wilhelm Röntgen of Germany for the discovery of X rays, which are often called Röntgen rays in his honor; for physiology or medicine, to Emil von Behring of Germany for the development of a diphtheria antitoxin; for chemistry, to Jacobus van't Hoff of the Netherlands for his presentation of laws of chemical dynamics and osmotic pressure in solutions; for literature, to René-François-Armand "Sully" Prudhomme of France; and for peace, to Henri Dunant of Switzerland and Frédéric Passy of France.

1897    On November 13, the first initially successful all-metal dirigible, designed by Hungarian David Schwarz, travels several miles from Berlin before crashing.
                                                                                                TECH

1898    Russian engineer Konstantin Tsiolkovsky states a group of mathematical laws that form the basis for the field of astronautics, or space flight.        ASTRO

1898    American astronomer William Pickering discovers Phoebe, a satellite of Saturn.
                                                                                                ASTRO

1898    Studying tobacco mosaic disease, Dutch botanist Martinus Beijerinck applies the term *virus* to a disease-causing microscopic agent that can pass through a bacteriological filter.        BIO

1898    German scientists Friedrich Loeffler and Paul Frosch prove that foot-and-mouth disease in livestock is caused by a virus.        BIO

1898    The International Congress of Zoology organizes an International Commission on Zoological Nomenclature to set up guidelines for naming animals. In 1981 an extensively revised edition will be published.        BIO

1898    French physicists Pierre Curie and Marie Curie discover radium and polonium. *See also* 1898, PHYS.        CHEM

1898    British chemists William Ramsay (who will win the 1904 Nobel Prize for chemistry) and Morris William Travers discover the elements krypton, neon, and xenon.        CHEM

1898    Scottish chemist James Dewar liquefies hydrogen.        CHEM

1898    Heroin is synthesized from morphine. At first deemed a useful painkiller, it proves to be more dangerous and addictive than morphine.        CHEM

1898    German chemist Johann Goldschmidt invents thermite, which proves useful in welding because it burns at high temperatures and leaves a residue of iron or chromium.        CHEM

1898    French physicist and chemist Marie Curie coins the word *radioactivity* to describe the radiation produced by uranium (*see* 1896, PHYS). She finds that other radioactive elements include thorium and the newly discovered radium and polonium (*see* 1898, CHEM). Curie postulates that radioactivity is a property of atoms rather than chemical compounds.        PHYS

1898    American psychologist William James introduces the term *pragmatism* to audiences at the University of California.        PSYCH

1898    English psychologist Havelock Ellis publishes the first of his seven volumes on sexuality, collectively entitled *Studies in the Psychology of Sex*. He is the first to discuss sex in psychological terms, and his work will be subject to controversy and litigation.                                                                          PSYCH

1898    The first magnetic wire recording device, called the Telegraphone, is patented by Danish engineer Valdemar Paulsen.                                                    TECH

1899    German art historian Robert Koldewy begins excavating in Babylon. For the next 18 years, these digs will uncover the remains of Nebuchadnezzar's palace, the Tower of Babel, and the Gate of Ishtar.                                              ARCH

1899    Using a spectroscope, astronomer William Wallace Campbell discovers that Polaris is a system of three stars.                                                       ASTRO

1899    French chemist André-Louis Debierne discovers the element actinium.   CHEM

1899    Scottish chemist James Dewar produces solid hydrogen, reaching the lowest temperature yet attained: 14 K, or 14 degrees Kelvin. (The degree symbol is conventionally omitted in the Kelvin scale.)                                          CHEM

1899    German mathematician David Hilbert publishes his *Foundations of Geometry*, proposing an improved set of axioms for geometry.                                    MATH

1899    British physicist J. J. Thomson calculates the charge of the electron, the particle he discovered in 1897. He also argues that ionization represents the splitting of electrons from the rest of the atom.                                          PHYS

1899    Franz Boas becomes the first professor of anthropology at Columbia University. He will found the "American historical school," which will stress the importance of historical research into folklore and belief.                             SOC

1900    British archaeologist Arthur John Evans excavates the palace of Knossos, the capital of Crete in the time of the legendary King Minos.                           ARCH

1900    German biologist Cal Correns, Dutch botanist Hugo de Vries, and Austrian botanist Erich Tachermak independently rediscover the significance of Gregor J. Mendel's 1866 work in pea pods. These scientists, reexamining this information in the light of British naturalist Charles Darwin's theory of evolution (*see* 1859, BIO), decide that the units of heredity found by Mendel provide a mechanism through which natural selection operates.                                 BIO

1900    German physicist Friedrich Ernst Dorn discovers the element radon.   CHEM

1900    Russian-born American chemist Moses Gomberg produces triphenylmethyl, the first known free radical, a molecule with an unattached carbon bond that can persist for an appreciable time.                                              CHEM

1900    German mathematician David Hilbert poses what will become known as Hilbert's second problem: Whether it can be proved that the axioms of arithmetic are consistent (i.e., will not lead to contradictions). Subsequent attempts to prove this problem are doomed by Gödel's theorem (*see* 1931, MATH). The problem is one of several in an address by Hilbert entitled "Mathematical Problems." *See also* 1934, MATH.                                                         MATH

1900    German physicist Max Planck, who will win the 1918 Nobel Prize for physics, states what will become known as Planck's radiation law, giving the distribution of energy radiated by a blackbody. This law, the foundation of quantum theory, introduces the concept that energy is radiated in discrete packets called quanta. With this law Planck also introduces the Planck formula and the Planck constant.                                                                       PHYS

1900    Two British physicists—J. W. Strutt, Lord Rayleigh, and James Hopwood Jeans—develop the Rayleigh-Jeans formula for predicting the distribution of energy radiated across long wavelengths by a blackbody. German physicist Wilhelm Wien, winner of the 1911 Nobel Prize for physics, works out the Wien formula for short wavelengths. Both are limiting cases of the Planck formula.                  PHYS

1900    Physicists measure the increase in the mass of electrons moving at speeds close to the speed of light, which confirms the Lorentz-FitzGerald contraction of 1895.                                                                                         PHYS

1900    French physicist Antoine-Henri Becquerel determines beta rays isolated in uranium radiation by Rutherford in 1897 are electrons. Becquerel will share the 1903 Nobel Prize for physics with the French physicists Marie Curie and Pierre Curie.                                                                                        PHYS

1900    French physicist Paul Ulrich Villard detects gamma rays—electromagnetic radiation of very short wavelength and high energy—in uranium radiation.          PHYS

1900    English physicist William Crookes discovers that a purified uranium compound in solution is only slightly radioactive but increases in radioactive intensity over time, suggesting that radioactivity is a form of change within the uranium atom.                                                                                PHYS

1900    British physicist Owen W. Richardson, who will win the 1928 Nobel Prize for physics, investigates the Edison effect of 1883 and discovers that heated metals tend to emit electrons.                                                            PHYS

1900    American sociologist Herbert Blumer is born (*d.* 1987). A student of American philosopher George Herbert Mead, he will found the sociological approach known as symbolic interactionism.                                                      SOC

1900    American Benjamin Holt invents the tractor.                                    TECH

1900    The Trans-Siberian Railway opens, linking Moscow and Irkutsk.          TECH

1900    Irish-American engineer John Phillip Holland designs the first modern submarine to be bought by the U.S. Navy. Named the *Holland*, it uses an internal combustion engine and an electric motor.                                               TECH

| | |
|---|---|
| 1900 | Eastman Kodak's one-dollar, six-exposure Brownie camera is introduced, to great success.                                                                      TECH |
| 1900 | The hamburger (ground lean beef on toast) is introduced by American businessman Louis Lassen, in Connecticut.                                                  TECH |
| 1900 | On July 2, Count Ferdinand von Zeppelin of Germany flies the first of his series of rigid-frame airships.                                                       TECH |
| 1901 | American astronomer Annie Jump Cannon introduces spectral subclasses into the Harvard Classification System for stars. *See* 1890, ASTRO.                       ASTRO |
| 1901 | Dutch botanist Hugo de Vries is the first to report gene mutations in plants.          BIO |
| 1901 | French chemist Eugène-Anatole Demarçay discovers the element europium.          CHEM |
| 1901 | French chemist Victor Grignard, who will share the 1912 Nobel Prize for chemistry with French chemist Paul Sabatier (*see* 1897, CHEM), discovers what will become known as Grignard reagents, a group of organometallic compounds of magnesium used as catalysts in organic synthesis reactions.          CHEM |
| 1901 | Indanthrene blue, the first synthetic vat dye, is produced.          CHEM |
| 1901 | F. A. Forel presents the first textbook on limnology, the intensive study of the physical and biological features of lakes.          EARTH |
| 1901 | Austrian pathologist Karl Landsteiner discovers three blood types, which he names A, B, and C (later renamed O). In the following year AB will be added to the list. His discovery that compatibility of blood types is governed by a simple set of rules will greatly improve the safety of blood transfusions. He will receive the 1930 Nobel Prize for physiology or medicine for his work. *See also* 1940, MED.          MED |
| 1901 | French physicist Pierre Curie measures the heat emitted by radium as 140 calories per hour. This energy flow becomes known as atomic energy because it is presumed to result from inside the atom rather than being caused by changes in chemical bonds to other atoms, as in the burning of wood.          PHYS |
| 1901 | The U.S. National Bureau of Standards is founded.          PHYS |
| 1901–1920 | French pioneer of the unconscious Pierre Marie Felix Janet does work on ego states via hypnosis. He is able to produce dissociative states, as are found in multiple personality disorder, and create a deeper understanding of the ego's role in personality.          PSYCH |
| 1901 | American Alfred Louis Kroeber founds the anthropology department at the University of California at Berkeley.          SOC |

1901     American businessman Patillo Higgins strikes oil at the Spindletop gusher in Beaumont, Texas, which becomes one of the nation's most productive oil sources. In a short time, Texas becomes the dominant petroleum-yielding state.     TECH

1901     The mercury-vapor electric lamp is developed, by American engineer Peter Cooper Hewitt.     TECH

1901     German auto maker Gottlieb Daimler introduces a new motor vehicle to the public. The car is named by its distributor, Austrian statesman Emil Jellinek, for his young daughter Mercedes.     TECH

1901     Polyunsaturated and unsaturated fatty acids are hydrogenated for the first time by English chemist William Normann. Incorporated in food to increase its shelf life, hydrogenated fatty acids will later be found to increase the risk of heart disease.     TECH

1901     The Multigraph, which prints by reproducing the written page, is produced by the American Multigraph Co.     TECH

1901     A usable electric vacuum cleaner is developed by British inventor Hubert Booth.     TECH

1901     On December 21, Italian electrical engineer Guglielmo Marconi broadcasts radio waves from England to Newfoundland, marking the invention of the radio. In 1909 he will share the Nobel Prize for physics with German physicist Ferdinand Braun (*see* 1897, TECH).     TECH

1902     The Code of Hammurabi, the oldest known legal code, dating from 1775 B.C., is discovered at Susa, Persia (Iran), by French archaeologists.     ARCH

c. 1902     While studying blood, American biochemist Lawrence Joseph Henderson introduces the nomogram (a chart of scaled lines for facilitating the calculating of variables) into biology.     BIO

1902     While studying beans, British geneticist William Bateson determines that variations in the beans caused by nutrition factors and the environment are not inherited by their offspring.     BIO

1902     French physicist Marie Curie makes a first determination of the atomic weight of radium as 225.93.     CHEM

1902     British physicist Ernest Rutherford and British chemist Frederick Soddy demonstrate that uranium and thorium break down in the course of radioactivity into a series of radioactive intermediate elements.     CHEM

1902     German-American chemist Herman Frasch perfects a method of removing sulfur from deep deposits by using superheated water to melt the substance. CHEM

1902    British-American engineer Arthur Edwin Kennelly and, independently, Oliver Heaviside discover what will become known as the Heaviside-Kennelly layer (or E-layer), a layer of the ionosphere (a part of Earth's atmosphere) that reflects medium-frequency radio waves.                    EARTH

1902    French meteorologist Léon-Philippe Teisserenc de Bort identifies the troposphere and stratosphere as two distinct layers of Earth's atmosphere.                    EARTH

1902    Denmark, Sweden, Norway, England, the Netherlands, and Russia establish the International Council for the Exploration of the Seas, which will still be functioning more than 90 years later.                    EARTH

1902    German mathematician Gottlob Frege refines and extends Boole's 1847 system of symbolic logic but is unable to account for an apparent self-contradiction noted by British mathematician Bertrand Russell.                    MATH

1902    French mathematician Henri Lebesgue redefines the theory of integration, introducing the Lebesgue measure.                    MATH

1902    German physicist Philipp Lenard, investigating the photoelectric effect (see 1887, PHYS), discovers that it results from the emission of electrons by certain metals in response to specific wavelengths of light.                    PHYS

1902    American psychologist Carl Ranson Rogers is born (d. 1987). He will take the focus off the psychoanalyst's couch and remove any authority the therapist has over an individual seeking help. Preferring the term *client* to *patient*, he will develop client-centered therapy, an introspective psychotherapy that concentrates on an individual's personal growth potential and self-healing capacity. Rogers will align with Abraham Maslow's later, related ideas of self-actualization.                    PSYCH

1902    The American Anthropological Association is founded.                    SOC

1902    The first automat, operated by the Horn & Hardart Baking Co., opens in Philadelphia. It dispenses food from a series of glass-walled compartments that open when money is deposited.                    TECH

1903    American geneticist Walter Sutton publishes a paper called "The Chromosomes Theory of Heredity," claiming that hereditary factors are located in chromosomes.                    BIO

1903    Norwegian chemists Kristian Birkeland and Samuel Eyde introduce the Birkeland-Eyde process for the fixation of nitrogen by passing air through an electric arc.                    CHEM

1903    Norwegian explorer Roald Amundsen becomes the first to sail through the Northwest Passage, thus reaching the Pacific Ocean from the Atlantic. He will also be the first to reach the South Pole. *See* 1911, EARTH.                    EARTH

1903        French physicist Marie Curie becomes the first woman to receive a Nobel Prize
            when she shares this year's Nobel Prize for physics with her husband, Pierre
            Curie, and Antoine-Henri Becquerel. The three French physicists are honored
            for their research on radioactivity. *See* 1896 and 1898, PHYS.                        PHYS

## Nobel Women

Since the Nobel Prizes for literature, peace, and science were instituted in 1901, more
than 400 people have received awards for science. Only 10 of them have been women.
The first woman to do so was the French physicist and chemist Marie Curie, born
Maria Sklodowska in Warsaw, Poland. She shared the 1903 Nobel Prize for physics
with her husband, French physicist Pierre Curie, and French physicist Antoine-Henri
Becquerel.

The three French physicists won the award, presented by the Royal Swedish
Academy of Sciences, for their research on radioactivity, a term coined in 1898 by
Marie Curie to describe radiation emitted by such elements as uranium and thorium.
In 1911 Marie Curie became the first person ever to win a second Nobel Prize when
she won the award for chemistry in recognition of her discovery of radium and polo-
nium in 1898.

The second woman to win a Nobel Prize for science was the French physicist and
chemist Irène Joliot-Curie, the daughter of Marie Curie and Pierre Curie. She shared
the 1935 Nobel Prize for chemistry with her husband, French physicist Frédéric Joliot.
The other eight women awarded Nobel Prizes for science are:

• Prague-born American biochemist Gerty Theresa Radnitz Cori (1947, physiology
or medicine; shared with her husband, Prague-born American biochemist Carl
Ferdinand Cori, and Argentine physiologist Bernardo Houssay).

• German-American physicist Maria Goeppert Mayer (1963, physics; shared with
German physicist J. Hans D. Jensen and Hungarian-American physicist Eugene P.
Wigner).

• Egyptian-born English chemist Dorothy Crowfoot Hodgkin (1964, chemistry).

• American biophysicist Rosalyn Sussman Yalow (1977, physiology or medicine;
shared with French-American physiologist Roger Guillemin and American biochemist
Andrew V. Schalley).

• American geneticist Barbara McClintock (1983, physiology or medicine).

• Italian-American neuroembryologist Rita Levi-Montalcini (1986, physiology or
medicine; shared with American biochemist and zoologist Stanley Cohen).

• American biochemist Gertrude B. Elion (1988, physiology or medicine; shared
with British pharmacologist James Black and American pharmacologist George H.
Hitchings Jr.).

• German biologist Christiane Nüsslein-Volhard (1995, physiology or medicine;
shared with American molecular biologist Eric F. Wieschaus and American geneticist
Edward B. Lewis).

1903        In his *History of Mechanics*, German physicist Ernst Mach critiques the concepts of absolute space and time in Newtonian physics in a work that will influence Albert Einstein in 1905.                    PHYS

1903        French mathematician Jules-Henri Poincaré argues that small discrepancies in initial conditions can result in large differences within a short period. This observation will become important to chaos theory in the 1970s and later.    PHYS

1903        Phenobarbital, a barbiturate and potent sedative, is first used.    PSYCH

1903        Austrian-born German chemist Richard Adolf Zsigmondy devises the ultramicroscope, a microscope that makes use of the Tyndall effect (*see* 1869, PHYS) to view colloidal particles. He will win the 1925 Nobel Prize for chemistry.    TECH

1903        Russian engineer Konstantin Tsiolkovsky publishes "The Investigation of Universal Space by Means of Reactive Devices," a seminal article in astronautics in which he suggests liquid hydrogen and liquid oxygen for use as propellants.    TECH

1903        American machinist Henry Ford and 12 other inventors found and incorporate the Ford Motor Co. in Detroit, Michigan, introducing their first automobile, the eight-horsepower, two-cylinder Model A.    TECH

1903        The Harley-Davidson motorcycle is developed in Milwaukee, Wisconsin, by William Harley, Arthur Davidson, Walter Davidson, and others.    TECH

1903        The Springfield rifle, which will be used widely in the armed forces, is developed at the U.S. arsenal in Springfield, Massachusetts.    TECH

1903        On December 17 at Kitty Hawk, North Carolina, Orville Wright flies 120 feet in 12 seconds, in a 12-horsepower biplane built by him and his brother Wilbur. It is the first successful flight of a heavier-than-air machine.    TECH

1904        American-Argentine astronomer Charles D. Perrine discovers the sixth satellite of Jupiter. An additional 10 Jovian moons will subsequently be discovered.
                                                                                      ASTRO

1904        German astronomer Johannes Franz Hartmann discovers spectral absorption lines that indicate the existence of interstellar clouds of gas and dust.    ASTRO

1904        German biologist Theodor Boveri, experimenting with sea urchins, decides on the necessity of a full set of chromosomes for an embryo's normal development. This discovery, coupled with Walter Sutton's discovery (*see* 1903, BIO), will form the basis for the Sutton-Boveri chromosome theory of inheritance.
                                                                                      BIO

1904        The local anesthetic novocaine, or procaine, is synthesized.    CHEM

| | |
|---|---|
| 1904 | German chemist Richard Abegg suggests that chemical reactions occur when electrons transfer from one atom to another. CHEM |
| 1904 | English chemist Frederick Stanley Kipping discovers silicones. CHEM |
| 1904 | Norwegian-American meteorologist Jacob Bjerknes publishes *Weather Forecasting as a Problem in Mechanics and Physics,* an influential study that takes a scientific approach to weather forecasting. EARTH |
| 1904 | German mathematician Ernst Zermelo formulates what will become known as Zermelo's axiom of choice, which states that in any set of mutually exclusive nonempty sets there exists at least one set that contains one, and only one, element in common with each nonempty set. Paul J. Cohen will show (*see* 1963, MATH) that the axiom of choice cannot be proved within set theory. MATH |
| 1904 and 1911 | Scottish paleontologist Robert Broom publishes papers on the Cretaceous and Triassic dinosaurs of South Africa which mark the beginning of sustained studies of dinosaurs in that region. PALEO |
| 1904 | British physicist Hertha Marks Ayrton becomes the first woman ever to address the Royal Society, reading a paper on the origin and growth of ripple marks in sand. She also becomes known about this time for her research on electric arcs. PHYS |
| 1904 | British physicist J. J. Thomson proposes a model for atomic structure in which electrons are embedded in a positively charged atom like raisins in a pound cake. PHYS |
| 1904 | Experimenting with the scattering of X rays, British physicist Charles G. Barkla discovers that the number of charged particles in an atom varies according to its mass. He also shows that X rays are transverse waves like light, confirming their electromagnetic nature. PHYS |
| 1904 | British physicist W. H. Bragg shows that the energies of alpha particles are emitted only within certain sharply defined ranges. PHYS |
| 1904 | American psychologist Edward B. Titchener establishes an organization of experimental psychologists who espouse the psychological viewpoint that will be referred to as existential psychology, or existentialism, which emphasizes subjectivity, personal decision, free will, and individuality. It will serve as a counterbalance to theories that stress the role of society and social groups. PSYCH |
| 1904–1905 | German sociologist Max Weber publishes a series of essays that will become his best-known work, *The Protestant Ethic and the Spirit of Capitalism*, which relates Protestant ideals to the development of capitalism. SOC |
| 1904 | The first electron radio tube, the diode thermionic valve, is developed by British engineer John Ambrose Fleming. TECH |

1904        Using hand-crafted muslin pouches, American shop proprietor Thomas
            Sullivan invents the teabag.                                      TECH

1904        On September 15 American aviator Orville Wright accomplishes the first air-
            plane maneuvers when he makes a turn with an airplane. On September 20
            his brother Wilbur makes the first complete circle.               TECH

1905        American astronomers Thomas Chrowder Chamberlin and Forest Ray
            Moulton formulate an early version of the planetesimal hypothesis, which
            states that the solar system was formed by accretion from smaller particles
            called planetesimals. This part of their theory will come to be accepted (*see*
            1944, ASTRO), though their belief that the solar system resulted from gaseous
            matter being pulled out of two stars as a result of a near collision will be dis-
            credited (*see* 1935, ASTRO).                                     ASTRO

1905        German chemist Richard Willstätter discovers the structure of chlorophyll.   CHEM

1905        German-born physicist Albert Einstein, who has been a Swiss citizen since 1901
            (having given up his German citizenship in 1896), proposes the theory of spe-
            cial relativity, which accounts for physical phenomena at constant (nonacceler-
            ated) velocities close to the speed of light. This theory assumes that the speed of
            light is a constant and shows that velocity has meaning only in that it is relative
            to the observer. It upholds the validity of Newton's laws of motion for subrela-
            tivistic speeds, or speeds far below the speed of light, and accounts for the
            Lorentz-FitzGerald contraction (*see* 1895, PHYS). *See also* 1916, PHYS.    PHYS

1905        Einstein deduces as a consequence of his theory of special relativity (*see* above)
            that the mass of a body is a measure of its energy content, according to the
            equation $E = mc^2$, where $E$ is energy, $m$ is mass, and $c$ is the velocity of light.
            This means that previous conservation laws (*see* 1789, CHEM, and 1847, PHYS)
            can be unified into a single law of the conservation of mass-energy.   PHYS

1905        Einstein explains the photoelectric effect (*see* 1902, PHYS) as a consequence of
            quantum theory (*see* 1900, PHYS). Einstein's treatment shows that light, then
            regarded solely as a wave, can also be seen as a particle in some respects. A unit
            of light regarded as a particle, or a quantum of electromagnetic radiation, will
            become known as a photon.                                         PHYS

1905        Einstein formulates an equation to describe Brownian motion (*see* 1827, PHYS)
            that will permit scientists to deduce the size of molecules and atoms.    PHYS

1905        American physicist P. W. Bridgman, who will win the 1946 Nobel Prize for
            physics, attains high-pressure conditions equal to 20,000 atmospheres.   PHYS

1905        French psychologist Alfred Binet publishes his first batteries of tests intended
            to measure intelligence, and introduces the term *intelligence quotient*, or *IQ*, for
            a score representing an individual's intelligence as measured against a stan-
            dard. *See also* 1911, PSYCH.                                     PSYCH

1906        Dutch astronomer Jacobus Cornelis Kapteyn begins a survey of the Milky Way
            in which he will calculate the galaxy's diameter as 23,000 light-years and its
            thickness as 6,000 light-years, closer to the truth than the 1785 results
            obtained by William Herschel but still too small.                          ASTRO

1906        American astronomer Percival Lowell begins searching for a planet beyond
            Neptune that would account for irregularities in the orbit of Uranus, but no
            such planet is discovered until after his death in 1916. *See* 1930, ASTRO.   ASTRO

1906        German astronomer Karl Schwarzschild theorizes that radiation is the princi-
            pal cause of heat transmission within stars.                               ASTRO

1906        British naturalist Henry Guppy publishes *Plant Dispersal*, the second volume of
            *Observations of a Naturalist in the Pacific Between 1896 and 1899*. In it Guppy
            tries to account for the current geographical distribution of plants and writes
            extensively about plant dispersal by flotation in seawater.                 BIO

1906        German chemist Richard Willstätter shows that chlorophyll molecules contain
            magnesium. In 1915 he will be awarded the Nobel Prize for chemistry.         BIO

1906        British biologist William Bateson coins the word *genetics* for the study of how
            physical, biochemical, and behavioral traits are transmitted from parents to
            children.                                                                   BIO

1906        Austrian chemist Carl Auer von Welsbach and French chemist Georges Urbain
            discover the element lutetium.                                              CHEM

1906        Russian botanist Mikhail Semenovich Tsvett develops the technique of chro-
            matography for separating complex mixtures.                                CHEM

1906        American millionaire Andrew Carnegie finances the foundation of the
            Geophysical Laboratory in Washington, D.C., for the experimental study of
            minerogenesis and petrogenesis.                                            EARTH

1906        British physicist J. W. Strutt, winner of the 1904 Nobel Prize for physics, dis-
            covers radioactivity in seawater.                                          EARTH

1906        The most deadly earthquake in American history strikes San Francisco on April
            18, devastating more than four square miles and killing more than 500 peo-
            ple. The earthquake will later be estimated to have measured 8.3 on the Richter
            scale (*see* 1935, EARTH).                                                  EARTH

1906        French mathematician Maurice Fréchet develops functional calculus.   MATH

1906        Russian mathematician A. A. Markov develops what will become known as
            Markov chains, or strings of linked probabilities.                          MATH

c. 1906     German physician Josef Schrudde develops a way to scrape fat off the body
            through small incisions in the skin. In the 1970s, Swiss physician Ulrich

Kesselring will add suction to the process of fat scraping, creating liposuction (suction lipectomy).                                                                    MED

1906    British physicist William Thomson disputes the developing theory of radioactive disintegration of atoms, suggesting that radium (discovered in 1898) is not an element but a molecular compound of lead and helium. Marie Curie will later prove him wrong. *See* 1910, PHYS.                                      PHYS

1906    British physicist Ernest Rutherford, who will win the 1908 Nobel Prize for chemistry, and German physicist Hans Geiger discover that alpha particles are related to helium atoms.                                                          PHYS

1906    British physicist Charles G. Barkla discovers a phenomenon characteristic of X rays: When they are scattered by diverse elements, more massive elements produce more penetrating X rays. Barkla will be awarded the 1917 Nobel Prize for physics.                                                                          PHYS

1906    German physical chemist Walther H. Nernst states the third law of thermodynamics, that all bodies at absolute zero would have the same entropy, though absolute zero can never be perfectly attained. In 1920 he will be awarded the Nobel Prize for chemistry.                                                        PHYS

1906    Upon the death of her husband, Pierre Curie, Marie Curie takes his place as chair of physics at the Sorbonne, becoming the university's first female professor.                                                                                       PHYS

1906    British physicist J. J. Thomson shows that a hydrogen atom has only one electron, contrary to other theories that had been predicting many more.        PHYS

1906    American psychiatrist Morton Prince is the key individual behind the establishment of the *Journal of Abnormal Psychology*.                              PSYCH

1906    British anthropologist W. H. R. Rivers introduces the genealogical method into social science research in his work *The Todas*.                         SOC

1906    The Audion, or three-electrode vacuum tube amplifier, is invented by American Lee De Forest. This device makes the further development of radio possible.                                                                                  TECH

1907    Biological productivity studies, measuring the amount and rate of production occurring in a given ecosystem in a certain time period, become popular. Biological productivity may apply to single organisms, populations, or entire communities and ecosystems. The concept helps scientists better understand food and fiber production, in addition to giving information about nonharvestable organisms.                                                                      BIO

1907    American zoologist Ross Granville Harrison becomes the first to culture tissues successfully.                                                                BIO

1907　　Dutch botanist Hugo de Vries publishes *Plant Breeding*, providing additional support for the importance of mutations in plant evolution.　　　BIO

1907　　Swiss chemist Jean-Charles de Marignac discovers the element ytterbium.　　　CHEM

1907　　Russian-born German mathematician Hermann Minkowski publishes *Time and Space*, in which he sets forth a mathematical treatment of a four-dimensional universe in which time is the fourth dimension. Einstein will make use of Minkowski's model in his 1916 general theory of relativity.　　　MATH

1907　　At Mauer, near Heidelberg, Germany, quarry workers discover a hominid mandible. Whether it represents *Homo erectus*, Neanderthal man, or some other intermediate form will remain unclear.　　　PALEO

1907　　American chemist Bertram Borden Boltwood proposes that lead is the final product of the radioactive decay of uranium and thorium, arguing that knowing the rate of decay will make radioactive dating possible.　　　PHYS

1907　　French physicist Pierre Weiss develops the theory that ferromagnetic substances consist of small magnetized regions called domains and that in strongly magnetized pieces the poles of the domains are aligned.　　　PHYS

1907　　American sociologist William Graham Sumner publishes *Folkways*, in which he analyzes the lasting impact of folkways and mores. He here originates the concept of ethnocentrism, or the belief in the superiority of one's own culture.　　　SOC

1907　　Nearly frictionless chrome and manganese-alloy ball bearings are developed by Swedish engineer Sven Gustav Wingquist.　　　TECH

1907　　The first all-in-one electric clothes washer, the Thor, is developed by the Hurley Machine Co. in Chicago. The Maytag washer is introduced in Iowa later this year.　　　TECH

1908　　The Archaeological Institute of America acknowledges the importance of conducting investigations in the New World when it establishes the School of American Archaeology in Santa Fe, New Mexico.　　　ARCH

1908　　American astronomer George Ellery Hale identifies magnetic fields in sunspots, leading to the discovery of the largely magnetic nature of sunspots.　　　ASTRO

1908　　A mysterious event near Tunguska, Siberia, creates huge craters, destroys a herd of deer, and levels trees for 20 miles. Later it will seem probable that the site was hit by a meteorite, possibly a small comet or a chunk of one.　　　ASTRO

1908　　Danish astronomer Ejnar Hertzsprung describes both the giant and dwarf stellar categories, also proposing the concept of absolute magnitude, or how to identify what the brightness of a star would be if all stars were seen at a standard distance.　　　ASTRO

1908		The first biological autoradiograph is taken when a frog is made radioactive and placed on a photographic plate, so that an image of the entire animal is formed.					BIO

1908		American zoologist Joseph Grinnell suggests that competition between species is a force that results in their adopting similar, yet separate habitats—distinct ecological niches.					BIO

1908		English scientist William Bayliss publishes his research on hormones in *The Nature of Enzyme Action*.					BIO

1908		English mathematician Godfrey H. Hardy and German obstetrician Wilhelm Weinberg independently formulate what will become known as the Hardy-Weinberg law, which states that population gene and genotype frequencies remain constant from generation to generation if mating is random and if mutation, selection, immigration, and emigration do not occur. This law is now a fundamental principle of population genetics.					BIO

1908		Dutch physicist Heike Kamerlingh Onnes liquefies helium at 4 K (four degrees Kelvin), reaching temperatures as low as 0.8 K, although he is unable to solidify helium.					CHEM

1908		German chemist Fritz Haber, who will win the 1918 Nobel Prize for chemistry, devises the so-called Haber process for producing ammonia by the reaction of nitrogen with hydrogen. It will become valuable in fixating nitrogen for fertilizers and explosives.					CHEM

1908		French mathematician Maurice Fréchet introduces abstract spaces.		MATH

1908		French physician Charles Mantoux invents a skin test to detect tuberculosis.		MED

1908		German paleontologist Friedrich von Huene publishes a thorough monograph on the Triassic dinosaurs of Europe, those dating from the earliest period of dinosaur evolution.					PALEO

1908		In Wyoming, paleontologists Charles H. Sternberg and his sons Charles M., George, and Levi discover the first known fossilized dinosaur skin, an impression of the skin of a duck-billed dinosaur.					PALEO

1908		Neanderthal remains are recovered in France at Le Mustier and La Chapelle-aux-Saints.					PALEO

1908		Using Einstein's equation based on Brownian motion (*see* 1905, PHYS), French physicist Jean Perrin, who will win the 1926 Nobel Prize for physics, calculates the approximate size of an atom as one hundred-millionth of a centimeter.					PHYS

1908		The International Conference on Electric Units and Standards adopts the international ampere as the basic unit of electric current.					PHYS

1908     German physicist Louis Paschen discovers what will become known as the Paschen series of lines, in the far-infrared region of the hydrogen spectrum.   PHYS

1908     British-American psychologist William McDougall writes a pioneering book on social psychology containing a controversial theory of instincts. This purposive psychology claims that instincts, like hunger, sex, escape, curiosity, and self-assertion, are the impetus to all behavior.   PSYCH

1908     Austrian psychoanalyst Abraham Arden Brill becomes Sigmund Freud's translator into English. Brill coins the psychoanalytic term *id* from Freud's German word for the unconscious, *Es* (*it*), by adopting the Latin word for *it*.   PSYCH

1908     American political scientist Arthur F. Bentley publishes *The Process of Government*, in which the concept of the group rather than the state is central.   SOC

1908     German physicist Hans Geiger invents the Geiger counter, a device to detect and measure the ionizing particles emitted by radioactive substances. *See also* 1928, TECH.   TECH

1908     American machinist and inventor Henry Ford introduces the only-in-black Model T, which at $850.50 becomes a nationally best-selling automobile. TECH

1908     The gyrocompass, which determines direction by the rotation of the earth, is developed by German engineer Hermann Anschutz-Kampfe.   TECH

1909     Dutch botanist Wilhelm L. Johannsen (1857–1927) suggests the term *genes* for the units of inheritance inside chromosomes.   BIO

1909     Austrian chemist Fritz Pregl develops techniques for analyzing minute amounts of organic chemicals. In 1923 he will be awarded the Nobel Prize for chemistry.   CHEM

1909     The system of pH numbers for quantifying acidity and alkalinity is developed by Danish chemist Søren Peter Lauritz.   CHEM

1909     Croatian geologist Andrija Mohorovičić notes a change in properties 32 kilometers (20 miles) beneath the earth's surface. The Mohorovičić discontinuity, or Moho, marks the junction between the earth's mantle and its crust.   EARTH

1909     Florence Bascom, the first woman awarded a Ph.D. from Johns Hopkins University, edits the U.S. Geological Survey's Philadelphia Folio, which will remain one of the most comprehensive geological descriptions of the area from Maine to Maryland.   EARTH

1909     On April 6, U.S. Navy engineer Robert E. Peary, his African-American dogsled driver Matthew Henson, and three Eskimos become the first to reach the North Pole. *See also* 1911, EARTH.   EARTH

1909          American neurosurgeon Harvey Cushing (1869–1939) discovers that
              acromegaly, a form of giantism, is due to an overproduction of the growth hor-
              mone in the pituitary gland. When Cushing surgically removes a portion of the
              oversized gland, the patient's symptoms disappear. This proves to have been an
              important step in understanding hormone activity.                         MED

1909          French physician Charles Nicolle (1866–1936), who will win the 1928 Nobel
              Prize for physiology or medicine, discovers that the body louse is responsible
              for transmitting epidemic typhus.                                         MED

1909          Paleontologist Earl Douglass of the Carnegie Museum in Pittsburgh discovers
              an almost complete skeleton of an *Apatosaurus* (*Brontosaurus*) at the Carnegie
              Quarry in what will become Dinosaur National Monument, Utah. Excavations
              from now through 1923 will uncover the largest known concentration of
              Jurassic dinosaurs, including the *Diplodocus, Stegosaurus, Antrodemus*
              (*Allosaurus*), and *Camptosaurus*.                                       PALEO

1909–1912     German paleontologists begin expeditions in Tendaguru, East Africa
              (Tanzania). Among the finds in the quarry of Jurassic dinosaurs is a skeleton of
              the largest known sauropod, *Brachiosaurus*.                              PALEO

1909          In the Canadian Rockies, American paleontologist Charles Walcott discovers
              the rich fossil site known as the Burgess Shale. This site contains the soft-bod-
              ied remains of creatures that lived 530 million years ago, just after the
              Cambrian explosion of multicellular organisms.                           PALEO

1909          Indian physicist Homi J. Bhabha is born (*d.* 1966). He will be known for his
              discovery of Bhabha scattering, the exchange of a virtual photon by an electron
              and a positron. *See also* 1945, MISC.                                    PHYS

1909          With the founding of Chicago's Juvenile Psychopathic Institute by psychiatrist
              William Healy, the first mental health facility specifically for children is estab-
              lished, marking the beginning of the child guidance movement.            PSYCH

1909          American physicist William David Coolidge develops a method of producing
              fine tungsten wires for use as filaments in light bulbs.                  TECH

1909          Bakelite becomes the first thermosetting plastic, one that once set does not
              soften under heat. Made of phenol and formaldehyde, it is invented by
              Belgian-American chemist Leo Hendrik Baekeland. Bakelite will be used first
              for electrical insulation, then later for a number of consumer products.   TECH

1909          On July 25, Louis Blériot becomes the first to fly across the English Channel,
              from France to England.                                                   TECH

1910s         As a pioneer of small-animal experiments, American biochemist Elmer Verner
              McCollum (1879–1967) establishes the first white-rat colony in the U.S. for
              nutrition research. He will subsequently identify vitamins A and B and prove

the necessity of trace elements. McCollum will also show that a calcium deficiency produces tetany (muscular spasm).                                                        BIO

1910s     This decade sees the foundation of the influential Norwegian, or Bergen, school of meteorology.                                                        EARTH

1910     American geneticist Thomas Hunt Morgan discovers the sexual differences in the inheritance of traits that will become known as sex-linked inheritance.

BIO

1910     William Burton uses thermal cracking to refine petroleum oil.                CHEM

1910     During work on celestial mechanics, French mathematician Jules-Henri Poincaré develops modern tide theories.                                                        EARTH

1910     British mathematicians Bertrand Russell and Alfred North Whitehead begin publication of their *Principia Mathematica*, the most definitive effort yet at rooting mathematics in logic and building it systematically.                MATH

1910     German bacteriologist and 1908 Nobel Prize winner Paul Ehrlich (who shared the prize for physiology or medicine with Russian embryologist and immunologist Ilya Metchnikoff) produces the arsenical compound salvarsan, the first drug effective in treating syphilis. This discovery marks the beginning of modern chemotherapy.                                                        MED

1910     American paleontologist Barnum Brown begins to excavate Cretaceous dinosaurs at the Red Deer River in Alberta, Canada. American paleontologists Charles H. Sternberg and sons will begin a separate dig at the same site in 1912. Brown's expeditions at this rich site will last through 1915, the Sternbergs' through 1917.                                                        PALEO

1910     British physicist J. J. Thomson confirms the existence of isotopes by using positive cathode rays to measure the atomic masses of two different isotopes of neon. Frederick Soddy (*see* 1913, CHEM) will further develop the concept of isotopes.                                                        PHYS

1910     French physicist Marie Curie shows conclusively that radium is an element, disproving William Thomson's argument. *See* 1906, PHYS.                PHYS

1910     University of Frankfurt psychologists Max Wertheimer, Kurt Koffa, and Wolfgang Köhler reject the prevailing associationism dominating German psychology and together found the school of Gestalt psychology, dealing mostly with perception processes and behavior. Gestalt therapists will seek to restore an individual's natural mental balance by heightening awareness, emphasizing present experiences rather than recollections of the past.                PSYCH

1910     Eric Berne, who will become known as the father of transactional analysis, is born in Canada (*d.* 1970). Berne will advocate this form of psychotherapy to help people exchange their feelings and thoughts more effectively. Among his popular books will be *Games People Play* (1964).                                    PSYCH

1910    Scottish anthropologist James George Frazer publishes *Totemism and Exogamy*, which will influence Viennese psychoanalyst Sigmund Freud in *Totem and Taboo* (1913), in which Freud will analyze the practices of "primitive" peoples to shed light on modern Western people's neuroses.                        SOC

1910    American sociologist Robert Merton is born. A student of Talcott Parsons (*see* 1937, soc), Merton will attempt to unite theoretical and empirical research.

SOC

1910    Steel begins to be used instead of wood in car bodies, for the Ford Model T. TECH

1910    The Pathé Gazette, a movie newsreel, is conceived by Frenchmen Charles and Émile Pathé. These newsreels are first presented in British and American markets.                                                                          TECH

1910    Cartoonist John Randolph Bray patents the cel process of animation. Nearly two decades later, in 1928, American filmmaker Walt Disney will license it to create the first Mickey Mouse cartoon, "Steamboat Willie."           TECH

1910    The first commercial flight of a dirigible, or airship, is made on June 22 in Germany. Piloted by Count von Zeppelin and carrying 20 passengers, the dirigible travels 500 miles, from Friedrichshafen to Düsseldorf, in nine hours. *See also* 1852, TECH.                                                                   TECH

1911    Using modern archaeological methods, Amadeo Maiuri begins the excavation of Pompeii and Herculaneum, destroyed by the eruption of Mount Vesuvius in 79.                                                                              ARCH

1911    Working independently, Danish astronomer Ejnar Hertzsprung and, in 1913, American astronomer Henry Norris Russell plot the magnitudes of stars against their colors and spectral classes. The resulting Hertzsprung-Russell diagram will provide important evidence for theories of stellar evolution.   ASTRO

1911    American geneticist Hermann J. Müller begins experimental breeding of the fruit fly (*Drosophil*). He will be best known for his discoveries of gene mutations caused by X rays and dire warnings concerning the effect of nuclear radiation on human genes.                                                     BIO

1911    American geneticists Thomas Hunt Morgan and Alfred Henry Sturtevant devise the first chromosome map, found by investigating the separation frequency of chromosome crossover from one gene to another.            BIO

1911    French physicist Marie Curie wins the Nobel Prize for chemistry, becoming the first person ever to win a second Nobel Prize. (She shared the 1903 Nobel Prize for physics with her husband, French physicist Pierre Curie, and French physicist Henri Becquerel.)                                       CHEM

| 1911 | Norwegian explorer Roald Amundsen reaches the Antarctic Pole on December 16, just ahead of British explorer Robert Falcon Scott, who will arrive on January 14, 1912. *See also* 1909, EARTH.                                    EARTH |

1911        Norwegian explorer Roald Amundsen reaches the Antarctic Pole on December 16, just ahead of British explorer Robert Falcon Scott, who will arrive on January 14, 1912. *See also* 1909, EARTH.                    EARTH

1911–1912   Australian explorer Douglas Mawson explores Adelie Land in Antarctica. EARTH

1911        The American Nurses' Association is established. Within 50 years it will be the largest professional women's organization in the world.                    MED

1911        British physicist Ernest Rutherford proposes a model of the atom in which the atom is mostly empty space. A massive, positively charged atomic nucleus is surrounded by outer regions of negatively charged electrons that leave the atom electrically neutral. This model makes it clear that alpha particles are helium nuclei, not helium atoms. *See also* 1913, PHYS.                    PHYS

1911        Scottish physicist C. T. R. Williams invents the cloud chamber, a device for studying the paths of particles of ionizing radiation.                    PHYS

1911        Dutch physicist Heike Kamerlingh Onnes, who will win the 1913 Nobel Prize for physics, discovers superconductivity, the absence of electrical resistance in certain substances at temperatures close to absolute zero.                    PHYS

1911–1920   Swiss psychiatrist Eugen Bleuler introduces the terms *schizophrenia* and *ambivalence* into psychiatry. Schizophrenia denotes what Bleuler considers a mental split from reality.                    PSYCH

1911        German neurologist Alois Alzheimer discovers the presenile dementia, or irreversible mental and intellectual deterioration, that will bear his name as Alzheimer's disease.                    PSYCH

1911        French psychologist Alfred Binet and French physician Théodore Simon develop the series of graded intelligence tests to measure a person's intelligence quotient, or IQ (*see* 1905, PSYCH), that becomes known as the Binet-Simon scale.                    PSYCH

1911        American ethnologist Alice Cunningham Fletcher publishes *The Omaha Tribe*, her most important monograph on American Indian culture.                    SOC

1911        American anthropologist Franz Boas publishes his influential work *The Mind of Primitive Man*.                    SOC

1911        C. F. Kettering, American businessman and owner of Dayton Engineering Laboratories (Delco), develops the electric self-starter, which will be widely used for automobile and truck engines.                    TECH

1911        The Chevrolet Motor Co. is founded by American automobile pioneer W. C. Durant. The company takes its name from Durant's partner, Swiss-American race-car driver Louis Chevrolet.                    TECH

1911      A sulfate process using alkali rather than acids improves paper production, resulting in—among other products—a sturdy brown material called Kraft paper.      TECH

1912      American archaeologist Hiram Bingham discovers the Inca strongholds of Machu Picchu and Vitcos near Cuzco, Peru.      ARCH

1912      Harvard astronomer Henrietta Leavitt discovers that the period of pulsation of a Cepheid variable star, one that varies in brightness in a regular way, increases with the star's luminosity or intrinsic brightness. This period-luminosity curve, as it is called, will become the basis for determining the distances of galaxies and distant stars. *See* 1914, ASTRO.      ASTRO

1912      Austrian-American physicist Victor F. Hess, who will share the 1936 Nobel Prize for physics with American physicist Carl D. Anderson (*see* 1932 and 1937, PHYS), discovers evidence of the existence of cosmic rays.      ASTRO

1912      Dutch physical chemist Peter Debye develops equations describing the behavior of polar molecules or dipoles, molecules with pairs of separated opposite electric charges. His work leads to the concept of the dipole moment, the product of the positive charge and the distance between charges.      CHEM

1912      German meteorologist and geophysicist Alfred Wegener proposes the theory of continental drift, arguing that the granite continents float on the basalt ocean floor, changing position over the ages. He claims that all the continents once formed a single landmass called Pangaea. At first rejected, his ideas bear similarities to the theory of plate tectonics (*see* 1960, EARTH), which holds that plates bearing continents, not the continents themselves, move.      EARTH

1912      For the next two years, Russian K. Silovski applies ultrasonic techniques to detect icebergs and submerged ice.      EARTH

1912      A compendium on the hydrology of the Mediterranean Sea, including numerous temperature and salinity stations, is published by N. Nielson.      EARTH

1912      Following the sinking of the *Titanic* on the night of April 14–15, underwater acoustics and communications becomes a rapidly growing area of study. EARTH

1912      British biochemist Frederick Gowland Hopkins and Dutch physician Christiaan Eijkman, who will share the 1929 Nobel Prize for physiology or medicine, establish the role of the accessory food factors called vitamins and are the first to pinpoint a dietary-deficiency disease. Polish-American biochemist Casimir Funk, theorizing that some diseases are caused by a lack of certain substances in the diet, coins the term *vitamin* for these missing substances.      MED

1912      English amateur archaeologist Charles Dawson claims to have discovered the missing link between apes and humans at Piltdown in Sussex, England. This so-called Piltdown man skull will fool paleontologists until 1953, when it will be shown to be a hoax, an artful combination of an ape's jaw and a modern human skull.                                                                    PALEO

1912        German physicist Max von Laue, who will win the 1914 Nobel Prize for physics, discovers how to use crystals for X-ray diffraction, permitting measurement of the wavelength of X rays.                                          PHYS

1912        British physicist J. J. Thomson studies canal rays (*see* 1886, PHYS), which he calls positive rays, since they are streams of positively charged atomic nuclei. He discovers also that there are at least two different varieties of neon atoms. PHYS

1912        French sociologist Émile Durkheim publishes his *Elementary Forms of Religious Life*, a major work in the development of cultural anthropology.          SOC

c. 1912     German chemist Carl Bosch improves the Haber process for making ammonia (*see* 1908, CHEM) and implements it in large industrial plants.           TECH

1912        German chemist Friedrich Bergius develops a coal hydrogenation process in which coal and heavy oil are treated with hydrogen to produce gasoline. In 1931 he will share the Nobel Prize for chemistry with German chemist Carl Bosch.                                                                           TECH

1912        American chemist Irving Langmuir finds that tungsten filaments in light bulbs filled with inert gases last longer than those in vacuum light bulbs.       TECH

1912        The first aeroboat, or amphibious aircraft, is developed, by German-American engineer Grover Loening.                                                TECH

1912        An effective high-vacuum tube, which amplifies electric current, is developed by American physicist H. D. Arnold.                                      TECH

1912        The Alpha Beta Food Market and Ward's Groceteria, two self-service grocery stores, the forerunners of supermarkets, open independently in California.   TECH

1913        Danish astronomer Ejnar Hertzsprung discovers that a Cepheid variable star of absolute magnitude −2.3 has a period of 6.6 days.                       ASTRO

1913        British chemist Frederick Soddy and, independently, Polish chemist Kasimir Fajans state the radioactive displacement law, which describes the loss of mass and electric charge incurred by radioactive atoms. Soddy, who will be awarded the Nobel Prize for chemistry in 1921, coins the term *isotopes* for atoms of the same element that have a differing mass and differing radioactive properties.

                                                                            CHEM

1913        American chemist Theodore William Richards finds that the atomic weight of lead varies according to the quantity of radioactive material in the lead ores, a finding that supports the concept of isotopes. In the following year he will win the Nobel Prize for chemistry.                                      CHEM

| | |
|---|---|
| 1913 | German chemists Leonor Michaelis and Maud Lenora Menten formulate the Michaelis-Menten equation describing the rate at which enzyme-catalyzed reactions take place.      CHEM |
| 1913 | French physicist Charles Fabry proves the existence of an ozone layer in the upper atmosphere.      EARTH |
| 1913–1917 | Large numbers of bottles are launched to study surface currents in the Sea of Japan.      EARTH |
| 1913–1919 | Dutch mathematician Luitzen Egbertus Jan Brouwer develops his theory of intuitionism, in which mathematics is considered to begin with a basic intuition of natural numbers. Truth claims are then made through constructivity rather than consistency, as advocated by German mathematician David Hilbert. *See* 1900, MATH.      MATH |
| 1913 | In a book on the Riemann surface, German mathematician Hermann Weyl develops new concepts and definitions that will be important to later research on manifolds.      MATH |
| 1913 | German surgeon A. Saloman develops the technique of mammography, an X-ray procedure to detect breast cancer.      MED |
| 1913 | Autopsies become legal in China.      MED |
| 1913 | American physicist Robert A. Millikan calculates the electric charge of a single electron. Ten years later he will be awarded the Nobel Prize for physics.      PHYS |
| 1913 | Danish physicist Niels Bohr applies quantum theory to the structure of the atom, describing electron orbits and electron excitation and de-excitation. PHYS |
| 1913 | German physicist Johannes Stark, who will win the 1919 Nobel Prize for physics, discovers what will become known as the Stark effect, a multiplication in spectral lines caused by strong electric fields.      PHYS |
| 1913 | Hans Geiger and Ernst Marsden provide supporting evidence for Rutherford's model of the atom (*see* 1911, PHYS) when they direct a beam of alpha particles at a piece of gold foil and note that a few of the particles are deflected.      PHYS |
| 1913 | Swiss psychiatrist Carl Gustav Jung breaks with his teacher, Sigmund Freud, and develops his own theories. He will classify people as introverts and extroverts and theorize that certain ideas, which he will call archetypes, are inherited from the distant past and are a part of all peoples' unconscious, referred to as the collective unconscious.      PSYCH |
| 1913 | American psychologist J. B. Watson explains his theory of behaviorism in an article entitled "Psychology as the Behaviorist Views It." For Watson, psychology is the "science of behavior," not the traditional "science of conscious expe- |

rience." Watson's aim is to predict and control behavior rather than describe and explain it. He also wishes to eliminate the traditional behavior distinction between humans and animals.     PSYCH

1913     Russian engineer Igor Ivan Sikorsky builds and flies the first multiengined aircraft.     TECH

1913     American physicist William David Coolidge invents what will become known as the Coolidge tube, a device for manufacturing X rays.     TECH

1913     Diesel-electric locomotives, invented by Rudolf Diesel, are put into operation in Sweden.     TECH

1913     In the United States, soap-laden steel wool pads are marketed under the name Brillo pads, by the Brillo Manufacturing Corp.     TECH

1913     On October 7, in Detroit, the Ford Motor Co. uses the assembly line system to build cars, which reduces assembly time from 12.5 to 1.5 hours.     TECH

1914     American astronomer Vesto Melvin Slipher discovers that 13 of 15 galaxies are receding from ours at hundreds of miles per second.     ASTRO

1914     American astronomer Walter Sydney Adams invents the technique of spectroscopic parallax for determining a star's distance by comparing its apparent magnitude with its absolute magnitude as derived from its spectral characteristics.     ASTRO

1914     American astronomer Harlow Shapley correlates the absolute magnitude and period of Cepheid variable stars, thereby providing a yardstick for determining galactic and stellar distances. *See also* 1912 and 1952, ASTRO.     ASTRO

1914     British physicist Henry Gwyn Jeffreys Moseley arrives at the concept of atomic number, a number representing the positive charge of the atomic nucleus. The periodic table of elements (*see* 1869, CHEM) is henceforth revised in order of atomic number, beginning with hydrogen at 1 and ending with uranium at 92.     CHEM

1914     British physicists W. H. Bragg and W. L. Bragg discover that sodium chloride and certain other compounds exist as groups of ions bound by electromagnetic interaction.     CHEM

1914     German chemists Paul Duden and J. Hess manufacture acetic acid synthetically.     CHEM

1914     German geologist Beno Gutenberg demonstrates the existence of a boundary in the lower depths of the earth's surface that he calls a discontinuity, because it causes an abrupt alteration in properties. Below the Gutenberg discontinuity is the earth's liquid core, above it the earth's mantle.     EARTH

1914        German mathematician Felix Hausdorff publishes his *Basic Features of Set Theory*, a systematic exposition that introduces the concept of Hausdorff topological spaces and marks the development of point set topology as a distinct discipline.                                                                                    MATH

1914        American biochemist Edward C. Kendall isolates thyroxin, the thyroid hormone, which will be used to treat thyroid insufficiencies. Kendall's work with hormones will lead him to isolate cortisone, an anti-inflammatory used widely in medicine. In 1950 Kendall, American physician Phillip S. Hench, and Polish-born Swiss chemist Tadeus Reichstein (*see* 1933, BIO) will jointly receive the Nobel Prize for physiology or medicine.                                   MED

1914        Renowned German paleontologist Friedrich von Huene endorses H. G. Seeley's 1887 classification of dinosaurs into two independent orders, *Ornithischia* and *Saurischia*. Von Huene's support promotes the general acceptance of this system.                                                                            PALEO

1914        British physicists W. H. Bragg and W. L. Bragg (father and son, respectively) show how to determine the wavelengths of X rays from diffraction by crystals. In the following year they will share the Nobel Prize for physics.                  PHYS

1914        British physicist James Chadwick shows that beta particles, unlike alpha particles, are emitted in a continuous range of energies.                             PHYS

1914        British physicist Ernest Rutherford gives the name *proton* to the positively charged nucleus of the hydrogen atom, which is now seen to be, in a sense, the fundamental atom, as Prout had suggested in 1815. Scientists of the time theorize that the nuclei of all other atoms are composed of a combination of protons and electrons, but this theory will be revised with the discovery of the neutron by Chadwick in 1932.                                               PHYS

1914        German physicists James Franck and Gustav Hertz, who will share the 1925 Nobel Prize for physics, perform experiments confirming Niels Bohr's model of electrons' orbits (*see* 1913, PHYS). Bohr will win the Nobel Prize for physics in 1922.                                                                              PHYS

1914        English anthropologist W. H. R. Rivers publishes his *History of Melanesian Society*, a classic work based on his research in the South Pacific.          SOC

1914        By the beginning of World War I, all the major European powers have submarines and continue to improve upon them, the research and development focusing only on their suitability for war purposes.                                           TECH

1914        The teletype machine is developed by German-American inventor Edward E. Kleinschmidt.                                                                          TECH

1914        The Panama Canal opens to traffic on August 31, carrying ships between the Atlantic and Pacific Oceans. Operating by a system of locks, the canal spans 50.7 miles.                                                                             TECH

1915        Scottish astronomer Robert Innes discovers Proxima Centauri, a faint companion to the double star Alpha Centauri. At 4.3 light-years away it is the closest star to Earth except the sun.                                    ASTRO

1915        American astronomer Walter Sydney Adams demonstrates that the star Sirius B is a white dwarf—extremely hot, dense, and small.                                    ASTRO

1915        Swedish biochemist Svante Arrhenius, winner of the 1903 Nobel Prize for chemistry, publishes the influential *Quantitative Laws in Biochemistry*.                                    BIO

1915–1917   Canadian bacteriologists Frederick William Twort and Felix Hubert d'Herelle separately discover bacteriophages, a type of virus able to infest and kill bacteria normally present in an organism.                                    BIO

1915        American geneticists Thomas Hunt Morgan, Calvin Bridges, Alfred Henry Sturtevant, and Hermann J. Müller publish *The Mechanism of Mendelian Heredity*, claiming that invisible genes within the chromosomes of the cell nucleus determine an offspring's hereditary traits.                                    BIO

1915        Japanese scientists K. Yamagiwa and K. Ichikawa identify the first of a long line of cancer-producing agents called carcinogens by exposing rabbits to coal tar for long periods of time.                                    MED

1915        American paleontologist Richard Swann Lull publishes his *Triassic Life of the Connecticut Valley* (revised 1953), a classic monograph on the ancient flora and fauna of that region. In it he revises an earlier study of fossil tracks (*see* 1848, PALEO), ascribing them not to birds but to dinosaurs.                                    PALEO

1915        German physicist Arnold Sommerfeld proposes that electrons travel in elliptical orbits. He combines quantum theory and relativity theory in revising the model of the atom, resulting in what is known as the Bohr-Sommerfeld atom. Sommerfeld and Danish physicist Niels Bohr work out the arrangement of electrons around an atom's nucleus, defining each in terms of three quantum numbers that vary in value. *See also* 1925, PHYS.                                    PHYS

1915        American physicist William Draper Harkins calculates that four hydrogen nuclei can fuse to form a helium nucleus, releasing a great deal of energy converted from the excess mass.                                    PHYS

c. 1915     American psychophysicist Stanley Smith Stevens formulates the power law of psychophysics. His developments in auditory scaling methods determine that physical continua usually conform to a psychophysical power law rather than Gustav Fechner's logarithmic law. The power function will prove controversial in psychophysics for more than 30 years.                                    PSYCH

1915        American neurologist and physiologist Walter B. Cannon presents a critique of the James-Lange theory of emotion in his book *Bodily Changes in Pain, Hunger, Fear, and Rage*. Cannon's substitute for the James-Lange theory, that a physical response precedes the appearance of emotion, will become known as the Cannon-Baird theory of emotion. It states that emotion is an emergency reaction to help humans cope in a crisis, not a follow-up to physical stimuli.                                                                    PSYCH

1915–1918      British anthropologist Bronislaw Malinowski does field research with the Trobriand Islanders. He will develop the theory of functionalism with his ethnographies in the 1920s and 1930s.                               SOC

1915           The first significant use of poisonous gas as a military weapon occurs when the Germans use chlorine gas against French troops at Ypres on April 22 during World War I. The Germans' first attempt at using tear gas, against the French in October 1914, was so ineffective that it was barely noticed by the troops.    TECH

1915           Heat- and shock-resistant borosilicate glass is developed in the United States by the Corning Glass Works and sold as cooking implements under the trade name Pyrex.                                                              TECH

1916           American astronomer Edward Emerson Barnard discovers what will become known as Barnard's Star, a star with the largest known proper motion.    ASTRO

1916           Physicist Albert Einstein's general theory of relativity (*see* 1916, PHYS) accounts for a long-standing anomaly about the orbital motion of the planet Mercury. *See* 1855, ASTRO.                                                               ASTRO

---

## Stellar Neighbors

About 1752, French astronomer Nicolas-Louis de Lacaille (1713–1762) discovered Alpha Centauri, which at a distance of 4.35 light-years was believed for a time to be the star nearest the solar system. Then in 1915 a faint star called Proxima Centauri was discovered at only 4.3 light-years distance, about 26 trillion miles.

As it turns out, Proxima Centauri and Alpha Centauri are part of one triple-star system called the Centauri System. Alpha Centauri is actually two stars revolving around each other every 80 years. Proxima, separated from its two siblings by a much greater distance, revolves around them once every million years.

Could any of these stars support life? The two stars of Alpha Centauri might. One is a type G2 star, about the size and color of our yellow sun; the other, a type K0, is a somewhat cooler orange star but might still be hot enough to support life. However, Proxima Centauri, which is a red dwarf (type M5), is probably too small and cool to sustain living things.

Despite its probable bleakness, Proxima Centauri is our nearest stellar neighbor. Even so, at the fastest rates of manned space travel yet achieved, it would take about 110,000 years to reach it, long enough for the crew's descendants and their relatives waiting back home to have evolved into separate species. Unless space travel becomes faster by several orders of magnitude, don't expect to see a visit to the nearest star any time soon.

---

1916        The Ecological Society of America is founded.      **BIO**

1916        American chemist Gilbert Newton Lewis shows how variations in the number of electrons in the outermost shell of an atom lead to the formation of chemical bonds, through electron transfer or sharing. He explains that the most stable elements are those in which the outer shell has either eight or two electrons, whereas elements with different arrangements are more or less reactive. Lewis's work explains the valences of elements. American chemist Irving Langmuir, who will win the 1932 Nobel Prize for chemistry, independently devises a similar model. *See* 1852, CHEM.      **CHEM**

1916        American feminist and nurse Margaret Sanger establishes the first American birth-control clinic, in Brooklyn, New York.      **MED**

1916        Physicist Albert Einstein, who had his German citizenship restored in 1914 (*see* 1905, PHYS), proposes his general theory of relativity, which extends his special theory of 1905 to systems moving at changing velocities relative to each other. The general theory accounts for gravitational interactions, arguing that mass generates a gravitational field that curves space. The predictions of Einstein's theory are close to those of Newton's, but with several differences that are testable, and subsequent tests support Einstein's theory. (*See* 1919 and 1925, PHYS.) In 1921 Einstein will be awarded the Nobel Prize for physics. **PHYS**

1916        Swiss linguist Ferdinand de Saussure publishes his compilation of lecture notes that he calls a *Course in General Linguistics*. The founder of structural linguistics, de Saussure argues that language sounds are arbitrary signs, while language itself is a systematic structure linking the sign and that which is signified.      **SOC**

1916        American radio station developer Lee De Forest broadcasts the first radio news report.      **TECH**

1917        American astronomer George Ellery Hale installs a reflecting telescope on Mount Wilson in California. Its lens of 100 inches (2.54 meters) makes it the largest until the one on Mount Palomar in 1948.      **ASTRO**

1917        German astronomer Karl Schwarzschild formulates equations based on Einstein's general theory of relativity that predict the existence of what will become known as black holes. *See* 1939 and 1960s, ASTRO.      **ASTRO**

1917        American embryologist Frank Rattray Lillie produces a major paper on questions of the origin, nature, and action of sex hormones, giving a thorough analysis of freemartinism, in which a sterile female calf is born as a co-twin with a normal bull calf.      **BIO**

1917        German physical chemist Otto Hahn and Austrian physicist Lise Meitner discover the element protactinium.      **CHEM**

| | |
|---|---|
| 1917–1919 | English mathematician G. H. Hardy collaborates at Cambridge with Indian colleague Srinivasa Ramanujan on number theory. <div align="right">MATH</div> |
| 1917 | Dutch-American physicist Peter Debye develops a technique for determining crystal structure by diffracting X rays through powdered crystalline solids. <div align="right">PHYS</div> |
| 1917 | French physicist Paul Langevin develops sonar (sound navigation and ranging), a system of echolocation used to detect underwater objects. <div align="right">TECH</div> |
| 1917 | Amplitude modulation for AM radio is pioneered through the development of a superheterodyne circuit by U.S. Army Signal Corps officer Edwin H. Armstrong. *See also* 1933, TECH. <div align="right">TECH</div> |
| 1918 | British archaeologist Leonard Woolley begins the excavations that will result in the discovery of Ur in ancient Mesopotamia (Iraq). *See* 1922, ARCH. <div align="right">ARCH</div> |
| c. 1918 | American astronomer Harlow Shapley makes the first fairly accurate estimate of the diameter of the Milky Way galaxy. He also calculates the position of the solar system relative to the Milky Way's galactic center. <div align="right">ASTRO</div> |
| 1918 | British scientists James Hopwood Jeans and Harold Jeffreys propose the tidal hypothesis as an explanation of the solar system's formation, arguing that angular momentum was imparted to the protoplanets by the gravitational pull of a star that nearly collided with our sun. *See also* 1905 and 1935, ASTRO. <div align="right">ASTRO</div> |
| 1918–1919 | A worldwide pandemic of Spanish influenza kills an estimated 21 million people, including 550,000 in the United States. No other pandemic in history has killed so many in so short a time. <div align="right">MED</div> |
| 1918 | In Germany, a laboratory test is introduced that determines the rate at which red blood cells settle in a test tube in the process called erythrocyte sedimentation. An increase or decrease in settling times indicates the presence of certain disease processes. <div align="right">MED</div> |
| 1918 | German mathematician Amalie (Emmy) Noether shows that every symmetry in physics implies a conservation law, and every conservation law a symmetry. <div align="right">PHYS</div> |
| 1918–1919 | Stanford University psychologist Lewis Madison Terman adapts the Binet-Simon Intelligence Scale for use in England, leading to the publication of the first standardized individual intelligence test in the United States, the Stanford-Binet Scale. <div align="right">PSYCH</div> |
| 1918 | Austrian psychologist and psychiatrist Alfred Adler publishes *The Theory and Practice of Individual Psychology*. Adler believes humans to be motivated by their expectations of the future and their striving toward three life goals: physical security, sexual satisfaction, and social integration. This conviction forms the basis of individual psychology. <div align="right">PSYCH</div> |

1918        Hungarian chemist Georg C. de Hevesy invents the technique of radioactive
            tracing, using a radioactive isotope of lead called radio lead.                  TECH

1918        A wood-cellulose bandage called Ceulcotton is developed to use with battle
            wounds, by German-American chemist Ernst Mahler for the U.S. company
            Kimberly & Clark. In 1921 it will be remarketed as a female sanitary napkin
            called Kotex. *See also* 1936, TECH.                                             TECH

1918        The pop-up toaster is patented by American inventor Charles Strite.             TECH

1919        American physicist Robert Goddard analyzes the mathematics of rocket
            propulsion, noting that these principles could be used to send a rocket to the
            moon.                                                                           ASTRO

1919        British chemist Francis W. Aston invents the mass spectrograph for distinguish-
            ing and studying ions with different masses. The device is used to determine rel-
            ative atomic masses and the relative abundance of isotopes. With it Aston shows
            that most stable elements occur in two or more stable isotopes, in which the
            nuclei have the same positive charge but a different mass or mass number. For
            this work he will receive the Nobel Prize for chemistry in 1922.                CHEM

1919        American astronomers George Ellery Hale and Walter Sydney Adams note the
            phenomenon of polarity inversion and realize that the true length of the solar
            cycle is 22 years rather than 11.                                               EARTH

1919        Belgian microbiologist Jules Bordet is awarded the Nobel Prize for physiology
            or medicine for his work in immunology. He has discovered that components
            in the blood act as complements to antibodies, making it possible for the anti-
            bodies to destroy bacteria—the basis for complement fixation—and maintain
            the body's defense against disease.                                             MED

1919        British physicist Ernest Rutherford produces the first artificial nuclear reactions
            when he uses subatomic bombardment to convert the nuclei of helium and
            nitrogen to nuclei of hydrogen and oxygen.                                      PHYS

1919        An expedition organized by British astronomer Arthur Stanley Eddington
            observes a total solar eclipse on May 29 from the island of Principe off West
            Africa. There, observations of starlight being bent by the sun's gravity confirm
            Einstein's general theory of relativity.                                        PHYS

1919        American physician C. F. Menninger and his sons, Karl and William, pioneer
            the development of community mental health programs when they open the
            Menninger Clinic in Topeka, Kansas. The resulting Menninger Foundation will
            become a nonprofit organization devoted to mental-illness prevention and
            treatment.                                                                      PSYCH

1919        American psychologist J. B. Watson, the founder of behaviorism (*see* 1913,
            PSYCH), publishes a book that will provide inspiration for experimental psy-
            chologists, *Psychology from the Standpoint of a Behaviorist.*                  PSYCH

1919        On June 15–16, British airmen John Alcock and Arthur Whitten Brown fly the first nonstop transatlantic flight, from Newfoundland to Ireland. *See also* 1927, TECH.

TECH

1919        On November 8 in Virginia, the American Telephone & Telegraph Co. introduces dial telephones.                                                          TECH

1920s       Soviet biochemist Alexander Ivanovich Oparin outlines his view that life gradually developed in the ocean of the ancient earth and that the primordial atmosphere consisted largely of ammonia, hydrogen, and methane. He will publish much of this theory in his 1936 book *The Origin of Life on Earth.* Stanley Miller and Harold C. Urey's later experiment (*see* 1952, CHEM) in creating amino acids will support Oparin's theory, though by the 1990s scientists will be raising doubts about Oparin's account of the early atmosphere.      BIO

1920s       The Russian school of chemistry proposes the first geochemical classification of the elements.                                                       EARTH

1920s       During this decade petroleum geology will develop.                        EARTH

1920s       The modern pastoral counseling movement, a psychologically sophisticated form of religious caring, begins as an alternative to traditional theological guidance, which is less practical, and early psychiatric treatment, which is less religious in its orientation.                                          PSYCH

1920s       The specialization of psychiatric social work comes into existence.       PSYCH

1920s       Austrian psychologist Otto Rank emphasizes the therapeutic importance of an individual's will. This "will" therapy is the forerunner of both assertiveness training and reality therapy.                                             PSYCH

1920s       Harvard and Columbia Universities take distinct approaches to anthropology, with Harvard specializing in archaeology and physical anthropology, Columbia in ethnology and linguistics.                                       SOC

1920        American astronomer Andrew Ellicott Douglass devises dendrochronology, a technique for dating objects based on the characteristic growth rings of trees in their given region.                                                   ARCH

1920        Using a stellar interferometer, American physicist Albert A. Michelson determines the diameter of Betelgeuse, the first such measurement for any star other than the sun.                                                          ASTRO

1920        On April 26, American astronomers Heber Doust Curtis and Harlow Shapley debate whether the Andromeda "nebula" is inside or outside the Milky Way galaxy. Curtis argues correctly (as verified by Edwin Hubble; *see* 1924, ASTRO) that it is outside.                                                      ASTRO

1920          Czech writer Karel Capek is the first to use the word *robot* as applied to
              mechanical people, in his play *R.U.R.* (*Rossum's Universal Robots*).          BIO

1920          English ornithologist H. Eliot Howard publishes his *Territory in Bird Life*, a
              study of how male birds fight over territory more often than over female birds
              and why they isolate themselves and maintain a certain distance from other
              birds.          BIO

1920          Austrian-born German zoologist Karl von Frisch conditions bees to go to cer-
              tain locations to gather nectar by conditioning them to certain colors. He also
              shows that bees can orient themselves in flight by using the direction of the
              sky's light polarization. More than half a century later he will be awarded the
              1973 Nobel Prize for physiology or medicine, which he will share with two
              other ethologists, the Austrian Konrad Lorenz (*see* 1966, BIO) and the Dutch-
              born Britisher Niko Tinbergen.          BIO

1920          German chemist Hermann Staudinger discovers the molecular structure of poly-
              mers, which will aid in the development of plastics. In 1953 he will be awarded
              the Nobel Prize for chemistry.          CHEM

c. 1920       Norwegian meteorologists Vilhelm Bjerknes and his son Jacob show that the
              atmosphere is made up of air masses differing in temperature, with sharp
              boundaries between them called fronts.          EARTH

c. 1920       Dutch physicist Willem Hendrik Keesom attains a temperature of 0.5 degrees
              above absolute zero, a new record low.          PHYS

c. 1920       Swiss psychologist Jean Piaget is involved in genetic epistemology, research that
              shows how intelligence varies qualitatively and quantitatively with age. He
              identifies stages in the development of a child's mental faculties, including rea-
              soning skills, and urges schools to provide developmentally appropriate chal-
              lenges for students.          PSYCH

1920          American psychologist Robert Sessions Woodworth develops the Woodworth
              personal data sheet, a screening device for military use to detect emotional
              instability that will become the prototype for future personality question-
              naires.          PSYCH

1920          Station KDKA in Pittsburgh becomes the first radio broadcasting station in the
              world. Developed by American engineer Frank Conrad for Westinghouse, its
              first broadcasts give the results of the November 2, 1920 U.S. presidential
              elelction, in which Ohio Republican Warren G. Harding is elected.          TECH

1921          Chinese records on the stellar explosion in 1054 are published, confirming
              supernova theory.          ASTRO

1921          British astronomer Edward Arthur Milne studies the sun's atmosphere, deter-
              mining the temperature of its layers and predicting the existence of the solar
              wind.          ASTRO

1921          British biochemist Frederick Gowland Hopkins isolates glutathione, a polypeptide important in physiological oxidations, from human tissues.     BIO

1921          In November, Canadian surgeon Frederick G. Banting and Canadian physiologist Charles Best isolate insulin. Within a few years it will be commercially produced and used for treatment of diabetes. Banting will share the 1923 Nobel Prize for physiology or medicine with Scottish physiologist John J. R. MacLeod.     MED

1921          Physicist Alfred Landé discovers half-integer quantum numbers. The previously known quantum numbers were assumed to be whole.     PHYS

1921          Swiss psychiatrist Hermann Rorschach devises what will be known as the inkblot test to assess personality and aid in the diagnosis of psychiatric disorders.     PSYCH

1921          American chemist Thomas Midgley Jr. discovers that tetraethyl lead serves as an antiknock additive in gasoline. He will sell this gasoline additive as Ethyl through his company, the Ethyl Corp.     TECH

1921          On December 1, the U.S. Navy dirigible the C-7 becomes the first to use helium as a lifting gas.     TECH

1922          British archaeologist Leonard Woolley excavates ancient Ur, discovering a great deal about the Sumerian civilization of ancient Mesopotamia (Iraq). He finds evidence of a great flood that swept Sumeria about 2800 B.C. and may have

---

## A Mathematician's Struggle

In 1922, German mathematician Amalie (Emmy) Noether (1882–1935) became the first woman to join the faculty of the University of Göttingen. Her career had been an uphill struggle since her days as a student at the University of Erlangen where, because of her gender, she was at first not allowed to matriculate and had to attend as an auditor only. Once graduated with a Ph.D. summa cum laude, she found it impossible to get a university appointment. The mathematician David Hilbert championed her cause and her achievements spoke for themselves, most notably her formulation of Noether's theorem in 1918, describing the relation between mathematical symmetries and physical conservation laws. Finally, in 1922, she was hired at Göttingen as an "unofficial associate professor."

Noether stayed at Göttingen until 1933, when she was dismissed for reasons that had nothing to do with her gender. Adolf Hitler had come to power and, along with other Jewish faculty, she received a notice saying, "I hereby withdraw from you the right to teach at the University of Göttingen." Noether emigrated to America, where she spent the last two years of her life lecturing and researching at Bryn Mawr College and the Institute for Advanced Studies at Princeton University.

given rise to the stories of the flood in the *Epic of Gilgamesh* and the Bible.

ARCH

1922    On November 4, British archaeologists George Herbert, Earl of Carnarvon, and Howard Carter discover the entrance to the tomb of Egyptian pharaoh Tutankhamen. Inside they find an abundance of ancient Egyptian artifacts that have lain untouched for thousands of years.

ARCH

1922    Russian mathematician A. A. Friedmann suggests that the universe has been expanding from an original dense core of matter. *See also* 1927, ASTRO.

ASTRO

1922    British bacteriologist Alexander Fleming isolates lysozyme from tears and mucus, finding it to have bacteria-killing properties, the first example of a human enzyme to have this capability.

BIO

1922    Polish mathematician Stefan Banach begins to introduce the concept of normed linear spaces. Completely normed ones will be called Banach spaces.

MATH

1922    American biochemist Elmer McCollum discovers a factor in fat that is essential to life—vitamin D. Like vitamin A, which McCollum had isolated in 1913, it is not soluble in water. Later, McCollum will contribute to the discovery of vitamin E.

MED

1922    Sickle-cell anemia is named. This hereditary blood disorder, a chronic anemia that afflicts primarily African-Americans, was first recognized by American physician James Herrick in 1910.

MED

1922    American paleontologist Roy Chapman Andrews begins a series of Central Asiatic Expeditions in the Gobi Desert of Mongolia. These excavations will unearth fossils of early mammals as well as Cretaceous dinosaurs.

PALEO

1922    At the Flaming Cliffs in Bain-Dzak, Mongolia, Andrews uncovers the remains of a primitive horned dinosaur, *Protoceratops*, and its fossilized egg shells. The next year many fossil eggs are found, the first dinosaur eggs ever to be discovered.

PALEO

1922    Dow Co. chemists William Hale and Edgar Britton discover what will become known as the Hale-Britton process, a cheaper, more efficient way to produce phenol. This method permits greater production of phenol-based pesticides and fungicides.

TECH

1923    British astronomer Arthur Stanley Eddington explains the relationship between a star's mass and its luminosity, describing the tension throughout its history between gravitational contraction and outward radiation. He speculates that white dwarf stars are made of degenerate matter.

ASTRO

1923    American physiologists Joseph Erlanger and Herbert S. Gasser develop a method to study the electric currents in nerves. They eventually determine the rate at which nerve fibers conduct impulses and ascertain that the velocity varies with the nerve fiber's thickness. In 1944 these scientists will share the Nobel Prize for physiology or medicine for their work.                    BIO

1923        Dutch physicist Dirk Coster and Hungarian chemist Georg C. de Hevesy discover the element hafnium. The latter will win the 1943 Nobel Prize for chemistry.                    CHEM

1923        American chemist Gilbert Newton Lewis formulates what will become known as the Lewis theory, which establishes a relationship between acid-base reactions and oxidation-reduction reactions. A Lewis acid is a substance that can accept a pair of electrons, a Lewis base one that can donate an electron pair.                    CHEM

1923        Dutch physical chemist Peter Debye and German chemist Erich Huckel develop their Debye-Huckel equations to explain the incomplete dissociation of certain compounds in solution.                    CHEM

1923        Danish chemist Johannes Nicolaus Bronsted develops the concept of acid-base pairs, in which an acid transfers a hydrogen ion, or proton, to a base.        CHEM

1923        Swedish chemist Theodor Svedberg invents the ultracentrifuge, a high-speed centrifuge for separating out small colloidal particles and macromolecules. Three years later he will win the Nobel Prize for chemistry.                    CHEM

1923        An effective vaccine against whooping cough, or pertussis, is produced in Copenhagen. Years later this vaccine will finally be replaced by the triple immunization (DPT) against diphtheria, pertussis, and tetanus.                    MED

1923        American physicist Arthur H. Compton discovers the phenomenon that will become known as the Compton effect: the reduction in the energy of the high-energy electromagnetic rays known as X rays and gamma rays when they are scattered by free electrons, and the corresponding gain in energy by the electrons. This effect demonstrates the particle aspects of an energetic wave. Compton coins the word *photon* for a unit of electromagnetic radiation considered as a particle. In 1927 he will share the Nobel Prize for physics with Scottish physicist C. T. R. Wilson (*see* 1896, PHYS).                    PHYS

1923        French physicist Louis de Broglie, who will win the 1929 Nobel Prize for physics, proposes that every particle should have an associated matter wave whose wavelength is inversely related to the particle's momentum. This theory leads to the consideration of the wave aspects of particles, as Compton (*see* above) drew attention to the particle aspects of waves. *See also* 1927, PHYS. PHYS

1923        Austrian psychologist Otto Rank publishes *The Trauma of Birth*, in which he claims that the phenomenon of birth, not the Oedipus complex as Freud teaches, is what gives rise to human anxiety. This clash in views will ultimately lead to a split between Rank and his mentor Freud. After 1934 Rank will live and lecture in the United States, where he will exert great influence on psy-chotherapy.                                                                          PSYCH

1923    American neurologist and physiologist Walter B. Cannon publishes his findings on the effects of "shell shock," as post-traumatic stress disorder was known following World War I.                                                                        PSYCH

1923    The sliding closure that will become known as the zipper is introduced by the B. F. Goodrich Co. Within five years, it will be used widely in men's and women's clothing. Accounts of the origin of the name vary, but the invention itself is a descendant of Whitcomb L. Judson's clasp locker of 1893.          TECH

1924    American astronomer Edwin Powell Hubble, working with the new Mount Wilson telescope, discovers that the Andromeda "nebula" is a separate galaxy outside the Milky Way.                                                                    ASTRO

1924    The Harvard Observatory completes the publication of the *Henry Draper* (or *Standard Draper*) *Catalogue* (1918–1924), listing 225,000 stars and their spec-tral types. American astronomer Annie Jump Cannon is the principal author. *See also* 1949, ASTRO.                                                          ASTRO

1924    German biochemist Otto Warburg discovers cytochrome oxidase, an enzyme of importance in biological oxidations.                                             BIO

1924    Russian-born British biochemist David Keilin discovers cytochrome, a pigment widely distributed in animals and plants that plays an important role in cellu-lar respiration.                                                                       BIO

1924    American biochemist Harry Steenbock shows that inactive precursors of vita-min D exist in food and that exposure to sunlight (irradiation) produces them and "adds" vitamin D to foods. After this discovery the irradiating of food will become common.                                                                    BIO

1924    Czech chemist Jaroslav Heyrovský devises the polarograph, a device that ana-lyzes chemical solutions of unknown composition through the use of mercury electrodes. In 1959 he will win the Nobel Prize for chemistry for his develop-ment of polarography.                                                         CHEM

1924    Australian-born South African anthropologist Raymond Arthur Dart discovers a fossil skull of *Australopithecus africanus*, a hominid or humanlike primate. Australopithecine fossils will later be found in East Africa as well. Living from 4 million to 1 million years ago, australopithecines were the earliest known hominids; they walked erect but had ape-sized brains. *See* 1974, PALEO.    PALEO

1924    Indian physicist Satyendra Nath Bose and German-born physicist Albert Einstein (who will become an American citizen in 1940) develop the Bose-Einstein statistics, a statistical method for handling the particles called bosons, for Bose. Bosons will come to be defined as elementary particles with an inte-gral spin. *See also* 1926, PHYS.                                            PHYS

| 1924 | Bell Laboratories is established by American Telephone & Telegraph and General Electric to carry out physics research. PHYS |
|---|---|

1924    Two U.S. Army Air Corps biplanes, flown by Lowell Smith and Erik Nelson, complete the first round-the-world flight.    ·    TECH

1924    The development of coal-based synthetic gasoline is studied at the I. G. Farben chemical complex in Germany.    TECH

1925    The Tennessee legislature passes the first state law prohibiting the teaching of Darwin's theory of evolution (*see* 1859, BIO).    BIO

1925    German chemist Walter Karl Noddack, his future wife Ida Eva Tacke, and Otto C. Berg together discover the element rhenium.    CHEM

1925    German oceanographers, using sonar, discover the Mid-Atlantic Ridge, an underwater mountain range running down the middle of the Atlantic Ocean. EARTH

1925    Russian scientist Pyotr A. Molchanov attempts the first radio-wave telemetry with balloon-borne instruments.    EARTH

1925    The first international symposium of the International Astronomical Union meets in Rome.    EARTH

1925    American astronomer Walter Sydney Adams discovers a red or Einstein shift in the spectral lines of white dwarf stars that is caused by the stars' massive gravity, as predicted by Einstein's general theory of relativity (*see* 1916, PHYS).    PHYS

1925    Pierre Auger discovers what will become known as the Auger effect, the ejection of an electron (an Auger electron) from an atom without the emission of a gamma-ray or X-ray photon, as a consequence of the de-excitation of an excited electron in the atom.    PHYS

1925    British chemist Francis W. Aston discovers the packing fraction, which is the algebraic difference between the relative atomic mass of an isotope and its mass number divided by the mass number. It refers to the energy change, also called the binding energy, produced by packing subatomic particles into a nucleus.    PHYS

1925    Austrian-born American physicist Wolfgang Pauli states what becomes known as the Pauli exclusion principle, that no two electrons in an atom can have the same set of quantum numbers, or numbers characterizing how electrons are arranged around their atomic nucleus. Pauli proposes that a fourth quantum number is needed in addition to the three identified by Bohr and Sommerfeld. *See* 1915, PHYS.    PHYS

1925    Dutch physicists George Eugene Uhlenbeck and Samuel Abraham Goudsmit define particle spin as the fourth quantum number.    PHYS

1925    Dutch physical chemist Peter Debye, who will win the 1936 Nobel Prize for chemistry, suggests using paramagnetic substances to lower the temperature of liquid helium to get it even closer to absolute zero than has yet been achieved. American chemist William F. Giauque makes the same suggestion independently.                                    PHYS

1925    German physicist Werner Heisenberg develops matrix mechanics, a mathematical technique for studying the energy levels of electrons.                                    PHYS

1925    Chinese psychologist Chen He-quin uses the diary, or journal, method to keep track of and study his child's physical and emotional development. This is the first official documentation of its kind in China.                                    PSYCH

1925    Psychologist Albert Bandura is born in Canada. He will be responsible for developing the theory of social learning, the view that the basic way humans learn new behavior is by watching and imitating—modeling—others.                                    PSYCH

1925    French anthropologist Marcel Mauss publishes *The Gift*, which analyzes the social bond of debt created by gift giving.                                    SOC

1925    American political scientist Charles E. Merriam publishes *New Aspects of Politics*, which argues that the discipline should make greater use of statistics in support of empirical observation. *See also* 1930, SOC.                                    SOC

1925    The Leica, a 35-millimeter camera with an adjustable lens and shutter speed, is developed by E. Leitz G.m.b.H. of Germany. This camera permits photos that vary in their depth of field and lighting.                                    TECH

1925    A deep-freezing process (patented in 1926) for cooked foods is developed by Americans Clarence Birdseye and Charles Seabrook.                                    TECH

1926    American astronomer Edwin Powell Hubble classifies galaxies by their structure, as elliptical, spiral, or irregular.                                    ASTRO

1926    While studying cytochromes, German biochemist Otto Warburg, who will be awarded the 1931 Nobel Prize for physiology or medicine, shows that they possess the same iron-containing heme—the complex molecular group that joins with protein to form hemoglobin—that hemoglobin does.                                    BIO

1926    American neurologist and physiologist Walter B. Cannon coins the term *homeostasis* to describe an organism's capacity to maintain its internal equilibrium.                                    BIO

1926    American biochemist James Batcheller Sumner crystallizes the first pure enzyme, jack bean urease, proving that enzymes are proteins and can act catalytically.                                    BIO

1926    Swiss-Norwegian geochemist Victor Goldschmidt codifies the isomorphism rule.                                    EARTH

1926  American researcher W. E. Brown detects slight irregularities in the motion of the earth.  EARTH

1926  On May 4, two Americans, Richard E. Byrd and Floyd Bennett, become the first to fly over the North Pole, in a 15-hour nonstop flight. *See also* 1929, EARTH.  EARTH

1926  American bacteriologist Thomas Rivers distinguishes between bacteria and viruses, thus establishing virology as a separate area of study.  MED

1926  German paleontologist Friedrich von Huene publishes a monograph on the late Cretaceous dinosaurs of Argentina.  PALEO

1926  German physicist Max Born develops the concept of the wave packet to describe the probabilistic aspect of electron waves. Along with German physicist Werner Heisenberg and Austrian physicist Erwin Schrödinger, he is considered one of the founders of quantum mechanics, a system of mechanics developed from quantum theory (*see* 1900, PHYS) to explain the properties of molecules, atoms, and subatomic particles. In 1954 Born will share the Nobel Prize for physics with German physicist Walther Bothe (*see* 1929, TECH).  PHYS

1926  Austrian physicist Erwin Schrödinger, who will share the 1933 Nobel Prize for physics with British physicist P. A. M. Dirac (*see* below), develops wave mechanics, a model of the atom in which the electron is regarded as a wave rather than a particle. This model accounts for hitherto unexplained aspects of electron orbits. Central to wave mechanics is the Schrödinger wave equation.  PHYS

1926  Italian physicist Enrico Fermi and British physicist P. A. M. Dirac develop the Fermi-Dirac statistics, which apply to all subatomic particles with half-integral spins, later called fermions. Such particles contrast with bosons, which have an integral spin and are treated with the Bose-Einstein statistics (*see* 1924, PHYS). Protons and electrons are fermions.  PHYS

1926  Irish physicist John Desmond Bernal invents what will become known as the Bernal chart, which assists researchers in using X-ray diffraction photographs to discover the structure of crystals.  PHYS

1926  Russian physiologist Ivan Pavlov, winner of the 1904 Nobel Prize for physiology or medicine, publishes his *Conditioned Reflexes*, an account of dogs salivating at the sound of a bell after being conditioned to eat after a bell rang. Pavlov's experiments will lend credence to the theory that human learning and behavior result, in part, from a conditioned reflex.  PSYCH

1926  The International Institute of African Languages is founded, to encourage linguistic and anthropological research in Africa.  SOC

1926    American chemist Waldo Lonsbury Semon develops an early synthetic rubber called Koroseal for B. F. Goodrich. This durable rubberlike material is based on the polymer polyvinyl chloride.    TECH

1926    In a demonstration, the first motion picture with synchronized sound is projected. The first sound feature will be *The Jazz Singer* (1927).    TECH

1926    The National Broadcasting Company (NBC), a network initially of nine radio stations, is founded by American David Sarnoff.    TECH

1926    On March 16, Robert H. Goddard launches the world's first liquid-fuel rocket, from a farm near Auburn, Massachusetts.    TECH

1927    Belgian astronomer Georges Lemaître, independently of Russian mathematician A. A. Friedmann (*see* 1922, ASTRO), theorizes that the universe has been expanding from an original small, dense core or "cosmic egg." This hypothesis is an early version of what will become accepted as the Big Bang theory.    ASTRO

1927    In Germany, the first rocket society is organized to promote rocket experimentation. Wernher von Braun (*see* 1932, ASTRO, and 1942, TECH) is involved, as is Rumanian Hermann Oberth, who envisions high-altitude research rockets, space travel, and space stations.    ASTRO

1927    Swedish astronomer Bertil Lindblad proposes that the Milky Way galaxy rotates, completing one cycle every 210 million years.    ASTRO

1927    American geneticist Thomas Hunt Morgan publishes *Experimental Embryology*.    BIO

1927    British biologist Charles Elton publishes *Animal Ecology*, a landmark study of ecology as scientific natural history.    BIO

1927    Cellulose acetate is developed.    CHEM

1927    Canadian anthropologist Davidson Black discovers a fossil molar of *Sinanthropus pekinensis*, or Peking man, in a cave near Peking (Beijing), China. This extinct relative of modern humans will come to be considered an example of *Homo erectus*.    PALEO

1927    Danish physicist Niels Bohr develops the concept of complementarity, which suggests that different but complementary models may be needed to explain the full range of atomic and subatomic phenomena.    PHYS

1927    German physicist Werner Heisenberg states the uncertainty principle or principle of indeterminism, that it is impossible to know both the position and the momentum of a subatomic particle with complete precision. *See also* 1812, PHYS.    PHYS

1927        American physicist Clinton J. Davisson and British physicist G. P. Thomson independently discover electron diffraction, which demonstrates the wave aspect of electrons (*see also* 1923, PHYS). Ten years later they will share the Nobel Prize for physics.                                                          PHYS

1927        German-American physicist Albert A. Michelson obtains a still more accurate value for the speed of light of 199,798 kilometers per second.                      PHYS

1927        German physicists Fritz Wolfgang London and Walter Heitler use quantum mechanics to explain the electron bond of the hydrogen molecule. Quantum mechanics will later prove a valuable aid in understanding chemistry.          PHYS

1927        American parapsychologist Joseph Banks Rhine begins research at Duke University to provide support for the existence of extrasensory perception (ESP), the ability to gain information unavailable to the five senses. His research will also provide evidence to support psychokinesis, clairvoyance, telepathy, and precognition.                                               PSYCH

1927        The first underwater passage from New York to New Jersey is put into use with the opening of the Holland Tunnel, named after its chief engineer, Clifton Holland.                                                                        TECH

1927        Transatlantic telephone service is offered between New York City and London.
                                                                              TECH

1927        Television is demonstrated for the first time in the United States by American Telephone & Telegraph executive Walter Gifford, who broadcasts images of Secretary of Commerce Herbert Hoover from Washington, D.C., to a group of viewers in New York City.                                                         TECH

1927        American inventors John and Mack Rust develop the all-mechanical cotton picker, which when popularized two decades later will lead African-Americans northward in search of jobs to replace those lost on southern farms.        TECH

1927        The American dairy company Borden begins to offer homogenized milk.      TECH

1927        On May 20, American aviator Charles Augustus Lindbergh sets off in *The Spirit of St. Louis* on the first solo nonstop transatlantic flight, from Roosevelt Field, Long Island, to Le Bourget Field outside Paris. The flight takes 33 hours, 29 minutes. *See also* 1919, TECH.                                                    TECH

1928        German chemists Otto Paul Hermann Diels and Kurt Alder discover what will become known as the Diels-Alder reaction, or diene synthesis, in which two compounds react in such a way as to yield a ring compound.               CHEM

1928        The first ultrasound echograph is developed.                              EARTH

1928        On December 20, the first flight over Antarctica occurs.                  EARTH

1928        Hungarian-born American mathematician John von Neumann develops the mathematical field of game theory.                                    MATH

1928        American neurosurgeons Harvey Cushing and W. T. Bowie introduce the process of surgical diathermy or blood vessel cauterization, reducing operating time and decreasing the risk of blood loss.                                    MED

1928        Greek-American pathologist George Papanicolaou develops the test for cervical cancer that will be referred to as the Pap smear. It will prevent thousands of cancer deaths through early detection of malignant change.                                    MED

1928        During an unrelated experiment, British bacteriologist Alexander Fleming isolates the mold called *Penicillium natatum*. This discovery is ignored until 1939, however, when British biochemist Ernst B. Chain and British pathologist Howard W. Florey take up Fleming's work and produce pure penicillin.                                    MED

1928        Indian physicist Venkata Raman, who will win the 1930 Nobel Prize for physics, discovers what will become known as the Raman effect, an inelastic scattering of electromagnetic radiation. This effect will be used in Raman spectroscopy to study the details of molecular structure.                                    PHYS

1928        Russian-born American physicist George Gamow, R. H. Fowler, and Lothar W. Nordheim develop the concept of electron tunneling.                                    PHYS

1928        German physicist Arnold Sommerfeld discovers that in a conductor electrons behave like a degenerate gas and only a few electrons with high energy participate in conducting electricity.                                    PHYS

1928        Hungarian-American physicist Eugene P. Wigner develops the concept of parity of atomic states.                                    PHYS

1928        In *Anthropology and Modern Life*, American ethnologist Franz Boas attacks theories of racial superiority that have been prevalent for decades.                                    SOC

1928        American anthropologist Margaret Mead publishes *Coming of Age in Samoa*, a landmark study of cultural traditions related to becoming an adult in Polynesian society.                                    SOC

1928        W. Müller produces an improved Geiger counter (*see* 1908, TECH), now known as a Geiger-Müller counter.                                    TECH

1928        American Amelia Earhart flies from Newfoundland to Wales, becoming the first woman to pilot an airplane alone across the Atlantic Ocean.                                    TECH

1928        A commercial bread-slicing machine is developed, by American inventor Otto Frederick Rohwedder. Two years later the Continental Bakery will sell the first loaves of prepackaged sliced bread, under the label Wonder Bread.                                    TECH

1929        American astronomer Edwin Powell Hubble confirms that the universe is expanding and formulates Hubble's law, that the increase in the velocity at which a galaxy is receding from ours is proportionate to the distance of that galaxy.                                                                        ASTRO

1929        American biochemist Edward A. Doisy and German chemist Adolf Butenandt independently isolate a female sex hormone, estrogen, from the estrogen group. Doisy will share the 1943 Nobel Prize for physiology or medicine with Danish biochemist Henrik Dam.                                                  BIO

1929        American geneticist Hermann J. Müller, who will be awarded the 1946 Nobel Prize for physiology or medicine, finds that the rate of gene mutation can be greatly increased by using X rays on genes.                                     BIO

1929        American chemist William F. Giauque discovers oxygen-16 to be the most common isotope of oxygen, with oxygen-17 and oxygen-18 less common.
                                                                                          CHEM

1929        Alexander Eugenevic Fersman formulates the concept of the geochemical migration of the elements.                                                         EARTH

1929        On November 28–29, American explorer Richard E. Byrd and his crew become the first to fly over the South Pole and back. See also 1926, EARTH.        EARTH

1929        German psychiatrist Hans Berger reports using electrodes, placed against the head, to record the electrical impulses called brain waves. The first electroencephalograph (EEG) is introduced to help in the diagnosis of neurological disorders.                                                                             PSYCH

1929        American psychologist Karl Spencer Lashley publishes Brain Mechanisms and Intelligence. In it he proposes the law of mass action, which states that the rate and accuracy of learning is proportionate to the amount of brain tissue available, and the principle of equipotentiality, that each part of the brain is as important as any other. Lashley will be known best for his work on brain function localization and research into how brain functions are generalized.                PSYCH

1929        German physicist Walther Bothe, who will share the 1954 Nobel Prize for physics with German physicist Max Born (see 1926, PHYS), invents the coincidence counter, a device for studying cosmic rays that machine-registers an event only when a particle passes through two detectors virtually simultaneously.                                                                             TECH

1929        British physicist John Cockcroft and Irish physicist Ernest Walton invent the voltage multiplier, the first particle accelerator, a device using electromagnetic fields to accelerate subatomic particles to high speeds.                        TECH

1929        The 16-millimeter movie camera, projector, and film are introduced to the public by Eastman Kodak.                                                       TECH

1929    On September 24, American pilot James Doolittle shows that instrument-guided flying is possible when he takes off and lands relying completely on instruments in the first "blind" flight.                                    TECH

1929    German automobile maker Fritz von Opel carries out the first flight powered by a rocket engine, traveling almost two miles in 75 seconds.                    TECH

1930s   Russian-American physicist George Gamow popularizes the theory of the Big Bang, the explosive origin of the universe.                                        ASTRO

1930s   While studying the competition of yeast and protozoa, Soviet biologist G. F. Gauze helps develop the principle that two species cannot simultaneously occupy the same ecological niche.                                          BIO

1930s   Evolutionist Theodosius Dobzhansky writes *Genetics and the Origin of the Species*, in which he shows that the facts of genetics are compatible with Darwinian natural selection, the chief cause of sustained changes in gene frequencies and a population's evolutionary changes.                           BIO

1930s   Several young Chinese mathematicians, including Shiing-Shen Chern, Wei-Lang Chow, and P. L. Hsu, go to the West to study. They will go on to become major researchers in their fields.                                           MATH

1930s   Chinese paleontologists M. N. Bien and Chung Chien Young excavate the so-called Lufeng series of late Triassic dinosaurs. Also discovered at this site are mammal-like reptiles related to the ancestors of modern mammals.      PALEO

1930s   American physicist J. Robert Oppenheimer demonstrates that bombardment by deuterons, the atomic nuclei of hydrogen-2, or deuterium (*see* 1931, CHEM), is equivalent to bombardment by neutrons.                                 PHYS

1930s   Bulgarians Ivan N. Stranski and Rostislav Kaïshev contribute to the formation of the molecular-kinetic theory of crystal growth.                              PHYS

1930s   Austrian researcher Manfred Sakel introduces insulin coma therapy, a somatic process to treat schizophrenia by insulin administration. Later discovered to be a potentially fatal procedure, this therapy will lose popularity in coming decades.                                                                       PSYCH

1930s   Australian-American psychologist Elton Mayo is responsible for the Hawthorne experiments in industrial psychology that identify work as a group activity and demonstrate that a sense of belonging, of identification with a group, is more important than the physical working conditions in sustaining worker morale. Much later research in social organization will be based on these findings.                                                              PSYCH

1930s   Keynesian economic theory, named for British economist John Maynard Keynes, calls for government intervention to boost a recessionary economy, encourages the development of econometrics, and provides a technique that combines theory with statistical and mathematical analysis in an attempt to improve the accuracy of economic forecasts. *See also* 1936, soc.                    soc

1930    Archaeologists translate the Edwin Smith papyrus, which dates from 1600 B.C. and was acquired by American Egyptologist Edwin Smith in 1862. This papyrus turns out to be an invaluable record of the surgical practices of ancient Egypt, based on an even older text written probably about 2500 B.C.    ARCH

1930    Russian-German optician Bernhard V. Schmidt invents the Schmidt telescope, which is free of the aberration known as coma.                    ASTRO

1930    Swiss-American astronomer Robert J. Trumpler shows that the Milky Way galaxy is about 60 percent smaller than in previous estimates, which had failed to account for interstellar dust that dims starlight and creates the illusion of greater distance.                    ASTRO

1930    French astronomer Bernard Ferdinand Lyot invents the coronagraph, a telescope for observing the inner corona of the sun in the absence of an eclipse.    ASTRO

1930    On February 18, American astronomer Clyde William Tombaugh discovers Pluto, the ninth planet from the sun and the last to be discovered.    ASTRO

1930    American biochemist John Howard Northrop crystallizes pepsin and shows it to be a protein.                    BIO

1930    British naturalist Henry Ridley writes on theories of long-distance plant dispersal in *The Dispersal of Plants Throughout the World*.                    BIO

1930    Russian botanist Trofim Lysenko, in an attempt to improve Soviet agriculture, turns to the earlier discovery of vernalization, the moisturizing and near-freezing of seed. He claims that winter wheat can be changed so that it can be sown in the spring and its offspring can be changed through vernalization so as to be sown as spring wheat. This concept contradicts theories of evolution that have been claiming that acquired characteristics are not inherited.    BIO

1930    English biologist Ronald Fisher publishes *The Genetic Theory of Natural Selection*, which reestablishes Darwin's theory of evolution (*see* 1859, BIO) by claiming that mutations occur by chance and natural selection controls the direction of evolution by weeding out harmful mutations and perpetuating useful ones.                    BIO

1930    American chemist Thomas Midgley Jr. discovers difluorodichloromethane, or Freon, which will be used as a coolant in refrigerators and air conditioners.    CHEM

1930    English biochemist William Thomas Astbury uses X-ray diffraction techniques to analyze the three-dimensional structure of proteins.                    CHEM

1930    The electrophoresis technique for studying particles in suspension is developed, for later use in the study of proteins.                    CHEM

1930            On June 11, American naturalists Charles Beebe and Otis Barton take a dive in the first bathysphere, built to study the ocean depths. The unmaneuverable, watertight sphere of metal is capable of withstanding intense water pressures.

                                                                                                EARTH

1930            British physicist P. A. M. Dirac, who will share the 1933 Nobel Prize for physics with Austrian physicist Erwin Schrödinger (*see* 1926, PHYS), proposes the existence of antiparticles, particles identical to known particles in mass but with a different charge. The antiproton has a negative charge, the antielectron (or positron) a positive one. Antiparticles make up a form of matter that will come to be called antimatter.

                                                                                                PHYS

1930            Superfluidity, the property of liquid helium at very low temperatures that allows it to flow without friction, is discovered.

                                                                                                PHYS

1930            In his *Principles of Quantum Mechanics,* British physicist P. A. M. Dirac develops a general mathematical theory in which wave mechanics and matrix mechanics represent special cases.

                                                                                                PHYS

1930            Austrian-American physicist Wolfgang Pauli, who will win the 1945 Nobel Prize for physics, proposes the existence of an electrically neutral, virtually massless particle to account for the energy missing in beta decay. In 1932 Italian physicist Enrico Fermi will give it the name *neutrino,* although proof of the existence of neutrinos will not be discovered until 1956. PHYS

1930            American political scientist Harold Lasswell publishes *Psychopathology and Politics,* which makes connections between politics and psychology. Lasswell and Charles E. Merriam (*see* 1925, SOC) are the leaders of the so-called Chicago school of political science, which emphasizes psychological factors and statistical analysis. It will influence the development of behaviorism in the late 1940s (*see* c. 1945, SOC).

                                                                                                SOC

c. 1930         Snorkels, short pipes with one end that goes into the mouth and the other above the water's surface, come into use to aid in underwater exploration. Rubber fins come into use in 1933.

                                                                                                TECH

1930            American physicist Ernest O. Lawrence, who will be awarded the 1939 Nobel Prize for physics, invents the cyclotron, a particle accelerator that speeds up subatomic particles by using a magnetic field to push them in spirals.

                                                                                                TECH

1930            American electrical engineer Vann Bush produces the first partially electronic computer, called a differential analyzer, capable of solving differential equations.

                                                                                                TECH

1930            British engineer Frank Whittle is the first to invent and patent a practical jet engine.

                                                                                                TECH

1930            The lightweight thermoplastic polymer trade-named Plexiglas is developed by Canadian research student William Chalmers.

                                                                                                TECH

| | |
|---|---|
| 1930 | The flash bulb for cameras is patented by German inventor Johannes Ostermeir. TECH |
| 1931 | The small planetoid Eros swings closer to Earth than any other celestial body except the moon, allowing astronomers to measure its parallax precisely and adjust their estimates of the scale of the solar system. ASTRO |
| 1931 | Karl G. Jansky of Bell Laboratories accidentally discovers radio radiation coming from the sky, thereby founding radio astronomy. *See also* 1933, ASTRO.  ASTRO |
| 1931 | German chemist Adolf Butenandt isolates the male sex hormone androsterone. BIO |
| 1931 | English bacteriologist William Joseph Elford discovers that viruses range in size from those of large protein molecules to those of tiny bacteria. BIO |
| 1931 | American pathologist Ernest Goodpasture devises a technique for culturing viruses in eggs. BIO |
| 1931 | British geneticist and biologist J. B. S. Haldane argues on July 2 that biology should be regarded as an independent science, being different in its orientation from the physical sciences. BIO |
| 1931 | American chemist Harold C. Urey discovers deuterium (heavy hydrogen, or hydrogen-2), an isotope of hydrogen. He will win the 1934 Nobel Prize for chemistry. CHEM |
| 1931 | American chemist Linus C. Pauling develops the concept of resonance, which uses quantum mechanics to explain electron sharing in organic compounds. CHEM |
| 1931 | Swiss physicist Auguste Piccard uses a sealed gondola attached to a balloon to ascend into the thin air of the stratosphere, reaching a height of 10 miles. EARTH |
| 1931 | Austrian mathematician Kurt Gödel develops the statement known as Gödel's proof, showing that in any system based on any set of axioms there will always be statements that cannot be proven or disproven on the basis of those axioms. *See also* 1900, MATH. MATH |
| 1931 | On Mount Carmel in Palestine and again in 1935 at Swanscombe, England, important discoveries of hominid skulls are made. The Swanscombe skull has characteristics similar to those of modern humans. PALEO |
| 1931 | Through his work on mental trauma, American neurologist and physiologist Walter B. Cannon discovers the hormone called sympathin, an adrenalinelike substance that works in the sympathic nervous system. PSYCH |
| 1931 | The 1,644-foot George Washington Bridge connecting New York and New Jersey opens. At the time it is the world's longest suspension bridge. Its cables |

were constructed by John A. Roebling's Sons, descendants of a designer of the Brooklyn Bridge.                                    TECH

1931          The refrigerant gas Freon 12 is used to replace less stable gases in refrigerators.          TECH

1932          Under scientists Walter Domberger and Wernher von Braun, the German Ordnance Corps begins researching rocket technology.                    ASTRO

1932          Using spectroscopy, astronomer T. Dunham discovers large quantities of carbon dioxide on Venus.                    ASTRO

1932          British biologist R. G. Canti takes some of the first pictures of cell division, using microcinematography.                    BIO

1932          German-born British biochemist Hans Krebs discovers the urea cycle, showing that when the amino acid arginine breaks down and is reconstituted it produces a urea molecule, the chief nitrogen-containing waste in humans.          BIO

1932          American geneticist Thomas Hunt Morgan argues on August 26 that genes can exert an influence outside the cells in which they are located, claiming that one of these extracellular gene activities results in hormone production.          BIO

c. 1932          The artificial respirator called the iron lung is invented.          MED

1932          American biochemist Charles King isolates and identifies vitamin C. *See also* 1933, BIO.                    MED

1932          American heart specialist A. S. Hyman develops the first clinical cardiac stimulator, calling it an artificial cardiac pacemaker.                    MED

1932          In India, Edward Lewis discovers a jaw fragment from *Ramapithecus*, which in 1981 will be shown to be an extinct primate that is probably an ancestor of the orangutan rather than the common ancestor of apes and hominids, as long believed.                    PALEO

1932          British physicist James Chadwick discovers the neutron, an electrically neutral particle that is a component of atomic nuclei.                    PHYS

1932          German physicist Werner Heisenberg, winner of this year's Nobel Prize for physics, develops the model of an atomic nucleus containing both protons and neutrons. He suggests that exchange forces (binding forces resulting from the interchange of particles) account for the stability of the nucleus. This model will be further developed by Japanese physicist Hideki Yukawa. *See* 1935, PHYS.                    PHYS

1932          Studying cosmic rays, American physicist Carl D. Anderson discovers the antielectron predicted by P. A. M. Dirac in 1930. Anderson calls it a positron.          PHYS

1932        British physicist John Cockcroft and Irish physicist Ernest Walton, who will share the 1951 Nobel Prize for physics, produce the first nuclear reaction to result from the bombardment of an element by artificially accelerated particles. They produce helium by bombarding lithium with hydrogen nuclei. PHYS

1932        Hungarian-born American physicist Leo Szilard grasps the possibility of a nuclear chain reaction, an idea that will not be realized for another decade. *See* 1942, PHYS.                                                                    PHYS

c. 1932     Hungarian psychiatrist Ladislas von Meduna invents metrazol shock therapy, one of the chemically induced forms of convulsive therapy used on schizophrenics.                                                                PSYCH

1932        British physiologist Charles S. Sherrington is awarded the Nobel Prize for physiology or medicine (shared with British physiologist Edgar D. Adrian) for his research on how neurons function. Since irregularities in neurotransmitter levels are linked to depression and schizophrenic disorders, Sherrington's work helps establish greater understanding of these mental illnesses and their treatment.                                                                PSYCH

1932        Austrian psychoanalyst Melanie Klein publishes *The Psychoanalysis of Children*, illustrating how anxieties affect a child's developing ego, superego, and sexuality and cause emotional disturbances. She also develops play therapy, in which children show and release their anxieties through playing with toys. PSYCH

1932        British psychologist Sir Frederic C. Bartlett publishes *Remembering*, in which he argues that all new learning builds on existing knowledge.         PSYCH

1932        American child development scientist Beth Wellman is the first to demonstrate that children's intelligence can diminish in deprived environments and increase in enriched ones.                                                      PSYCH

1932        German electrical engineer Ernst Ruska invents the electron microscope. For this invention and his subsequent work in electron optics, he will share the 1986 Nobel Prize for physics with German physicist Gerd Binnig and Swiss physicist Heinrich Rohrer.                                                     TECH

1932        The first synthetic light-polarizing film, Polaroid film, is developed by American inventor Edwin Herbert Land.                                        TECH

1933        In the first discovery of radio astronomy, Karl G. Jansky detects powerful radio-wave radiation coming from the center of the Milky Way. *See also* 1931 and 1937, ASTRO.                                                              ASTRO

1933        German-American astronomer Walter Baade and others develop the theory of neutron stars, arguing that a star larger than 1.4 solar masses would collapse into small, dense objects composed of neutrons.                           ASTRO

1933    Polish-born Swiss chemist Tadeus Reichstein successfully synthesizes vitamin C. In 1950 he will share the Nobel Prize for physiology or medicine with American biochemist Edward C. Kendall and American physician Phillip S. Hench (*see* 1914, MED).

BIO

1933    American geneticist Thomas Hunt Morgan receives the Nobel Prize for physiology or medicine for proving that chromosomes carry hereditary traits.    BIO

1933    American chemist Gilbert Newton Lewis discovers deuterium oxide, or heavy water.

CHEM

1933    Hubert James and Albert Sprague Coolidge apply quantum mechanics to deduce the strength of the covalent bond. Their highly accurate results lend support to quantum theory.    CHEM

1933    A prototype defibrillator, to electrically restore an irregular heartbeat to normal, is introduced.    MED

1933    German-American physicist Otto Stern, who will win the 1943 Nobel Prize for physics, demonstrates the wave aspects and magnetic characteristics of molecular beams: streams of molecules, atoms, or ions traveling at low pressure in the same direction.    PHYS

1933    American chemist William F. Giauque uses the magnetic techniques he and P. A. M. Dirac had earlier developed (*see* 1925, PHYS) to reach a new record low temperature of 0.25 K (degrees Kelvin).    PHYS

1933    Walther Meissner discovers what will become known as the Meissner effect, that there is a falling off of magnetic flux in the interior of a superconducting material when it is cooled below a critical temperature.    PHYS

1933    Wiley Post becomes the first pilot to circle the globe solo.    TECH

1933    American electrical engineer Edwin H. Armstrong refines the process of frequency modulation (FM), a method of transmitting radio waves without static. *See also* 1917, TECH.    TECH

1933    The walkie-talkie, a portable pair of short-distance radios, is invented by the U.S. Army Corps of Engineers and Motorola employee Paul Galvin.    TECH

1933    The first milk fortified with vitamin D is sold commercially by the Borden Co.    TECH

1934    German-American astronomer Walter Baade and Swiss astronomer Fritz Zwicky identify the differences between novae and supernovae, and suggest that neutron stars are the product of supernova eruptions. *See* 1968, ASTRO.

ASTRO

1934    Dutch botanist F. W. Went demonstrates the existence of plant hormones.    BIO

1934        German chemist Adolf Butenandt isolates progesterone, a female hormone vital to successful gestation. He will share the 1939 Nobel Prize for chemistry with Swiss chemist Leopold Ružička.                                                              BIO

1934        The pH meter for electronically measuring acidity and alkalinity is invented by Arnold O. Beckman.                                                              CHEM

1934        Russian mathematician Aleksandr O. Gelfond publishes Gelfond's theorem, which solves Hilbert's seventh problem. *See* 1900, MATH.                                   MATH

1934        Sodium pentothal is introduced for use as an intravenous anesthetic.       MED

1934        French physicist Irène Joliot-Curie (the daughter of Marie Curie and Pierre Curie) and her husband, French physicist Frédéric Joliot, are the first to achieve artificial radioactivity when they create the radioactive isotope phosphorus-30 by bombarding the nucleus of an aluminum atom with alpha particles. In the following year the Joliot-Curies (as they call themselves) will be awarded the Nobel Prize for chemistry.                                              PHYS

1934        Italian physicist Enrico Fermi studies the effects of bombarding uranium atoms with neutrons.                                                                     PHYS

1934        Enrico Fermi develops the concept of the weak interaction, the fundamental event that accounts for the beta decay of particles and atomic nuclei.               PHYS

1934        Soviet physicist Pavel Cherenkov discovers what will become known as Cherenkov radiation, the wake of light produced by particles moving faster than light in a medium other than a vacuum, a phemomenom explained by Soviet physicists Igor Tamm and Ilya Frank. This form of radiation proves useful in calculating the speed of very fast particles. All three Soviets will share the 1958 Nobel Prize for physics.                                                          PHYS

1934        British physicist James Chadwick, who will be awarded next year's Nobel Prize for physics, and Austrian-American physicist Maurice Goldhaber determine the mass of the neutron.                                                                 PHYS

c. 1934–     American psychiatrist Harry Stack Sullivan serves as director of the William
1943        Alanson White Foundation, during which time he contributes his theory of interpersonal relations, that personality development and adjustment are determined by the results of interactions with significant others in a person's life (family, friends, peers, spouse), not solely by biological and sexual factors.
                                                                                      PSYCH

1934        German-born psychologist Erich Fromm immigrates to the United States, where his prolific work will highlight the importance of social influences— especially alienation—on individual personalities. His publications will include *Escape from Freedom* (1941).                                                    PSYCH

1934            American anthropologist Ruth Benedict publishes *Patterns of Culture*, an account of her work with the Zuni and Hopi Native American peoples. This book contributes to the development of the field of cultural psychology.   soc

1935            American astronomer Henry Norris Russell shows that the catastrophic theories of solar system formation advanced by Chamberlin and Moulton in 1905 and Jeans and Jeffreys in 1918 violate the laws of conservation. Future studies by astronomers will further disprove the idea that the solar system could have resulted from a near collision of stars. *See* 1944, ASTRO, for the currently accepted theory of the solar system's origin.   ASTRO

1935            German astronomer Rupert Wildt discovers methane and ammonia on the large planets such as Jupiter and Saturn that will become known as gas giants.   ASTRO

1935            American biochemist Wendall Stanley isolates a virus in crystals, proving it is protein in nature, much the same way John Howard Northrop crystallized pepsin (*see* 1930, BIO). In 1946 Stanley, Northrop, and James Batcheller Sumner (*see* 1926, BIO) will share the Nobel Prize for chemistry.   BIO

1935            German biologist Hans Spemann is awarded the Nobel Prize for physiology or medicine for his discovery of the organizer effect in embryo development.   BIO

1935            German bacteriologist Gerhard Domagk publishes his discovery of the antibacterial effects of Prontosil, a red dye that becomes the first sulfa drug. His resulting 1939 Nobel Prize for physiology or medicine will go unclaimed, as the Nazis will forbid him to accept it. The Nazi refusal stems from anger over the granting of the 1935 Nobel Peace Prize to imprisoned German pacifist Carl von Ossietzky.   BIO

1935            The Richter scale for measuring the severity of earthquakes is developed by American seismologist Charles Richter.   EARTH

1935            Swedish biochemist Ulf von Euler discovers prostaglandins, a group of fatty acids made naturally in the body that act much like hormones. Von Euler finds them in semen, but they will also be discovered in many different bodily tissues and found to vary in their chemical structure. He will share the 1970 Nobel Prize for physiology or medicine with American biochemical pharmacologist Julius Axelrod and German-born British biophysicist Bernard Katz.   MED

1935            Canadian-American physicist Arthur Jeffrey Dempster discovers uranium-235, an isotope of uranium that will be used in producing the first sustained nuclear chain reaction. *See* 1942, PHYS.   PHYS

1935            Japanese physicist Hideki Yukawa, who will win the 1949 Nobel Prize for physics, develops the theory of the strong interaction, the fundamental interaction that binds particles in the atomic nucleus. He proposes the existence of an exchange particle binding the nuclear particles or nucleons. Later called a *mesotron* or *meson*, such a particle is predicted to be intermediate in mass between electrons and protons.                                                               PHYS

1935    American chemist William F. Giauque, winner of the 1949 Nobel Prize for chemistry, cools helium to a new record low temperature of 0.1 K (degrees Kelvin).                                                                            PHYS

1935    In Akron, Ohio, stockbroker Bill W. and surgeon Dr. Bob S. found Alcoholics Anonymous (AA), the first self-help fellowship. By the 1990s, AA will have inspired numerous other programs centered on mutual support to recover from addiction.                                                                    PSYCH

1935    Psychologists Henry A. Murray and Conway Lloyd Morgan introduce a projective test to study personality. Their Thematic Apperception Test consists of 30 pictures and one blank, to and around which the patients assign stories. Information about the patient's personality and motivation is then obtained from the stories and later used for diagnostic purposes.                        PSYCH

1935    Portuguese neurologist Egas Moniz, who will share the 1949 Nobel Prize for physiology or medicine with Swiss physiologist Walter R. Hess, develops the lobotomy, a form of brain surgery, to relieve some forms of mental disturbance. The procedure involves cutting fibers in the brain that connect the frontal lobes with the anterior ones. In time, the lobotomy's benefits will prove inconclusive and its side effects dangerous, so that it will be discontinued after the mid-1950s.                                              PSYCH

1935    American anthropologist Margaret Mead publishes *Sex and Temperament*, which examines gender-based social expectations in three cultures.                    SOC

1935    American pollster George Gallup founds the American Institute of Public Opinion in Princeton, New Jersey. His correct prediction of the outcome of the 1936 presidential election will confirm the validity of sampling as a method for determining public opinion. Scientific polling will become an important tool of social scientists, politicians, and market researchers.                     SOC

1935    Robert H. Goddard becomes the first to fire a liquid-fuel rocket faster than the speed of sound.                                                                     TECH

1935    Scottish physicist Robert Alexander Watson-Watt invents the first practical system of radar (radio detection and ranging). Radar uses microwaves (shortwavelength radio waves) to locate and track objects. Radar systems are developed independently in several countries in the 1930s.                          TECH

1935    The Rural Electrification Administration is established by American president Franklin D. Roosevelt to subsidize, through loans and underwriting, the development of electrical service for rural areas of the country.              TECH

1935    The U.S. Army demonstrates the B-17 bomber, an all-metal, four-engine monoplane that will be used extensively in World War II.                               TECH

1935     Polyethylene, a plastic made of polymerized ethylene that will eventually have many uses, is invented in Britain by Imperial Chemical Industries.     TECH

1935     Kodachrome color film, using a three-color process, is introduced for 16-millimeter movie cameras by Eastman Kodak. In Hollywood, the first three-color Technicolor feature, *Becky Sharp*, is released.     TECH

1936     The quartz clock becomes a standard part of astronomic instrumentation.     ASTRO

1936     André Leallemard, of the Strasbourg and Paris observatories, invents the first electronic telescope accessory—the image-intensifying tube—which becomes important in the study of faint objects.     ASTRO

1936     British geologist Arthur Holmes begins to use the uranium-lead absolute dating method on Precambrian minerals.     EARTH

1936     French-American surgeon Alexis Carrel, winner of the 1912 Nobel Prize for physiology or medicine, collaborates with American navigator Charles Lindbergh to invent the first artificial heart or cardiac pump.     MED

1936     Hungarian-born American physicist Eugene P. Wigner introduces the concept of the nuclear cross section in developing the mathematics of neutron absorption by atomic nuclei. In 1963 he will share the Nobel Prize for physics with German physicist J. Hans D. Jensen and German-American physicist Maria Goeppert Mayer (*see* 1948, PHYS).     PHYS

1936     British economist John Maynard Keynes publishes *The General Theory of Employment, Interest, and Money*, in which he argues for government intervention in the market and deficit spending as a remedy for recession. Keynes becomes one of the chief architects of modern macroeconomic theory.     SOC

1936     The Houdry catalytic cracking process of producing gasoline from oil is employed by Socony-Vacuum and Sun Oil. This process, developed by French-American engineer Eugène Houdry, works at a lower pressure and temperature than previous refining processes.     TECH

1936     Southwestern U.S. states are provided with inexpensive electric power after the completion of the 726-foot-high Boulder Dam, known after 1947 as Hoover Dam.     TECH

1936     Douglas Aircraft debuts the DC-3, an early example of the commercial passenger plane. This two-engine vehicle transports up to 21 passengers.     TECH

1936     An electronic television system is set up by the British Broadcasting Company (BBC).     TECH

1936     The tampon, developed by American physician Earl Haas, is produced commercially for the first time by Tampax, Inc. *See also* 1918, TECH.     TECH

1937        The first radio telescope, with a 9.4-meter (31-foot) dish, is installed by American astronomer Grote Reber in Illinois.                    **ASTRO**

1937        British plant pathologist Frederick Charles Bawden shows that the tobacco mosaic virus is not all protein but also contains small amounts of RNA (ribonucleic acid). It will eventually be discovered that viruses contain either RNA or DNA (deoxyribonucleic acid).                    **BIO**

1937        German-born British biochemist Hans Krebs discovers the citric acid cycle, later called the Krebs cycle. This series of chemical body reactions is the main pathway of terminal oxidation in the process of utilizing carbohydrates, fats, and proteins. He will share the 1953 Nobel Prize for physiology or medicine with German-American biochemist Fritz Lipmann.                    **BIO**

1937        American biochemist Conrad Arnold Elvehjem discovers that nicotinic acid (niacin) and nicotinamide (niacinamide) are vitamins that prevent and cure pellagra.                    **BIO**

1937        Swedish chemist Arne Tiselius introduces electrophoresis, the movement of charged colloidal particles through a medium in which they are dispersed as a result of changes in electrical potential. The process, which will earn him the 1948 Nobel Prize for chemistry, will quickly become important in biochemistry, microbiology, immunology, and chemistry.                    **BIO**

1937        Italian physicists Emilio Segrè and Carlo Perrier discover the element technetium. With no stable isotope, it is the first of many elements to be manufactured rather than discovered in nature.                    **CHEM**

1937        German scientists develop polyurethane.                    **CHEM**

1937        British mathematician Alan Mathison Turing describes a "Turing machine," a hypothetical device that can solve any computable problem. Turing's work will contribute to the development of digital computers in the 1940s.                    **MATH**

1937        American physician D. W. Gordon Murray introduces heparin, a complex organic acid that prevents blood clotting, into general medical practice.                    **MED**

1937        Physicists H. A. Jahn and Hungarian-born American Edward Teller predict what will become known as the Jahn-Teller effect, a distortion of the structure of nonlinear molecules or ions that would be likely to have degenerate orbitals.                    **PHYS**

1937        American physicist Carl D. Anderson, who shared last year's Nobel Prize for physics with Austrian-American physicist Victor F. Hess (see 1912, **ASTRO**) discovers a particle that is at first believed to be a meson (see 1935, **PHYS**) and is thus called a mu-meson. But when it is shown that this particle does not behave like a meson, it is renamed a muon and placed in a class of particles called leptons, which interact by the electromagnetic and weak interactions.                    **PHYS**

1937       Physicist H. A. Kramers develops the concept of charge conjugation, a property that determines the difference between a particle and its antiparticle.    PHYS

1937       Italian physicians Ugo Cerletti and Lucio Bini pioneer electric shock treatment for the symptomatic relief of schizophrenia. Such electroconvulsive treatment (ECT), although controversial, will become standard for forms of depression until the introduction of antipsychotic drugs in the 1950s.    PSYCH

1937       German-American psychiatrist and psychoanalyst Karen Horney publishes *The Neurotic Personality of Our Time*, in which she explores the concept of basic anxiety.    PSYCH

1937       American psychologist Gordon Willard Allport publishes *Personality: A Psychological Interpretation*, with which he begins to make his mark as a specialist in personal dispositions, functional autonomy, and the mature personality. Both this book and his 1961 *Pattern and Growth in Personality* will be professionally well received.    PSYCH

1937       Yale University begins recording its Human Relation Area Files (HRAF), a compilation of ethnographic data for the statistical comparison of cultures.    SOC

1937       American sociologist Talcott Parsons publishes *The Structure of Social Action*, which bridges the gap between American and European schools of sociology. Parsons will become known for his structural-functional theory, a grand unifying theory of nearly every aspect of society.    SOC

1937       Russian-American sociologist Pitirim A. Sorokin publishes *Social and Cultural Dynamics*, in which he expounds influential theories of social process and the typology of cultures.    SOC

1937       Canadian physicist James Hillier invents the first electron microscope, which improves vastly on optical microscopes. His device reaches a magnification of 7,000 times, in contrast to the 2,000-fold magnification of the best optical microscope.    TECH

1937       German-born American physicist Erwin Wilhelm Mueller invents the field-emission microscope, which allows direct observation of atoms.    TECH

1937       While working for Du Pont, American chemist Wallace Hume Carothers patents nylon, the first fully synthetic fiber. Nylon will be used in many products previously made with silk and wool.    TECH

1937       The explosion of the German dirigible *Hindenburg* at Lakehurst, New Jersey, killing 36 people, marks the end of the use of hydrogen-borne dirigibles for air travel.    TECH

1937       Great Britain and other European countries, as well as parts of South America and Asia, adopt the phone number 999 as a universal distress signal for police and firefighters. In 1968 New York will become the first U.S. state to adopt a universal emergency phone number, in this case 911.                    TECH

1937    Using principles of electrostatics and photoconductivity, American student Chester Carlson develops a dry-copy process he calls Xerography, which revolutionizes office technology.                    TECH

1937    American grocery store owner Sylvan Goodman develops the first large-sized grocery shopping cart, from folding chairs and hand-held shopping baskets.                    TECH

1937    On July 4, German pilot Hanna Reitsch is the first to fly a helicopter successfully. Her FW-61 helicopter was built by Heinrich Focke.                    TECH

1938    German physicists Hans A. Bethe, who will win the 1967 Nobel Prize for physics, and Carl F. von Weizsäcker independently develop the theory that stars are powered by thermonuclear fusion.                    ASTRO

1938    British physicist P. A. M. Dirac links the Hubble constant, which concerns the rate at which galaxies recede from each other, to constants describing subatomic particles.                    ASTRO

1938    American physicist J. Robert Oppenheimer and George Volkoff predict the existence of neutron stars rotating at a rapid rate. *See* 1967, ASTRO.                    ASTRO

1938    A coelacanth, a species of fish believed extinct for 70 million years, is discovered alive in the waters off South Africa.                    BIO

## Nylon Day

Although it was 1937 when American chemist Wallace Hume Carothers patented the strong polymeric fiber built of diamines and dicarboxylic acids that would rival silk, the substance now known as nylon did not reach the market until 1938, in the form of toothbrush bristles and later, more dramatically, women's hosiery.

The first toothbrush with nylon bristles, the Dr. West's Miracle Tuft Toothbrush, was first sold in 1938, but because of the toughness of the nylon it met with only limited success. To introduce its nylon hosiery the manufacturer and patent holder Du Pont orchestrated a more wide-ranging plan: a nationwide launch called Nylon Day. On May 15, 1940, when the first nylon hosiery was sold, women lined the streets in front of stores in anticipation, and the product took off.

From 1941 until the end of World War II, in 1945, nylon would be used primarily for parachutes and other military purposes. But starting on Nylon Day, nylon came to dominate people's lives more intimately. Despite wartime shortages, the synthetic fiber virtually replaced silk as the standard for sheer in stockings, even providing the basis for their new name, nylons.

1938    In South Africa, Scottish–South African paleontologist Robert Broom discovers hominid fossils he classifies as a new genus and species, *Paranthropus robustus*. The species is of a heavier, more robust build than *Australopithecus africanus* (*see* 1924, PALEO). Some later paleontologists will classify Broom's find as *Australopithecus robustus*. The remains date from 2 million to 1.5 million years ago.

PALEO

1938    Austrian-born American physicist I. I. Rabi, who will win the 1944 Nobel Prize for physics, develops the technique of magnetic resonance, which allows measurement of the energies absorbed and emitted by the particles of a molecular beam.

PHYS

1938    The law of baryon conservation is proposed. Baryons include protons and neutrons.

PHYS

1938    Bulgarian physicist Georgi Nadjakov formulates the photoelectric state of matter.

PHYS

1938    On December 18, German physical chemist Otto Hahn achieves nuclear fission, the splitting of an atomic nucleus into two parts, when he bombards uranium-235 with neutrons. The results will be announced in the following month (*see* 1939, PHYS).

PHYS

1938    American behaviorist B. F. Skinner publishes the results of his first experiments with the Skinner box, a simple piece of laboratory equipment that makes possible a series of systematic experiments in operant conditioning on rats and pigeons. His work will be considered different from, but equal in value to, Ivan Pavlov's earlier conditioned response experiments with dogs.

PSYCH

1938    Dutch physicist Frits Zernike, who will be awarded the 1953 Nobel Prize for physics, invents the phase-contrast microscope, which uses light diffraction to improve resolution.

TECH

1938    The rear-engined Volkswagen ("people's car"), also known as the Beetle, is first produced in Germany, designed by Austrian engineer Ferdinand Porsche.

TECH

1938    Fiberglass is developed by Owens-Illinois and Corning. Able to be woven or spun, this material will prove useful in many applications.

TECH

1938    American chemist Roy Plunkett invents the material trade-named Teflon (polytetrafluoroethylene) while working for Du Pont. It is originally sold in Britain, under the trade name Fluon.

TECH

1938    Hungarian brothers Ladislao and George Biro patent the first ballpoint pen.

TECH

1939    American physicist J. Robert Oppenheimer theorizes that a star greater than 3.2 solar masses will collapse from its own weight into a single point, an object that will become known as a black hole.

ASTRO

1939          Indian astronomer Subrahmanyan Chandrasekhar, who will share the 1983
              Nobel Prize for physics with American physicist William A. Fowler, determines
              what will become known as the Chandrasekhar limit, the maximum possible
              mass for a star prevented from collapsing by degeneracy pressure. For a white
              dwarf star the limit is 1.4 times the mass of the sun; heavier stars may become
              neutron stars. *See also* 1934, ASTRO.                                        ASTRO

---

# Racism and Blood Transfusion

All human blood looks the same, but medical attempts at blood transfusion before the 20th century showed, sometimes fatally, that there were differences. Some patients were helped by receiving blood from outside donors, but others died more quickly as a result. Not until 1901 did Austrian pathologist Karl Landsteiner show that human blood occurs in classes (originally named A, B, and C; C was later renamed O, and AB was discovered in the following year) and that a simple set of rules governed which class was compatible with which (O could be given to any receiver, AB only to AB receivers, A to A and AB, and B to B and AB). Mixing incompatible blood types could result in clumping of red blood cells, blocked vessels, and death.

The difficulty was in finding donor blood of the type needed when and where it was needed, often under the most pressing emergency conditions. Conscious of this problem, African-American physician Charles Drew, a medical professor at Howard University, became interested in the idea of storing blood in "blood banks" for use in transfusions. However, blood storage had many problems, because refrigeration extended blood's therapeutic benefits for only a few hours and freezing destroyed red blood cells. But the imminence of war in Europe at the end of the 1930s made it important to find a solution. While working with blood chemist John Scudder at Columbia University in 1938–1940, Drew discovered that blood plasma (the yellowish fluid part of blood in which cells are suspended) could be stored for long periods and was effective in treating blood loss and burn victims.

Drew therefore proposed that plasma banks be set up for massive wartime programs. He directed the Plasma for France and Plasma for Britain projects and, in 1941, was named medical director of the American Red Cross's National Blood Bank program. He directed the preparation of liquid plasma and researched ways to prepare frozen and dried plasma.

Then Drew found out that the U.S. military would accept blood only from Caucasians. If African-American blood was accidentally accepted, it had to be isolated and transfused only to African-Americans. Unlike the segregation of blood types that had made transfusions possible, this segregation was a result purely of racial prejudice and had no medical basis. Drew protested but was told that the whites-only policy was required to ensure the general population's cooperation with blood drives.

Drew resigned and returned to Howard University to continue teaching. In 1949 he took a position as surgical consultant to the U.S. Armed Forces, at a point when the military no longer distinguished between blood donated by whites and blood donated by African-Americans.

1939    Russian-born British biochemist David Keilin further demonstrates the existence and importance of essential minerals (essential trace elements) with his discovery that zinc is necessary to life.                                    BIO

1939    DDT, a hydrocarbon pesticide, is developed by Swiss chemist Paul Müller for the Geigy Co. It is first used in Switzerland, then, due to its efficacy and low cost, becomes widely popular. Because of its effectiveness in controlling typhus fever and malaria during World War II, DDT will be deemed so successful that Müller will be awarded the 1948 Nobel Prize for physiology or medicine.                                    BIO

1939    French physicist Marguerite Perey discovers the element francium.                                    CHEM

1939    The French survey of mathematics called *Éléments de mathématique* begins publication under the name of Nicolas Bourbaki, a pseudonym for a group of mathematicians. The work emphasizes logical structure and an axiomatic approach.                                    MATH

1939    French-born American microbiologist René Jules Dubos isolates the antibiotic substance tyrothricin. Though not very effective, it is for a time one of the few tools physicians have against infection.                                    MED

1939    British pathologist Howard W. Florey and British biochemist Ernst B. Chain produce pure penicillin, the first powerful antibiotic. They wil share the 1945 Nobel Prize for physiology or medicine with British bacteriologist Alexander Fleming (*see* 1928, MED).                                    MED

1939    World War II begins on September 1, when German dictator Adolf Hitler invades Poland, prompting Great Britain and France to declare war on September 3. The war will spur the development of numerous technologies, including the atomic bomb.                                    MISC

1939    Hungarian-born American physicist Leo Szilard hears of Otto Hahn's discovery of nuclear fission (*see* 1938, PHYS) and observes that it can be used to produce a nuclear chain reaction (*see* 1932, PHYS), which can be put to destructive use in a bomb.                                    PHYS

1939    Swiss-born American physicist Felix Bloch calculates the magnetic moment (a measure of magnetic strength) of the neutron. This discovery indicates that the electrically neutral particle is made up of smaller charged particles. Independently, American physicist Edward M. Purcell makes the same discovery.                                    PHYS

1939    Under pressure from Hungarian-born American physicists Leo Szilard, Edward Teller, and Eugene P. Wigner, German-born physicist Albert Einstein, who will become an American citizen in the following year, sends a letter to President Franklin D. Roosevelt urging him to develop an atomic bomb before the Germans do.                                    PHYS

| | |
|---|---|
| 1939 | W. C. Herring discovers a way of calculating the properties of substances from quantum principles, a technique he applies in explaining the properties of beryllium. **PHYS** |
| 1939 | German physical chemist Otto Hahn's achievement of nuclear fission (*see* 1938, PHYS) is announced in a paper dated January 26 and published by his colleague Austrian physicist Lise Meitner and her nephew Otto Robert Frisch (also a physicist). For his discovery of nuclear fission, Hahn will be awarded the 1944 Nobel Prize for chemistry. Interned in England at the time, he will not be able to accept the prize until 1946, when he will have been allowed to return to a defeated Germany. **PHYS** |
| 1939 | American psychologist Myrtle B. McGraw demonstrates the swimming reflex in infants. She will later pioneer in studying twins. **PSYCH** |
| 1939 | American psychologist David Wechsler introduces the Wechsler-Bellevue Adult Intelligence Scale, which measures verbal, numerical, social, and perceptuo-motor abilities. **PSYCH** |
| 1939 | Fluorescent lighting is developed by General Electric. **TECH** |
| 1939 | Pan American Airways introduces commercial passenger airline service with a four-engine Boeing craft traveling from Port Washington, New York, to Marseilles, France. **TECH** |
| 1939 | FM radios are sold commercially for the first time. **TECH** |
| 1939 | The Germans fly the first turbojet, a Heinkel He-178 plane powered by a Heinkel S3B turbojet engine. **TECH** |
| 1940s | Ecological studies show that unproductive land can often be reclaimed and made productive again by correcting its nutritional deficiencies and not over-fertilizing it. **BIO** |
| 1940s | American geneticists George W. Beadle and Edward L. Tatum provide one of the first important clues as to how chromosomes and their genes copy exactly from cell to cell when they find that genes direct enzyme formation through the units called polypeptides that make them up. **BIO** |
| 1940s | Norwegian-American meteorologist Jacob Bjerknes identifies the jet stream, a narrow, eastward wind current above the lower troposphere. **EARTH** |
| 1940s | I. Efremov and other Soviet paleontologists in Mongolia discover fossils of *Syrmosaurus*, an armored dinosaur that links the plated dinosaurs of the Jurassic period with the armored dinosaurs of the Cretaceous. **PALEO** |
| 1940s | Psychologists at the University of Minnesota develop the Minnesota Multiphasic Personality Inventory (MMPI) test to measure more than one personality dimension at a time. It will be used more than any other personality |

test and be considered reliable in indicating psychological pathology. *See also* 1989, PSYCH.

PSYCH

1940s    Austrian-born Canadian endocrinologist Hans Selye, studying the results of injecting rats with hormones, discovers the general adaptation syndrome, in which intense physiological changes in body organs occur in response to stress. The body changes consist of the alarm reaction, resistance, and exhaustion. Selye will continue to study the effects of stress, publishing major works in 1950 and 1976.

PSYCH

1940s    American child psychologist and pediatrician Arnold Gesell researches childhood stages of mental and emotional growth. The Gesell developmental scale comes into wide use during this decade. He is the first psychologist to observe patients/clients through one-way mirrors.

PSYCH

1940s    At McGill University, neurosurgeon William Penfield and his colleague Herbert Jasper give one of the first demonstrations in neuroscience of the human brain's information storage and retrieval capabilities. With electrodes, they stimulate exposed temporal brain lobes on neurological patients under a local anesthetic, a process that evokes vivid memories of isolated and insignificant past events.

PSYCH

1940    Italian physicist Emilio Segrè, Dale Corson, and K. R. Mackenzie discover the element astatine.

CHEM

1940    American physicists Edwin M. McMillan and Philip Hauge Abelson discover neptunium, the first known transuranium element, or element with an atomic number greater than 92, the number of uranium.

CHEM

1940    American physicist Glenn T. Seaborg and his colleagues discover the element plutonium.

CHEM

1940    Canadian-American biochemist Martin David Kamen discovers carbon-14, an isotope of carbon with a half-life of about 5,700 years. *See* 1947, ARCH.

CHEM

1940    The British Association Seismological Committee begins to publish the seismic wave travel timetables known as the Jeffreys and Bullen tables. They will be superseded in 1991 by timetables based on tomography (CT scanning).

EARTH

1940    Nine-year-old Milton Sirotta is the first to use the googol, the number 10 raised to the 100th power, or 1 followed by 100 zeroes.

MATH

1940    Rockefeller Institute scientists, including Austrian-American pathologist Karl Landsteiner, discover the Rhesus factor in blood. This subgrouping of blood types, which determines if a person is Rh positive or Rh negative, will make blood transfusions safer.

MED

1940    Paintings dating from the Cro-Magnon era some 17,000 years ago are discovered in September on the walls of a cave in Lascaux, France.

PALEO

1940        American physicist Philip Hauge Abelson proposes a process for enriching ura-
            nium, by accumulating the rare isotope uranium-235 in quantities sufficient
            for use in an atomic bomb. The process involves the evaporation and gaseous
            diffusion of the liquid uranium hexafluoride.                            PHYS

1940        American physicist Donald William Kerst invents the betatron, an accelerator
            that pushes electrons to speeds close to that of light, now making electron
            bombardment practical.                                                  PHYS

1940        British anthropologist Edward Evan Evans-Pritchard publishes *The Nuer*, the
            first work in an influential trilogy that will include *Kinship and Marriage Among
            the Nuer* (1951) and *Nuer Religion* (1956).                             SOC

1940        American coal executive Carson Smith and engineer Harold Silver invent a
            deep-cutting, continuous-digging machine able to carve a nearly 20-foot-wide
            tunnel. Upon the rights to the machine being purchased by manufacturer
            Joseph Joy, the implement becomes known as the Joy machine.              TECH

1940        The Soviet MiG-1 fighter plane is introduced.                           TECH

1940        The four-cylinder general purpose field vehicle called the jeep is developed by
            American engineer Karl Pabst for the Bantam Car Co. More than 600,000 jeeps
            will be produced for use in World War II.                                TECH

1940        The first American superhighway with tunnels, the 160-mile Pennsylvania
            Turnpike, opens.                                                        TECH

1940        Nylon stockings are sold for the first time in the United States.        TECH

1941        Previously unexplained lines in the spectrum of the solar corona, known as
            coronium lines, are found to be produced by iron, calcium, and nickel ion-
            ized by the corona's intense heat of about 1 million° C.                ASTRO

1941        Arnold O. Beckman invents the spectrophotometer, a device for measuring a
            material's chemical composition based on reflected wavelengths of light. CHEM

1941        American physician Dickinson W. Richards, German physician Werner
            Forssmann, and French-American physician André Cournand develop a pro-
            cedure in which a tiny plastic tube or catheter is passed into the heart through
            a blood vessel to withdraw blood samples, to test cardiac output and blood
            pressure. This technique of cardiac catheterization will advance the diagnosis
            of heart disease and heart defects. The three scientists will share the 1956
            Nobel Prize for physiology or medicine.                                 MED

1941        On December 6, one day before the Japanese attack on Pearl Harbor will
            bring the United States into World War II, President Franklin D. Roosevelt
            signs a secret directive ordering the development of a nuclear fission bomb
            in an operation known as the Manhattan Project.                         PHYS

| | |
|---|---|
| 1941 | Terylene, a polyester fiber composed of terephthalic acid and ethylene glycol, is developed by British chemist John Rey Whinfield. It will become known in the United States as Dacron and be sold by Du Pont.            TECH |
| 1942 | Radio waves from the sun are detected in England.            ASTRO |
| 1942 | American astronomer Grote Reber makes the first radio maps of the universe. Among his discoveries is the first known radio galaxy, Cygnus A, some 700 million light-years away.            ASTRO |
| 1942 | American biochemist Vincent du Vigneaud isolates vitamin H (biotin). BIO |
| 1942 | Italian-born American microbiologist Salvador Edward Luria is able to photograph bacteriophages with the magnifying aid of the electron microscope, the first time a virus has been recorded as something greater than a speck.            BIO |
| 1942 | Curare, a substance used for centuries by South American Indians as a poison, is introduced as a muscle relaxant for patients in surgery.            MED |
| 1942 | On December 2, on a converted squash court at the University of Chicago, Italian-born American physicist Enrico Fermi, winner of the 1938 Nobel Prize for physics, achieves the first sustained nuclear chain reaction. This uranium-235–based fission reaction, produced in a structure called an atomic pile, will lead to the development of nuclear weapons and nuclear power.            PHYS |
| 1942 | American psychologist William Herbert Sheldon (1899–1977) publishes his constitutional theory of personality, claiming that body structure alone determines personality. In his system, body type is classified in terms of three components: ectomorphy, endomorphy, and mesomorphy.            PSYCH |
| 1942 | American theoretical physicist John V. Atanasoff and his assistant Clifford Berry build the first computer that successfully uses vacuum tubes to perform calculations. The machine is called the Atanasoff Berry Computer, or ABC.            TECH |
| 1942 | In the first U.S. jet plane flight, Robert Stanley flies the Bell XP-59 *Airacomet* on October 1 at Muroc Army Base, California.            TECH |
| 1942 | On October 3, Wernher von Braun and other scientists in Peenemünde, Germany, successfully launch the world's first ballistic missile, the 12-ton AS-4 rocket that is the predecessor of the V-2 rockets that will wreak havoc on London in 1944 and 1945.            TECH |
| 1943 | Chinese-born American biochemist Choh Hao Li isolates a hormone that stimulates the adrenal cortex to produce and release cortical hormones. It is called adrenocorticotropic hormone (ACTH).            BIO |
| 1943 | Swiss chemist Albert Hoffman synthesizes lysergic acid diethylamide, or LSD, which will become widely used as a hallucinogen in the 1960s.            CHEM |

1943      Swiss chemists produce xylocaine (lidocaine) for use as a local anesthetic. It is faster acting and longer lasting than procaine, making it a popular choice of physicians.       **MED**

1943      Austrian-born psychiatrist Leo Kanner is the first to describe infantile autism, a brain disease of childhood characterized by withdrawal, language disturbance, mutism, fear of change, emotional detachment, and repetitive rhythmic movements.       **PSYCH**

1943      Neurophysiologist Warren McCulloch and mathematician Walter Pitts show that the human brain's fundamental mechanisms can be described in terms of symbolic (Boolean) logic. They find that electrical impulses pass along the axon and trigger chemical processes that cause adjoining neurons either to fire or not fire. The discovery suggests that human thought mechanisms may be reproducible on complex computer systems.       **PSYCH**

1943      Construction on the Pentagon, the largest office building in the world at 6.5 million square feet, is completed at a cost of $83 million. The five-sided building will house the Department of Defense, which will be created in 1949, on 34 acres in Arlington, Virginia, across the Potomac River from Washington, D.C.       **TECH**

1944      German astronomer Carl F. von Weizsäcker reexamines Pierre Simon de Laplace's nebula hypothesis of the origin of the solar system (*see* 1796, ASTRO). His elaborations on it, as well as later revisions by Swedish astrophysicist Hannes Alfvén and British astronomer Fred Hoyle, establish the theory that the planets formed from the coalescing of smaller particles called planetesimals, which in turn arose from eddies in an original planetary nebula. *See also* 1905, 1918, and 1935, ASTRO. Alfvén will share the 1970 Nobel Prize for physics with French physicist Louis Néel.       **ASTRO**

1944      German-American astronomer Walter Baade distinguishes two populations of stars: population I (younger stars found in the spiral arms of galaxies) and population II (older stars found in galactic cores).       **ASTRO**

1944      Dutch astronomer Hendrik van de Hulst predicts that interstellar hydrogen emits radiation with a 21-centimeter wavelength, a prediction later verified by Ewen and Purcell. *See* 1951, ASTRO.       **ASTRO**

## Mummification

In 1944 the oldest complete mummy was found in Saqqâra, Egypt. The preserved body was that of a court musician named Wati from about 2400 B.C.

Egyptian mummification is believed to have begun about 2600 B.C., during the fourth dynasty. The oldest known mummy fragment is the skull of a woman from about that time, found near the Great Pyramid of Cheops, or Khufu, at Giza in 1989.

1944        American astronomer Gerard P. Kuiper discovers that there is an atmosphere
            on the Saturnian moon Titan.                                          ASTRO

1944        Astronomer Carl Seyfert discovers several spiral galaxies with compact nuclei
            radiating enormous quantities of energy at all wavelengths, which will become
            known as Seyfert galaxies. *See* 1957, ASTRO.                          ASTRO

1944        American geneticist T. M. Sonneborn explains that genes cannot operate
            except in the presence of other substances he calls primers, which as yet
            remain unidentified.                                                   BIO

1944        Canadian bacteriologist Oswald Theodore Avery proves that deoxyribonucleic
            acid (DNA) is the fundamental substance that determines heredity.      BIO

1944        American paleontologist George Gaylord Simpson becomes a leading figure in
            evolutionary thought with such as works as his *Tempo and Mode in Evolution*.   BIO

1944        American physicist Glenn T. Seaborg and his colleagues discover the elements
            americium and curium.                                                 CHEM

1944        British biochemists Archer Martin and Richard Synge invent paper chro-
            matography, a technique for analyzing mixtures using absorbent paper. They
            will share the 1952 Nobel Prize for chemistry.                        CHEM

1944        American chemists R. B. Woodward, who will win the 1965 Nobel Prize for
            chemistry, and William E. Doering synthesize quinine, an antimalarial drug.
            *See also* 1819, MED.                                                 CHEM

1944        American mathematicians John von Neumann and Oskar Morgenstern pub-
            lish *The Theory of Games and Economic Behavior*, a major work in the develop-
            ment of game theory.                                                  MATH

1944        Dutch physician Willem Kolff produces the first kidney machine, to cleanse
            the blood of people whose own kidneys have failed.                    MED

1944        Austrian-born psychoanalyst Helene Deutsch publishes *The Psychology of Women*,
            which corroborates many Freudian ideas. She was the first female psychoanalyst
            to be analyzed by Freud.                                              PSYCH

1944        German-American psychologist Kurt Lewin becomes the director of the
            Research Center for Group Dynamics at the Massachusetts Institute of
            Technology (MIT). His work will focus on motivation problems in groups and
            individuals, child development, and personality characteristics. He will estab-
            lish what will become known as field theory, a method for analyzing causal
            relations and building scientific constructs.                         PSYCH

1944        The first nuclear reactors, built to convert uranium into plutonium for atomic
            bombs, begin operation in Washington State.                           TECH

1944      At the California Institute of Technology (Caltech), research begins on U.S. high-altitude rockets.                                                    TECH

1944      At Harvard University, the Harvard-IBM Automatic Sequence Controlled Calculator is developed under the direction of Howard Hathaway Aiken. It contains more than 750,000 parts and takes a few seconds to complete simple arithmetic calculations.                                         TECH

1944      The U.S. National System of Interstate Highways is set up by a congressional federal highways act designating the construction of 40,000 miles of highway across the country.                                               TECH

1944      Kodacolor negative film is developed for color snapshots by Eastman Kodak.   TECH

1944      In June, the German Messerschmitt Me-163B Komet becomes the first rocket-engined fighter plane to go into production.                          TECH

1944      The first V-2 rockets, developed by German rocket engineer Wernher von Braun, are fired on September 7 by the Germans at London.               TECH

1945      British science-fiction writer Arthur C. Clarke arrives at the concept of communication satellites in geosynchronous orbits (a stationary orbit above a particular longitude on Earth) to provide worldwide communication. *See* 1962, TECH.                                                              ASTRO

1945      American microbiologists Salvador Luria and Alfred Hershey show that bacteriophages mutate, which will explain why flu and the common cold are difficult to develop an immunity against.                              BIO

---

# Alamogordo, July 16, 1945

At 5:30 A.M. on July 16, 1945, the first atomic bomb was exploded in a test site at Alamogordo Air Base, New Mexico. The U.S. War Department issued a release about the test, saying in part:

> Mounted on a steel tower, a revolutionary weapon destined to change war as we know it, or which may even be the instrumentality to end all wars, was set off with an impact which signaled man's entrance into a new physical world. . . . At the appointed time there was a blinding flash lighting up the whole area brighter than the brightest daylight. A mountain range three miles from the observation point stood out in bold relief. Then came a tremendous sustained roar and a heavy pressure wave which knocked out two men outside the control center. Immediately thereafter, a huge multicolored surging cloud boiled to an altitude of over 40,000 feet. Clouds in its path disappeared. Soon the shifting substratosphere winds dispersed the now gray-mass. . . . The test was over, the project a success.

1945        Cambridge University geneticists J. F. Danielli and D. G. Catchside report
           being the first to have witnessed the process whereby genes influence cellular
           activity.                                                                    BIO

1945        The fluoridation of water supplies to retard tooth decay begins, in Grand
           Rapids, Michigan.                                                            MED

1945        Indian physicist Homi J. Bhabha founds and becomes first director of the Tata
           Institute of Fundamental Research, a multidisciplinary scientific institute in
           Bombay.                                                                      MISC

1945        American physicist Edwin M. McMillan and, independently, Soviet physicist
           Vladimir I. Veksler invent the synchrocyclotron, an accelerator that produces
           particle energies in excess of 20 million electron volts.                    PHYS

1945        American anthropologist Ralph Linton publishes *Cultural Background of
           Personality*, which develops an interdisciplinary approach to the study of cul-
           ture and personality.                                                        SOC

c. 1945     Inspired by the Chicago school (*see* 1930, soc), behaviorism becomes domi-
           nant in political science from the late 1940s to the 1960s. Behaviorists attempt
           to explain and predict political behavior across cultures and throughout his-
           torical periods, using empirical methodologies previously employed by other
           social sciences. *See* 1960s, soc.                                           SOC

1945        A pilot is killed in Germany in the first attempt at manned rocket flight.  TECH

1945        At the end of World War II, more than 120 German scientists, including
           rocket designers Wernher von Braun and Walter Domberger, surrender to
           the United States and begin working for their former enemies. The Soviets
           recruit their own German scientists, and both the Soviets and Americans
           capture rocket equipment.                                                    TECH

1945        The White Sands Proving Ground in New Mexico is established for rocket
           research, launching its first captured V-2 rocket in 1946.                   TECH

1945        Frozen orange juice, a concentrate of fresh juice, is developed in the United
           States.                                                                      TECH

1945        Five thousand American homes now have television sets—three years from
           now the number will be 1 million. By 1968, Americans will own 78 million
           television sets.                                                             TECH

1945        The first atomic bomb is detonated just before dawn on July 16 in a secret test
           at Alamogordo, New Mexico. Its force is equivalent to about 20,000 tons of
           conventional high explosives.                                                TECH

1945          On August 6, Hiroshima, Japan, is devastated by an American atomic bomb based on uranium-235 in the first public display and wartime use of nuclear weapons.                                                                                      TECH

1945          On August 9, Nagasaki, Japan, is destroyed by a plutonium-based atomic bomb. The surrender of Japan to the Allies will follow five days later (August 14), ending World War II.                                                                   TECH

1946          The radio source Cygnus A is discovered. In 1951, astronomers at Mount Palomar will identify it with a distant cluster of galaxies in the constellation Cygnus, making Cygnus A the first radio galaxy.                                           ASTRO

1946          German-born American microbiologist Max Delbrück and American microbiologist Alfred Hershey show that the genetic material of different virus strains can be combined to form a new strain. In 1969 they will share the Nobel Prize for physiology or medicine with Italian-born American microbiologist Salvador Luria (*see* 1942 and 1945, BIO).                                                 BIO

1946          American chemist Vincent Schaefer creates the first artificially induced precipitation when he seeds clouds with dry ice, resulting in a snowstorm. Later, seeding clouds with other chemicals, he will succeed in producing rain.   EARTH

1946          In *Foundations of Algebraic Geometry*, French mathematician André Weil develops a theory of polynomial equations in any number of indeterminates and with coefficients in an arbitrary field. Weil makes a number of conjectures concerning algebraic topology that are eventually proven true. *See* 1974, MATH. MATH

1946          The Atomic Energy Commission releases radioisotopes (radionuclides) for medical use. Nuclear medicine will explore their diagnostic, therapeutic, and investigative uses.                                                                        MED

1946          Swiss-born American physicist Felix Bloch and American physicist Edward M. Purcell, who will share the 1952 Nobel Prize for physics, independently develop the technique of nuclear magnetic resonance (NMR), for determining nuclear moments and measuring magnetic fields. This process will be the basis for the medical technique of magnetic resonance imaging (MRI), a noninvasive way of producing images of the body's interior.                                    PHYS

1946          American physicist Willis E. Lamb Jr. discovers what will become known as the Lamb shift, a small energy difference between two levels in the hydrogen spectrum. This discovery will contribute to the development of quantum electrodynamics (QED) (*see* 1948, PHYS). Lamb will share the 1955 Nobel Prize for physics with American physicist Polykarp Kusch.                               PHYS

1946          Abraham Pais and C. Moller coin the term *lepton* to describe particles such as electrons and muons that are not affected by the strong force.   PHYS

c. 1946       The field of artificial intelligence (AI) begins as the first computers are developed. AI—often defined as a multidisciplinary field encompassing computer science, neuroscience, philosophy, psychology, robotics, and linguistics—will attempt to reproduce with machines the methods and results of human reasoning and brain activity. PSYCH

1946    American pediatrician Benjamin Spock writes *The Common Sense Book of Baby and Child Care*, encouraging parents to show more affection toward their infants and be less structured in their feeding habits. Later retitled *Baby and Child Care*, Spock's book will become an all-time bestseller on the subject of child rearing. PSYCH

1946    In psychoanalysis the idea of brief therapy begins to evolve. Any form of this therapy is goal specific and has relatively limited and delineated objectives. Brief therapy may be completed in a limited number of sessions, whereas traditional psychoanalysis can go on indefinitely. Further, brief therapy strives to focus on the present problem and work on modifying current variables. PSYCH

1946    American anthropologist Ruth Benedict publishes *The Chrysanthemum and the Sword*, a classic study of Japanese culture and society. SOC

1946    American postwar testing of nuclear weapons begins in the Pacific Ocean. TECH

1946    At the California Institute of Technology (Caltech), it is found that a liquid polysulfide polymer is an effective propellant for space vehicles. TECH

1946    The first automatic electronic digital computer, the ENIAC, is constructed at Harvard University by electrical engineers John Presper Eckert and John William Manchly in consultation with John Atanasoff. This electronic numerical integrator and computer contains radio tubes and runs by electrical power to perform hundreds of computations per second. TECH

1946    The word *automation* is used for the first time, by Ford Motor Co. engineer Delmar Harder to describe the 14-minute process by which Ford engines are produced. TECH

1947    American chemist Willard F. Libby, who will win the 1960 Nobel Prize for chemistry, invents the technique of carbon dating, in which the radioactive isotope carbon-14 discovered by Martin Kamen in 1940 is used to determine the age of archaeological objects dating as far back as 45,000 years. ARCH

1947    The Dead Sea Scrolls are discovered in earthenware jars in a cave near Khirbet Qumran on the northwestern shore of the Dead Sea. These scrolls contain religious texts offering insights into ancient Judaism and early Christianity. More than 10 more caves with other such scrolls will be discovered in the 1950s and 1960s. ARCH

1947    Prague-born biochemists Gerty Theresa Radnitz Cori and Carl Ferdinand Cori share the Nobel Prize for physiology or medicine for discovering the series of steps by which the human body converts glycogen into glucose and back again to glycogen, the process that will become known as the Cori cycle. BIO

1947        From a group of molds, American microbiologists isolate chloramphenicol, the first broad-spectrum antibiotic.                                            BIO

1947        American chemists J. A. Marinsky, L. E. Glendenin, and C. D. Coryell discover promethium, the last element from the periodic table to be identified.   CHEM

## Bizarre Treatments for Mental Illness

Along with shock therapy and straitjackets, here are some of the more arcane, barbaric, and bizarre treatments for mental illness offered throughout history:

Apples.  Those suffering from madness were once allowed to eat nothing but apples for 30 days.

Bad news.  In this 16th-century therapeutic technique, manic patients were given frequent unpleasant and depressing news.

Bleeding.  Blood was let out of the patient's body, based on the belief that too much hot blood caused insanity.

Branking.  Starting in 16th-century Scotland, the insane were put in iron or leather headpieces (branks), complete with a mouth gag.

Carbon dioxide therapy.  As recently as the 1940s, neurotics were prescribed carbon dioxide inhalation to the point of coma.

Diamonds.  As early as 1582, diamonds were worn to cure depression and prevent nightmares.

Fish.  Citerrochen fish, which are naturally charged with electricity, were placed on patients' foreheads as a shocking apparatus.

Human skin belts.  Skin belts made from human corpses were worn by those suffering from hysteria.

Malaria.  In early 20th-century America, intentional infection with malaria was used as a form of therapy for the general paralysis of the insane.

Peas.  Head wounds were inflicted on patients, then stuffed with dried peas. The peas were supposed to produce a counterirritation to combat the mental irritation in the brain.

Withholding afternoon tea.  In early 19th-century England, hot drinks were believed to cause suicide.

This list by no means exhausts the questionable practices undertaken in the name of restoring mental health. Since antiquity water has been a favorite therapy, usually cold and dropped from heights onto patients' heads, or used to immerse them to the point of drowning. Whipping was once a form of shock therapy, and sneezing powder was popular. Perhaps the most radical treatment was one recommended by an American neurologist in 1877: rest, seclusion, good food, and a massage.

c. 1947        Researchers abandon the idea that the oceans' floors are flat and produce arguments in favor of the continental drift hypothesis.                                    EARTH

1947           Nikolai Vasilevich Belov develops his theory concerning maximum ionic density.                                    EARTH

1947           American geobiologist Maurice Ewing conducts systematic studies of the North Atlantic, using depth probes, and determines the existence there of a huge abyssal plain.                                    EARTH

1947           British physicist Cecil F. Powell discovers the subatomic particle called a pi-meson or pion, the first true meson to be discovered (*see* 1935 and 1937, PHYS). Three years later he will win the Nobel Prize for physics.                                    PHYS

1947           Hungarian-British physicist Dennis Gabor, who will win the 1971 Nobel Prize for physics, develops the theory of holography, though full implementation of it will await the invention of the laser in 1960.                                    PHYS

1947           Austrian-born psychoanalyst Anna Freud, daughter of Sigmund Freud, founds the London Hampstead Child-Therapy Clinic. In 1936 she had introduced the theory of ego defense mechanisms such as repression. She becomes well known for her work in the psychoanalysis of children.                                    PSYCH

1947           The Institute of Sex Research is founded as an affiliate of Indiana University by Alfred C. Kinsey. In the next five years, Kinsey and three colleagues will publish the results of research on male and female sexuality known as the *Kinsey Reports*. These findings will be used to correct prevalent misconceptions about female sexual arousal, childhood sexuality, and homosexuality.                                    PSYCH

## First Microwave Oven

In 1947, when the American manufacturer Raytheon introduced the first microwave oven, the company predicted it would revolutionize cooking. They had reason to be confident. After all, it had been five years since Raytheon scientist Percy Spencer had discovered that the magnetron, or electronic tube, he was testing would manipulate food molecules into a heated, cooked state, and the company had used the war years to develop a working prototype.

But the public test of the Radarange was a failure. Since the oven lacked modern-day browning devices, the tested foods, unlike those in the publicity photos, were pale and rubbery. Worse, at the size of a standard oven and a cost of $3,000, the first microwave was hardly an affordable convenience. Not until Amana Refrigeration introduced a $495 table-top microwave oven in 1967, built with a smaller electron tube developed by Japanese engineer Keishi Ogura, did the product sell. Until then the microwave oven was as popular as the 1957 giant meant to revolutionize its own industry, the Ford Edsel.

1947        The first self-sealing tubeless automobile tires are sold, by B. F. Goodrich. TECH

1947        The eight-engine *Spruce Goose*, then the largest aircraft in the world, is intro-
            duced by its designer Howard Hughes in a one-mile flight in Long Beach
            Harbor, California.                                                        TECH

1947        The first Levittown suburban housing development is erected, on Long Island,
            by American construction designer Abraham Levitt and his sons. Over the next
            decade, thousands of these moderately priced homes will be mass-produced
            on Long Island, in Pennsylvania, and in New Jersey.                        TECH

1947        The Radarange, the first microwave oven for commercial use, is built and
            demonstrated by the Raytheon Co., but it is not an immediate success.      TECH

c. 1947     Diesel-electric trains are replacing steam locomotives on U.S. railroad
            lines.                                                                     TECH

1947        On October 14, in the first piloted supersonic airplane flight, Capt. Charles E.
            Yeager flies the Bell X-1 rocket-powered plane *Glamorous Glennis* (named for
            his wife) faster than the speed of sound at Muroc Air Force Base in California.
                                                                                       TECH

# The War on Fungi

As long as there have been people there have been fungi to torment them. Neither
plants nor animals, fungi such as mold, yeast, and mushrooms are a distinct kingdom
of living things that absorb nutrients directly from the environment, either from dead
organic matter or parasitically from living hosts. Some varieties have long been useful—yeast in baking and brewing, mushrooms as food—but others are more inclined
to use people as food. Fungi spoil granaries, rot books, and cause maddening itches in
warm, moist body crevices.

In 1948 two American women discovered the first safe fungicide for human use.
Nystatin, named in honor of New York State, which funded the scientists' work, was
the discovery of microbiologist Elizabeth Hazen and chemist Rachel Brown. Hazen
and Brown were convinced, based on previous research, that an antifungal organism
existed in certain soils. While vacationing in Virginia, Hazen collected a soil sample
from a friend's cow pasture and sent it to Brown in New York for analysis. Brown then
isolated an antitoxin agent in the soil. The two used it to develop nystatin, the first
broadly effective antifungal antibiotic.

Since then, nystatin has appeared in the form of oral and vaginal tablets, ointments,
powder, and liquid medication. Horticulturists also use nystatin to combat Dutch elm
disease. Its mold-destroying ability prevents spoilage in everything from bananas to
zebra feed. In 1966, when the Arno River overflowed in Italy, nystatin was used to stop
mold from ruining priceless paintings and books damaged by flood waters.

1948        Austrian-born astronomers Hermann Bondi and Thomas Gold theorize that
            the universe is expanding but has no beginning or end. New matter, they say,
            is constantly being created from nothing. This model of what becomes called
            the steady-state universe will be popularized by British astronomer Fred Hoyle.
                                                                              **ASTRO**

1948        American researchers at White Sands, New Mexico, launch a monkey named
            Albert in a V-2 rocket's nose cone.                                **ASTRO**

1948        The Hale telescope at Mount Palomar, California, named for American
            astronomer George Ellery Hale, becomes and remains the largest reflecting
            telescope in the world. Its lens is 200 inches (5.08 meters) in diameter. *See also*
            1897, **ASTRO**.                                                   **ASTRO**

1948        American astronomer Gerard P. Kuiper discovers Miranda, the fifth known
            moon of Uranus.                                                    **ASTRO**

1948        American microbiologist John F. Enders, along with American virologist
            Thomas H. Weller and American physician Frederick C. Robbins, develops a
            technique to study viruses within living cells. Using chicken eggs, Enders grows
            viruses in the developing embryos, then adds penicillin to prevent bacteria
            growth without destroying the viruses. This method becomes useful in finding
            ways to battle viral diseases. The three American scientists will share the 1954
            Nobel Prize for physiology or medicine.                           **BIO**

1948        After years of study on mice, American geneticist George D. Snell locates the
            specific gene sites (histocompatibility genes) concerned with the acceptance or
            rejection of tissue transferred from one organism to another. He will be award-
            ed the 1980 Nobel Prize for physiology or medicine, which he will share with
            Venezuelan-American geneticist Baruj Benacerraf and French biologist Jean
            Dausset, who in the late 1950s will have identified the first human histocom-
            patibility protein.                                               **BIO**

1948        Soviet geneticist and biologist I. V. Michurin's alternative theory of genetics
            wins out over neo-Mendelism in the Soviet Union, where Michurin becomes
            heralded as a great and original thinker. His fundamental theory is that hered-
            ity can be altered by changing the environment.                   **BIO**

1948        American biochemists Stanford Moore and William H. Stein invent starch
            chromatography. They will share the 1972 Nobel Prize for chemistry with
            American biochemist Christian Anfinsen.                           **CHEM**

1948        Geologists O. F. Tuttle and Norman L. Bowen develop the first petrogenetic
            grid and apply it to metamorphosis and serpentinization.          **EARTH**

1948        Swiss physicist Auguste Piccard builds the first bathyscaphe, an improvement
            upon the bathysphere for deep-sea dives. Piccard tests, rebuilds, and continues
            to improve upon the craft.                                        **EARTH**

1948    American mathematician Norbert Weiner publishes *Cybernetics*, a landmark investigation of the mathematics of computer-controlled systems.    MATH

1948    Chinese mathematician Wu Wenjun discovers what will become known as the Wu classes ($W_i$), a family of characteristic classes in topology.    MATH

1948    American botanist Benjamin Duggar isolates and introduces Aureomycin, a tetracycline, which proves second only to penicillin in combating infection.    MED

1948    The United Nations establishes the World Health Organization, stating that "the health of all peoples is fundamental to the attainment of peace and security."    MED

1948    German-born American physicist Maria Goeppert Mayer and, independently, German physicist J. Hans D. Jensen advance the shell model of the atomic nucleus, introducing the concept of magic numbers for the numbers of protons or neutrons that produce the most stable structures. They will share the 1963 Nobel Prize for physics with Hungarian-born American physicist Eugene P. Wigner (*see* 1936, PHYS).    PHYS

1948    American physicist Richard P. Feynman develops the theory of quantum electrodynamics (QED), the study of the properties of electromagnetic radiation and its interaction with charged matter in terms of quantum mechanics. American physicist Julian S. Schwinger and Japanese physicist Shin'ichiro Tomonaga independently develop the theory. All three will share the 1965 Nobel Prize for physics.    PHYS

c. 1948    At Cornell University, the first major attempt is made to put together a pseudobrain out of electrical circuits.    PSYCH

1948    In *Male and Female: A Study of the Sexes in a Changing World*, American anthropologist Margaret Mead argues that many aspects of gender identity are determined by cultural practices.    SOC

1948    American economist Paul Samuelson, a Keynesian, publishes *Economics*, which will long remain a standard textbook. Samuelson will win the 1970 Nobel Memorial Prize in economic sciences for his role in developing the mathematical basis of economics.    SOC

1948    The transistor, which will greatly reduce the size of electronic devices, is developed for Bell Laboratories by American physicists William Shockley, John Bardeen, and Walter H. Brattain. The three will share the 1956 Nobel Prize for physics.    TECH

1948    The long-playing vinyl phonograph record is developed and demonstrated by CBS engineer Peter Goldmark. The 12-inch record runs at a speed of 33 1/3 revolutions per minute and plays for about 45 minutes.    TECH

1948        Chinese-American computer scientist An Wang develops the concept of the ferrite core memory (patented 1955), which will be crucial to the development of pre-1970 computers.                                        TECH

1948        Japanese computing pioneer Hideo Yamashita finishes a calculating machine based on binary logic. Yamashita will found the Information Processing Society of Japan in 1960.                                          TECH

1948        Hexachlorophene, a bacteria-killing compound, is an active ingredient in Dial, the first deodorant soap.                                          TECH

1949        Russian-born American physicist George Gamow predicts that if there was a Big Bang at the creation of the universe, there should be a homogeneous background of radio radiation indicating an average temperature of the universe of about 5° K. *See* 1964, ASTRO.                                          ASTRO

1949        German-American astronomer Walter Baade discovers the asteroid Icarus.        ASTRO

1949        Astronomer Ralph Belknap Baldwin theorizes that meteoritic impacts account for lunar features.                                          ASTRO

1949        American astronomer Fred L. Whipple theorizes that comets are "dirty snowballs" composed of ice and dust.                                          ASTRO

1949        American astronomer Gerard P. Kuiper discovers Nereid, one of Neptune's two satellites.                                          ASTRO

1949        Publication of the *Henry Draper Extension*, begun in 1925, is completed. Based primarily on the work of American astronomer Annie Jump Cannon, the *Henry Draper Extension* and the *Henry Draper Catalogue* of 1918–1924 (*see* 1924, ASTRO) together catalog some 350,000 stars.                                          ASTRO

1949        British anatomist P. B. Medawar develops a technique leading to the reduction of problems associated with tissue transplants. Working with mouse embyros, Medawar discovers they have not yet developed an immunological system to reject foreign proteins, so that when these embyros begin independent life and form antibodies they do not treat injected foreign cells as invaders. In 1960 he will share the Nobel Prize for physiology or medicine with Australian immunologist Macfarlane Burnet.                                          BIO

1949        The first photograph of genes, the units that transmit physical characteristics from one generation to the next, is taken by Daniel Chapin Pease and Richard Baker at the University of Southern California.                                          BIO

1949        American physicist Glenn T. Seaborg and his colleagues discover the element berkelium.                                          CHEM

1949        British chemist Derek H. R. Barton begins to study complex organic molecules. He will eventually demonstrate the high dependence that chemical properties have on molecular shape. In 1969 Barton will share the Nobel Prize for chemistry with Norwegian chemist Odd Hassel.                                        CHEM

1949    British biochemist Dorothy Crowfoot Hodgkin is the first to enlist the aid of an electronic computer in discovering the structure of an organic compound, in this case penicillin.                                        CHEM

1949    V-2 rockets are used by the United States to explore the upper atmosphere.    EARTH

1949    Chinese engineer J. S. Lee becomes known for his work as the founder of geomechanics, which approaches tectonics through experiments with clay- and paraffin-structured models under rotational and tilting forces.               EARTH

1949    Psychopharmacologist J. F. J. Cade publishes the first report on lithium's antimanic effects. Although lithium was discovered in 1817 by Swedish chemist Johan August Arfwedson as a naturally occurring salt, it will not be approved in the United States for the treatment of affective disorders until 1970.    PSYCH

1949    British writer George Orwell coins the term *brainwashing* in his novel *Nineteen Eighty-Four*.                                                          PSYCH

1949    The Wechsler Intelligence Scale for Children (WISC) is published as an intelligence test for children 5 to 15 years old.                               PSYCH

1949    French anthropologist Claude Lévi-Strauss publishes his first major work, *Elementary Structures of Kinship*. In this and later titles, including *Structural Anthropology* (1958) and *The Savage Mind* (1962), he expounds his theory of structuralism, which will become important not only to anthropology but to literary studies.                                                     SOC

1949    A rocket-testing site is founded at Cape Canaveral, Florida, the future site of U.S. space launches.                                                 TECH

1949    American engineers build and launch the first multistage rocket, made up of a smaller rocket on top of a V-2.                                        TECH

1949    On August 29 the Soviet Union explodes its first atomic bomb, a plutonium-based fission weapon. U.S. president Harry S. Truman will inform the American public of the Soviet test on September 23, one day after the American Naval Research Laboratory will have reported that it had detected evidence of the explosion.                                           TECH

1950s   Metal-shadowcasting and freeze-drying techniques are developed in the microscopic study of viruses.                                                 BIO

1950s   American biochemists Edmond H. Fischer and Edwin G. Krebs discover a cellular regulatory mechanism used to control a variety of metabolic processes important to life. In 1992 they will share the Nobel Prize for physiology or medicine for their work in this area.                                   BIO

1950s    Researchers find that they can increase plant root systems by inoculating them with soil fungi or mycorrhizae. The fungi colonize and extend down to the root system, providing more root-surface area for water and nutrient absorption.   BIO

1950s    Some insect pests are successfully controlled biophysically, by sterilizing males with radiation.   BIO

1950s    Prompted by studies of paleomagnetism in the late 1940s, geologists begin to accept the concept of continents moving relative to magnetic poles and to one another. This change represents the beginning of significant studies of continental drift, sea-floor spreading, and plate tectonics. *See also* 1912 and 1960, EARTH.   EARTH

1950s    Several nations, including Iceland and New Zealand, begin to access geothermal energy from water that is naturally heated in volcanic and earthquake areas, where molten rock is close to the surface and hot springs and geysers plentiful.   EARTH

1950s    Meteorological and climatological services in China expand and improve in quality. By 1960, surface weather-observing stations in China number about 400 and upper-air stations 60 or 70.   EARTH

1950s    During this decade, scientific researchers in every discipline learn to make use of computers. Computers will be used in calculating planetary orbits, weather patterns, molecular structures, population trends, and more.   MISC

1950s    Two main classes of elementary particles, fundamental units smaller than the atom, are identified. Hadrons, including nucleons (protons and neutrons), mesons, and hyperons interact by the force known as the strong interaction and are found to have a complex internal structure. Leptons, including electrons, neutrinos, and muons, interact either by the weak or electromagnetic interactions (or forces) and have no apparent internal structure.   PHYS

1950s    American psychiatrist Nathan Kline introduces reserpine as the first major tranquilizer.   PSYCH

1950s    German psychoanalyst Ludwig Binswanger formulates a mode of therapy from the ideas of existentialism. He claims that neurosis must be explained by its meaning to the patient, not in terms of its origin or etiology. Existential therapy in all its forms becomes not a "school" of psychology but rather an understanding and therapy based on personal values, concrete experience, and respect for each patient.   PSYCH

1950s    German-American psychologist Stanley Milgram experiments with human obedience, finding that people will go to the extent of torturing others in order to obey authority figures. Milgram suggests that this tendency helps to explain people's compliance with Nazi brutality during World War II.   PSYCH

1950s        American research psychologists Neal Miller and John Dollard introduce the frustration-aggression hypothesis, that frustration always causes a certain amount of aggression. This hypothesis will play a part in evaluating and diagnosing mental health problems like depression.        **PSYCH**

1950s        Between now and the 1960s, the monoamine (MAO) inhibitors and tricyclic antidepressants are discovered and developed.        **PSYCH**

1950         Dutch astronomer Jan Hendrik Oort suggests that comets originate in a vast cloud of material revolving around the sun far beyond Pluto, a region that becomes known as the Oort Cloud. American astronomer Gerard P. Kuiper will posit the existence of a disk-shaped belt of comets just beyond Pluto.        **ASTRO**

1950         American biochemist William Cumming Rose conclusively establishes the protein-building role of the essential amino acids: isoleucine, leucine, lysine, methionine, phenylalanine, threonine, tryptophan, valine, and histidine.        **BIO**

1950         Using the electron microscope, Belgian cytologist Albert Claude discovers the structural network of membranous vesicles in cells' cytoplasm called the endoplasmic reticulum. He will share the 1974 Nobel Prize for physiology or medicine with Belgian cytologist Christian de Duve (*see* 1955, **BIO**) and Romanian-born American physiologist George E. Palade (*see* 1956, **BIO**).        **BIO**

1950         German-born American biochemist Konrad Bloch, using stable carbon-13 and radioactive carbon-14 as tracers, shows detailed changes occurring with the buildup of the cholesterol molecule in the body. Bloch will share the 1964 Nobel Prize for physiology or medicine with German biochemist Feodor Lynen (*see* 1951, **BIO**).        **BIO**

1950         American physicist Glenn T. Seaborg and his colleagues discover the element californium. For their discovery of this and other transuranium elements (*see* 1940, **CHEM**), Seaborg and American physicist Edwin H. McMillan will share the 1951 Nobel Prize for chemistry.        **CHEM**

1950         In Paris, an international meeting agrees to adopt a new astronomical unit, the ephemeridical unit, to measure time. It is based on Earth's movement around the sun.        **EARTH**

1950         A French team led by Maurice Herzog climbs Annapurna, the first 8,000-meter peak (26,000 feet) ever scaled, in north-central Nepal.        **EARTH**

1950         The World Meteorological Organization, an international technical organ of the United Nations, is founded on March 23, replacing the International Meteorological Organization begun in 1873.        **EARTH**

1950         The Chinese Academy of Sciences is established.        **MISC**

| | |
|---|---|
| 1950 | German-French physicist Alfred Kastler, who will be awarded the 1966 Nobel Prize for physics, develops an optical pumping system, which uses electromagnetic waves to excite atoms and is a precursor to the laser. **PHYS** |
| c. 1950 | American psychiatrist Jacob L. Moreno develops psychodrama (therapy involving role playing to bring about emotional catharsis) and a social psychology methodology called sociometry, a technique for measuring attraction and repulsion among people. **PSYCH** |
| 1950 | German-born psychoanalyst Erik H. Erikson writes his first work on the developmental stages in humans. The stages he elaborates are trust vs. mistrust, autonomy vs. doubt, industry vs. inferiority, identity vs. diffusion, intimacy vs. isolation, generativity vs. stagnation, and integrity vs. despair. **PSYCH** |
| 1950 | American social psychologist Stanley Schachter begins developing a psychological cognitive theory of emotion, claiming that humans cannot discriminate emotions unless they have some cognitive indication of what their feelings relate. **PSYCH** |
| 1950 | American psychology professor James J. Gibson writes *The Perception of the Visual World*, a book on human perception that explores the role human senses take in selecting information from stimuli. His theory comes to be called psychophysical correspondence. **PSYCH** |
| 1950 | British mathematician Alan Mathison Turing proposes the "Turing Test" for determining whether a machine thinks: If a person communicating with a computer cannot tell whether its responses come from a human or a machine, the computer can be considered intelligent. **TECH** |
| 1950 | Orlon, a polymerized acrylonitrile fiber that will be widely employed in clothing, is introduced by Du Pont, as developed in consultation with William Hale Church. **TECH** |
| 1950 | The first Xerox machine is built by the Haloid Co. of New York. **TECH** |
| 1951 | Dutch-American astronomer Dirk Brouwer is the first to use a computer to calculate planetary orbits. **ASTRO** |
| 1951 | Using spectroscopic analysis, American astronomer William Wilson Morgan shows that the Milky Way galaxy has a spiral structure like that of its neighbor the Andromeda galaxy (M31). **ASTRO** |
| 1951 | American astronomers Harold Irving Ewen and American physicist Edward M. Purcell discover radio emissions from hydrogen clouds in interstellar space (*see* 1944, **ASTRO**). Their 21-centimeter-wavelength radiations will allow astronomers to map the structure of the galaxy and confirm that the galaxy rotates once every 200 million years. **ASTRO** |

1951        Henrietta Lacks, a 31-year-old cervical cancer patient, dies in Baltimore. Cells from her cervical tumor are preserved and, when multiplied, become the first continuously cultured strain of cancer cells, called HeLa cells.        BIO

1951        German biochemist Feodor Lynen is the first to isolate acetyl coenzyme A, a compound important in biochemical functions and as an intermediate in the Krebs cycle, the cycle of intracellular chemical reactions by which organisms convert food chemicals to energy. Lynen will share the 1964 Nobel Prize for physiology or medicine with German-born American biochemist Konrad Bloch (*see* 1950, BIO).        BIO

1951        Following a major outbreak of locusts this year, China mobilizes what may be the most extensive pest management program in history. By the 1970s, aerial dusting with insecticides will have given way to more successful, ecologically oriented efforts.        BIO

1951        Insect sterilization by irradiation is determined to be an effective method for lowering insect levels, according to findings by U.S. Department of Agriculture entomologist Edward Knipling.        BIO

1951        Drawing on quantum theory, American physicist John Bardeen develops an explanation of superconductivity.        PHYS

1951        The Chrysler Corporation introduces power steering in their high-end automobiles. Eventually, power steering will be installed in other Chrysler models, as well as those made by other automobile companies.        TECH

1951        The Univac computer is introduced for business use by Remington Rand.        TECH

1951        Chinese-American scientist An Wang founds Wang Laboratories, which will introduce a desktop calculator in 1964 and a word processor in 1971.        TECH

1951        Color television programming is transmitted for the first time, by CBS, though color TV sets will not be marketed commercially until 1954.        TECH

1951        In December an experimental reactor in Idaho generates the first electricity from nuclear power.        TECH

1952        Archaeologists discover signs of human settlement at a site near Clovis, New Mexico, dating from 11,500 years ago. At the time, the remains of these "Clovis people," believed to have migrated from Asia about 12,000 years ago, are the earliest undisputed evidence of human settlement in the Americas. *See also* 1970, ARCH.        ARCH

1952        Michael Ventris deciphers the ancient Cretan language known as Linear B.        ARCH

1952        German-American astronomer Walter Baade discovers an error in the Cepheid luminosity scale (*see* 1912 and 1914, ASTRO), based on differences between

Cepheids in population I and population II stars. As a result, he determines that other galaxies are about twice as far away as previously thought.    ASTRO

1952    Astronomers Adriaan Blaauw and Georg Herbig discover evidence of ongoing star formation in the Milky Way galaxy, while Martin Schwarzschild investigates signs of stellar evolution in globular clusters.    ASTRO

1952    American biologists Robert William Briggs and Thomas J. King successfully transplant living nuclei from blast cells to enucleated frog's eggs.    BIO

1952    After working with implanted tumors in chick embryos, Italian embryologist Rita Levi-Montalcini shows the nerve growth factor to be a soluble substance that the tumor releases, which hastens nerve growth.    BIO

1952    American biophysicist Rosalyn Sussman Yalow develops the radioimmune assay, a method for detecting and following antibodies and other minute biologically active proteins and hormones present in the body.    BIO

1952    American physicist Albert Ghiorso and his colleagues discover the element einsteinium.    CHEM

1952    American chemists Stanley Miller and Harold C. Urey demonstrate that simple chemical compounds such as water, hydrogen, ammonia and methane—like those believed to have composed the earth's early atmosphere and ocean—can interact with electrical discharges to produce more complex organic compounds and even amino acids. This experiment supports the theory that life originated from simpler, nonliving substances.    CHEM

1952    British biochemist Arthur J. P. Martin develops gas chromatography.    CHEM

1952    American chemist William Gardner Plann develops zone refining, a technique for reducing the impurities in metals, alloys, semiconductors, and other substances.    CHEM

1952    An earthquake at Tehachapi in Kern County, south central California, registers 7.7 on the Richter scale—the largest earthquake in the state since the 1906 San Francisco earthquake. Eleven people die. *See also* 1992, EARTH.    EARTH

1952    German physician and philosopher Albert Schweitzer receives the Nobel Peace Prize for his work with the sick in Africa.    MED

1952    American biochemists discover an antibacterial called isoniazid, which will be used in the long-term treatment of tuberculosis.    MED

1952    In the United States erythromycin, an antibiotic used to treat skin, chest, throat, and ear infections is isolated.    MED

1952    Fossil remains of a giant extinct ape, called *Gigantopithecus*, are discovered in Asia.    PALEO

1952    Polish physicists Marian Danysz and Jerzy Pniewski discover the K meson or kaon, about 0.5 the mass of a proton, and the lambda particle, approximately 1.2 times the mass of a proton. Particles more massive than protons will eventually be grouped as hyperons.                    PHYS

1952    The American Psychiatric Association (APA) publishes its first *Diagnostic and Statistical Manual of Mental Disorders* (DSM-1), which will serve as a mental disorders classification system and be the diagnostic standard of the APA. *See also* 1979, PSYCH.                    PSYCH

1952    Martinique-born psychiatrist Frantz Omar Fanon examines the significance of racism and cultural prejudice in his book *Black Skin, White Masks*, in which he argues that racial oppression can cause debilitating mental illness. In 1953, while practicing psychiatry in Algeria, Fanon will join with the Algerian Liberation Movement and attempt to overthrow French rule. He will eventually call for violent revolution to end colonial tyranny.                    PSYCH

1952    British anthropologist Alfred Reginald Radcliffe-Brown publishes *Structure and Function in Primitive Society*, on his theory of structural functionalism.    SOC

1952    Japanese company Sony produces the first pocket-sized transistor radio.    TECH

1952    Japan's first 35-millimeter single-lens reflex camera, the Asahiflex, is manufactured by Asahi Optical Co. of Japan.                    TECH

1952    Using a De Havilland Comet, British Overseas Airways (BOAC) institutes on May 2 the first jetliner service, between London, England, and Johannesburg, South Africa.                    TECH

1952    The "club" of nuclear-armed nations, which already includes the U.S. and U.S.S.R., continues to grow as Britain explodes its first fission bomb, on October 3, on the Monte Bello Islands, off the western coast of Australia. France (1960) and China (1964) will also soon produce nuclear weapons.                    TECH

1952    On November 1, the U.S. detonates the world's first hydrogen bomb, at Eniwetok Atoll in the South Pacific. This fusion bomb releases energy equivalent to 10.4 million tons of high explosives, about 700 times the force of the Hiroshima fission bomb.                    TECH

1953    Superclusters of galaxies—clusters of clusters—are discovered.    ASTRO

1953    American biologists Robert William Briggs and Thomas J. King succeed in growing tadpoles from eggs whose nucleus has been replaced with one from a partly differentiated cell of a developing embryo.                    BIO

1953    British molecular biologist Francis Crick, British biophysicist Maurice H. F. Wilkins, and American molecular biologist James D. Watson discover the double-helix structure of the deoxyribonucleic acid (DNA) molecule. The three will share the 1962 Nobel Prize for physiology or medicine.                    BIO

1953    American physicist Albert Ghiorso and his colleagues discover the element fermium.                    CHEM

1953    German chemist Karl W. Ziegler and Italian chemist Giulio Natta, who will share the 1963 Nobel Prize for chemistry, develop isotactic polymers, non-branching, uniformly ordered polymer chains useful in industry. These chains employ catalysts that combine monomers into polymers in a regular way.
                    CHEM

1953    American geologists Maurice Ewing and Bruce Charles Heezen discover an underwater canyon running the length of the Mid-Atlantic Ridge (*see* 1925, EARTH). In 1956, they will propose the existence of a world-girdling formation of mountains called the Mid-Oceanic Ridge, accompanied by a formation of canyons called the Great Global Rift. The discovery that the rift separates the earth's crust into plates contributes to the developing theory of plate tectonics (*see* 1960, EARTH).                    EARTH

1953–1979    It is official U.S. Weather Service policy to use women's names for tropical cyclones, hurricanes, and typhoons. Afterward, names of both genders will be used to designate storms.                    EARTH

1953    New Zealander Edmund P. Hillary and his Nepalese guide, Tenzing Norkay , reach the top of Mount Everest on May 29. At 8,848 meters (29,028 feet) high, this mountain on the Tibet–Nepal border is the world's tallest. Tibetans call Everest *Chomolungma*, "Goddess Mother of the World."                    EARTH

1953    The world's first successful open-heart surgery, using American surgeon John Gibbon Jr.'s newly developed heart-lung machine, is performed.                    MED

1953    American physician and epidemiologist Jonas Edward Salk begins preliminary testing of a poliomyelitis vaccine he developed in 1952. By 1955 this polio vaccine will be used worldwide to dramatically reduce the incidence of this disease. Virologist Albert Bruce Sabin will develop an oral polio vaccine in 1957.                    MED

1953    American physicist Donald A. Glaser invents the bubble chamber, a device for detecting ionizing radiation. He will be awarded the 1960 Nobel Prize for physics.                    PHYS

1953    American physicist Murray Gell-Mann investigates the property of the elementary particles called kaons and hyperons that is known as strangeness, the tendency to decay slowly by way of the weak interaction even though the particles are subject to the strong interaction. To these particles Gell-Mann assigns a quantum number $s$ (for strangeness number) that has an integral value and does not equal zero.                    PHYS

1953    A regularly recurring sleep stage characterized by rapid eye movement (REM) is discovered. This REM stage, which occurs spontaneously about every 90 minutes during sleep, is considered to indicate the presence of dreams in humans.    PSYCH

1953    African-American psychologist Mamie Phipps Clark assists in the preparation of a social science brief addressing self-awareness and self-esteem in black children. This brief will form the basis of the 1954 U.S. Supreme Court decision *Brown v. Board of Education of Topeka*, which will mandate the desegregation of public schools.    PSYCH

1953    American political scientist David Easton publishes *The Political System*, which develops the approach known as systems analysis. Drawing metaphors from physics and biology, Easton treats the political system as one part of an over-all social system. His strategy provides a framework for many topics of study, including the interaction of elites, interest groups, and political parties.    SOC

1953    On August 12 the Soviet Union becomes the second nation to explode a hydrogen (fusion) bomb. *See also* 1949 and 1952, TECH.    TECH

1953    American physicist Charles H. Townes and, independently, Soviet physicists Aleksandr Prokhorov and Nikolai Basov, invent the maser (microwave amplification by stimulated emission of radiation), a device for producing a coherent beam of microwave radiation. All three will share the 1964 Nobel Prize for physics.    TECH

1953    International Business Machines introduces the IBM 701, its first computer for scientific and business use.    TECH

1953    The plastic valve for aerosol cans that will bear his name is developed by American inventor Robert Abplanalp.    TECH

1953    The Ziegler process, a catalytic technique used in making low-cost polyethylene plastics, is developed by German chemist Karl W. Ziegler.    TECH

1953    Cinemascope, a film process that widens the view projected on the screen, is used for the first time in the film *The Robe*.    TECH

1953    *Bwana Devil*, the first three-dimensional, or 3-D, film is shown in theaters. The technique requires special viewing devices to appreciate the 3-D effects.    TECH

1954    In Egypt, Kamal el-Malakh and his colleagues discover two chambers near the base of the Great Pyramid of Cheops, or Khufu, at Giza. In one they find a 142-foot boat probably meant to transport the deceased pharaoh to the next world.    ARCH

1954    American biochemist Vincent du Vigneaud synthesizes the hormone oxytocin, the first naturally occurring protein to be synthesized with the exact makeup it has naturally in the body.    BIO

1954        Polish-American biochemist Daniel Israel Arnon obtains intact chloroplasts
           from spinach-leaf cells that have been disrupted and shown their ability to
           photosynthesize extracellularly.                                    BIO

1954        Russian-born American physicist George Gamow proposes the existence of a
           multinucleotide genetic code.                                       BIO

1954        The microprobe is invented for use in experimental mineralogy.      EARTH

1954        The National Geographic Society and the Mount Palomar Observatory togeth-
           er publish the *National Geographic–Palomar Sky Atlas*.              EARTH

1954        On February 15, off the coast of West Africa, two French naval officers
           descend to a depth of 4,050 meters (13,300 feet) in Auguste Piccard's bathy-
           scaphe. *See* 1948, EARTH.                                          EARTH

1954        Scientists at the University of California build the bevatron, a particle acceler-
           ator capable of accelerating protons to energies of 5 billion or 6 billion electron
           volts.                                                              PHYS

1954    ⌐   CERN, the European Organization for Nuclear Research, is founded in
           Geneva, Switzerland.                                                PHYS

1954        Chinese-American physicist Chen Ning Yang and American physicist Robert
           Mills develop the mathematics of Yang-Mills gauge-invariant fields, concern-
           ing symmetry at the level of fundamental interactions.              PHYS

1954        Dutch-born American physicist Abraham Pais coins the term *baryon* to
           describe particles such as protons and neutrons that are affected by the strong
           force. This definition will be applied later to hadrons (*see* 1962, PHYS). Baryons
           will be understood to be a subclass of hadrons, those with a half-integral spin.
           Nucleons (protons and neutrons) will be considered a subclass of baryons.
                                                                              PHYS

1954        In March, chlorpromazine is approved for use as an antipsychotic in the
           United States, under the trade name Thorazine.                     PSYCH

1954        In the United States the Durham Rule, named for defendant Monte Durham,
           becomes law in June. It states that a criminal is not guilty if his unlawful
           behavior is the result of "mental defect or disease."               PSYCH

1954        Silicon transistors are introduced by Texas Instruments.           TECH

1954        The oxygen steel-manufacturing furnace, already popular in Europe, is intro-
           duced to the United States in a steel mill in Detroit.              TECH

1954        BHA, butylated hydroxyanisole, is approved by the U.S. Food and Drug
           Administration for use as a preservative in foods.                  TECH

1954        American salesman Raymond Kroc purchases the franchise rights to the
            California-based McDonald brothers' hamburger chain and begins to develop it
            into the largest fast-food restaurant chain in the world.                    TECH

1954        The U.S.S. *Nautilus*, the first submarine powered by an onboard nuclear reac-
            tor, is launched. It will remain in service until 1980.                     TECH

c. 1955     The Schwarzschild radius, named for the German astronomer Karl
            Schwarzschild, is identified. It is the radius that must be exceeded for light to
            escape an object of a given mass, and it marks the event horizon of a black
            hole. *See* 1960s, ASTRO.                                                   ASTRO

1955        The United States and the Soviet Union initiate separate satellite programs,
            with the Soviets being the first to launch a satellite, two years later.    ASTRO

1955        British astronomer Martin Ryle invents the radio interferometer, a device that
            improves the resolution of radio telescopes.                               ASTRO

1955        American astronomers detect radio emissions from Jupiter.                   ASTRO

1955        Measuring the polarization of light, Dutch astronomer Jan Hendrik Oort con-
            firms a 1953 hypothesis by I. S. Shklovskii that radio emission from the Crab
            Nebula is the result of synchrotron radiation.                             ASTRO

1955        British biochemist Frederick Sanger uses paper chromatography to show that
            the protein hormone insulin consists of 50 amino acids along two intercon-
            nected chains. He also shows their exact order on each chain. In 1958 he will
            win the Nobel Prize for chemistry for his discovery of the complete structure
            of the insulin molecule, among those of other proteins.                     BIO

1955        Belgian cytologist Christian de Duve discovers and names lysosomes.He will
            share the 1974 Nobel Prize for physiology or medicine with Belgian cytologist
            Albert Claude (*see* 1950, BIO) and Romanian-born American physiologist
            George E. Palade (*see* 1956, BIO).                                         BIO

1955        American physicist Albert Ghiorso and his colleagues discover the element
            mendelevium.                                                                CHEM

1955        Using high temperatures and pressures and with chromium as a catalyst, sci-
            entists produce the first synthetic diamonds out of graphite.               CHEM

1955        French mathematician Henri Cartan and American mathematician Samuel
            Eilenberg develop homological algebra, an innovation that unites abstract alge-
            bra and algebraic topology.                                                 MATH

1955        At the urging of American social activist Margaret Sanger, American biologist
            and endocrinologist Gregory Pincus develops the first successful birth-control
            pill, based on his discovery that the hormone norethindrone is effective in
            preventing conception.                                                      MED

1955        Italian-American physicist Emilio Segrè, who became an American citizen in
            1944, and American physicist Owen Chamberlain discover the first known
            antiprotons, negatively charged particles that have the mass of protons. Segrè
            and Chamberlain will share the 1959 Nobel Prize for physics.          PHYS

1955        Two types of K mesons with differing modes of decay are detected, the tau and
            the theta.                                                            PHYS

1955        American clinical psychologist Albert Ellis develops Rational-Emotive Therapy
            (RET), which emphasizes how the holding of unrealistic expectations and irra-
            tional thinking and beliefs can cause and perpetuate human misery. RET works
            to overcome problems created by false beliefs and to correct the human ten-
            dency toward irrational thought.                                      PSYCH

1955        American clinical psychologist George Kelly publishes his two-volume work
            *The Psychology of Personal Constructs*. Kelly's personal construct theory is based
            on the idea that the most important determinant of human behavior is the
            individual's own conception of the world and the people he or she meets. Kelly
            will be the first to found a psychological clinic for training in a theory.   PSYCH

1955        American anthropologist Julian Steward publishes his *Theory of Culture
            Change: The Methodology of Multilinear Evolution*, a study of cultural evolution.
                                                                                  SOC

1955        American physicist Erwin Wilhelm Mueller invents the field ion microscope,
            which is capable of magnifications of more that a million times. It is the first
            device that can yield images of individual atoms.                      TECH

1955        The IBM 752, International Business Machines' first computer designed exclu-
            sively for business use, is produced.                                 TECH

1956        Microwave radiation detected on Venus indicates that its surface temperature
            is as high as 600° F.                                                 ASTRO

1956        American biologists T. T. Puck, S. J. Cieciura, and P. I. Marcus grow clones of
            human cells successfully in vitro.                                    BIO

1956        Using the electron microscope, Romanian-born American physiologist George
            E. Palade discovers that microsomes (small bodies in cell cytoplasm) contain
            ribonucleic acid (RNA), and renames them ribosomes. They are later found to
            be the protein-manufacturing site in the cell. In 1974 he will share the Nobel
            Prize for physiology or medicine with Belgian cytologist Albert Claude (*see*
            1950, BIO) and Christian de Duve (*see* 1955, BIO).                    BIO

1956        Chinese-born American biochemist Choh Hao Li isolates the human growth
            hormone from the pituitary gland. He also studies the structure of ACTH and
            the melanocyte-stimulating hormone (MSH).                             BIO

1956        American biochemist Earl W. Sutherland Jr., who will win the 1971 Nobel
            Prize for physiology or medicine, isolates cyclic adenosine monophosphate
            (AMP), a molecule that stimulates enzymatic activity, affecting the functioning
            of certain hormones.                                                    BIO

1956        The United States Congress passes the Water Pollution Control Act, one of the
            early modern attempts at marine protection.                             BIO

1956        American biochemist Mahlon Bush Hoagland discovers small RNA molecules
            in cytoplasm and shows that each variety has the capacity to combine with a
            particular amino acid.                                                  BIO

1956        Halothane, a colorless liquid inhaled as a vapor to induce and maintain gen-
            eral anesthesia, is introduced.                                         MED

1956        American physicists Frederick Reines and Clyde Lorrain Cowan Jr. dis-
            cover proof of the existence of neutrinos, subatomic particles whose exis-
            tence had been proposed by Wolfgang Pauli in 1930. In 1995 Reines will
            share the Nobel Prize for physics with American physicist Martin L. Perl
            (see c. 1974, PHYS).                                                     PHYS

1956        Chinese-American physicists Chen Ning Yang and Tsung-dao Lee show that
            the property called parity, the quality of being odd or even, is conserved in the
            strong and electromagnetic interactions but not in the weak interaction. In the
            following year they will share the Nobel Prize for physics.              PHYS

1956        Antineutrons, differing from neutrons in the orientation of their magnetic
            field, are discovered.                                                  PHYS

1956        American anthropologist Gregory Bateson formulates the double-bind theory
            of communication patterns, which argues that schizophrenia develops in chil-
            dren whose parents give contradictory messages, often sent subliminally. PSYCH

1956        The Dartmouth Summer Research Project on Artificial Intelligence (AI) is held
            to explore the idea that intelligence can be described so precisely that a
            machine can be made to simulate it. Among those attending the conference
            are neurologist and mathematician Marvin Minsky, Herbert Simon, and Allen
            Newell. Minsky and conference host John McCarthy will found the AI lab at
            the Massachusetts Institute of Technology (MIT), Simon and Newell the one
            at Carnegie Mellon in Pittsburgh.                                       PSYCH

1956        Americans William C. Boyd and his wife, Lyle Boyd, identify 13 "races" of
            Homo sapiens, based on blood groups.                                    SOC

1956        American sociologist C. Wright Mills publishes The Power Elite, an analysis of
            American class structure, in which he critiques the highly theoretical sociolo-
            gy of Talcott Parsons and others, contending that they neglect such issues as
            group conflict and social change. Three years later he will publish The
            Sociological Imagination, which will become a classic in its field.      SOC

1956        The first nuclear-powered jet engine is tested in Idaho. It is meant to power a bomber that can fly for months without refueling, but the project will be canceled in 1961 after intercontinental ballistic missiles make it obsolete.     TECH

1956        Dutch-born American physicist Nicolaas Bloembergen invents the continuous maser. In 1981 he will share the Nobel Prize for physics with American physicist Arthur L. Schawlow and Swedish physicist Kai Siegbahn.     TECH

1956        A videotape-recording machine is shown publicly by the U.S.-based Ampex Corporation.     TECH

1957        British astronomer Martin Ryle, who whill share the 1974 Nobel Prize for physics with British astronomer Antony Hewish, argues that energy fluctuations in Seyfert galaxies (see 1944, ASTRO) result from the ejection of matter at near-light speed.     ASTRO

1957        On October 4, the Soviet Union launches *Sputnik I*, the world's first artificial satellite, into orbit. The United States reacts by beginning a space race with the Soviet Union.     ASTRO

1957        On November 3, the Soviet Union launches the satellite *Sputnik II*, which carries a dog named Laika.     ASTRO

1957        British biochemist John C. Kendrew solves for the first time the three-dimensional structure of a protein. He will share the 1962 Nobel Prize for chemistry with Austrian-born British molecular biologist Max Perutz (*see also* 1967, BIO).     BIO

1957        American organic biochemist Melvin Calvin, who will be awarded the 1961 Nobel Prize for chemistry, discovers and isolates all the details of plant photosynthesis.     BIO

1957        Gibberelins—plant hormones used to increase plant size, especially in wine and table grapes—are isolated from a fungus of the genus *Gibberella*.     BIO

1957        English-born American zoologist G. Evelyn Hutchinson defines the concept of the ecological niche as the place or function of a given organism within its ecosystem, which is a collection of living things and the environment in which they live.     BIO

1957        The lightweight plastic called polypropylene is invented.     CHEM

1957        American seismologist Charles Richter establishes a new relationship between the magnitude and the energy produced by an earthquake.     EARTH

1957        Researchers in 70 countries engage in the systematic, coordinated study of the earth and its atmosphere during the International Geophysical Year. One study, for instance, focuses on measuring the flattening of the earth at its poles.     EARTH

| 1957 | British bacteriologist Alick Isaacs discovers interferon, an antiviral protein produced by the body in response to viral infections. It will be used against a wide variety of drug-resistant viral diseases and in cancer research. MED |
|---|---|

1957      Amniocentesis, the study of amniotic fluid extracted from the amniotic sac with a needle and syringe, is developed to test for genetic disorders. MED

1957      American physicists John Bardeen, Leon N. Cooper, and J. Robert Schrieffer advance the so-called BCS theory to explain the phenomenon of superconductivity. All three will share the 1972 Nobel Prize for physics. PHYS

c. 1957      American researchers Allen Newell, Herbert Simon (who will win the 1978 Nobel Memorial Prize in economic sciences), and J. C. Shaw develop their Logic Theorist program, one of the first artificial intelligence programs. PSYCH

1957      American social psychologist Leon Festinger presents his cognitive dissonance theory, having to do with the relationship among cognitive elements such as self-knowledge, behavior, and environment. Festinger's research concentrates on the discrepancies between attitude and behavior and on the consequences of decisions. PSYCH

1957      In his work *Syntactic Structures*, American linguist Noam Chomsky proposes the revolutionary theory of transformational-generative grammar, in which he argues that innate structures in the mind are the basis for human languages. This theory seeks to uncover the underlying structure and rules that govern the production of sentences. SOC

1957      Hungarian mathematician L. Kalmár, known for developing the theoretical foundation of a formula-contolled computer, launches a university course in computer science in Szeged. TECH

1957      The Wankel rotary engine, an improved internal combustion engine, is developed by German engineer Franz Wankel for use in automobiles and other types of machinery. TECH

1957      Soy protein foodstuffs are more easily created with the development of an improved spinning process to mix soy flour and alkaline liquids. TECH

1958      American physicist Eugene N. Parker discovers the solar wind, a flow of charged, subatomic particles emanating from the sun. ASTRO

1958      American astrophysicist Herbert Friedman discovers X rays emanating from the sun that are probably produced in its corona. ASTRO

1958      On January 31, the United States launches its first satellite, *Explorer I*. In addition, American scientists discover Earth's Van Allen radiation belt and launch four more satellites and three lunar probes. *See also* 1957, ASTRO. ASTRO

1958    On October 7, the United States announces its Project Mercury, the first American manned space program.                                ASTRO

1958    British molecular biologist Francis Crick writes on a principle he calls the central dogma of molecular biology: "Once 'information' has passed into protein it cannot get out again. . . . The transfer of information from nucleic acid to nucleic acid may be possible, but transfer from protein to protein, or from protein to nucleic acid, is impossible."                                BIO

1958    Russian scientist Ilya Darevsky discovers the first known example of an all-female vertebrate species, a lizard species in Soviet Armenia that reproduces without male fertilization.                                BIO

1958    In China, entomology and plant pathology combine in the applied discipline of plant protection, heavily supported by the government as an aid to agricultural production.                                BIO

1958    American geneticists Joshua Lederberg, George W. Beadle and Edward L. Tatum share the Nobel Prize for physiology or medicine, Lederberg for his work on genetic mechanisms, Beadle and Tatum for discovering how genes transmit hereditary characteristics.                                BIO

1958    American physicist Albert Ghiorso and his colleagues discover the element nobelium.                                CHEM

1958    Scottish physician Ian McDonald pioneers the use of high-frequency sound waves (ultrasound) as a diagnostic and therapeutic tool. Ultrasound is used to destroy diseased tissue and restore damaged tissue.                                MED

1958    German physicist Rudolf L. Mössbauer discovers what will become known as the Mössbauer effect, a sharp narrowing of the energy spread (range of wavelengths) of gamma-ray emission from atoms in certain solids possessing a lattice configuration. Atoms of the same crystal will absorb only gamma rays of the same energy spread. *See also* 1960, PHYS.                                PHYS

1958    The Council of Mental Health of the American Medical Association validates the therapeutic use of hypnosis.                                PSYCH

1958    American psychologist Arnold Lazarus is the first to use the term *behavior therapy* in describing certain mental illness treatment strategies.                                PSYCH

1958    American experimental and comparative psychologists Harry F. Harlow and John Bowlby experiment with baby monkeys, mannequin mothers, and the idea of maternal deprivation. They prove that behavioral disturbances and detachment occur when adequate interaction, holding, and bonding do not take place between mother and child.                                PSYCH

1958    Social psychologist Fritz Heider publishes what will become a classic treatise on social psychology, *The Psychology of Interpersonal Relations*. In it Heider presents his balance theory and explains why individuals strive for cognitive consistency—how an individual organizes beliefs and perceptions in a consistent, organized way.                                                                 PSYCH

1958    The Boeing 707, the first U.S. jet for passenger service, is put into operation by Boeing Aircraft.                                                              TECH

1958    A saccharin-based artificial sweetener is introduced to the American market under the brand name Sweet 'n Low. *See also* 1983, TECH.                        TECH

1959    The United States experimentally launches two monkeys and a chimpanzee into space.                                                                        ASTRO

1959    The Soviet unmanned spacecraft *Luna 1* makes the first flyby of the moon.    ASTRO

1959    The Soviet unmanned spacecraft *Luna 2*, launched on September 12, is the first vehicle to reach the moon, where it crashes on September 14.              ASTRO

1959    On October 7, the Soviet unmanned craft *Luna 3* takes the first pictures of the moon's far side.                                                             ASTRO

1959    American biochemist Christian Anfinsen, who will share the 1972 Nobel Prize for chemistry with American biochemist Stanford Moore and William H. Stein (*see* 1948, CHEM), publishes *The Molecular Basis of Evolution*, based on his work with enzymes.                                                                        BIO

1959    Spanish-American biochemist Severo Ochoa and American biochemist Arthur Kornberg are awarded the Nobel Prize for physiology or medicine for discoveries related to compounds within chromosomes that play a role in heredity.  BIO

1959    The U.S. satellite *Vanguard II* becomes the first to transmit weather information to Earth.                                                                    EARTH

1959    The Antarctic Treaty is signed by 12 nations, including the U.S. and U.S.S.R. The treaty freezes territorial claims and establishes freedom of scientific activity.                                                                    EARTH

1959    The U.S. satellite *Explorer 6* takes the first television pictures of Earth's cloud cover as seen from space.                                               EARTH

1959    British anthropologists Louis and Mary Leakey discover fossils of *Zinjanthropus boisei* in the Olduvai Gorge of what will become Tanzania, Africa. Believed to have lived 2.5 million to 1 million years ago, this thick-boned hominid with large back teeth is now classified either as *Australopithecus robustus* or *Australopithecus boisei*.                                                    PALEO

1959    Japanese physicists Saburo Fukui and Shotaro Miyamoto invent the spark chamber, a device to selectively detect ionizing particles.                       PHYS

1959        German-born psychiatrist Viktor Frankl publishes *Man's Search for Meaning,* a popular mental-health book emphasizing the importance of free will.    PSYCH

1959        The drug haloperidol, or Haldol, is first synthesized, for use with psychotic disorders.    PSYCH

1959        The first ground-based nuclear rocket engine is tested, with the goal of designing nuclear rockets to fly into space. After many experiments the program will be scrapped in 1973 as the space effort is scaled back.    TECH

1959        The microchip, an integrated circuit made of a single silicon wafer, is invented by American engineers Jack Kilby of Texas Instruments and Robert Noyce of Fairchild Semiconductors.    TECH

1959        Sony introduces the first transistorized television set.    TECH

1959        The first pantyhose are developed, by Glen Raven Mills in North Carolina.    TECH

1960s       American physicist John Archibald Wheeler coins the term *black hole* for a collapsed star whose surface gravity is so great that nothing, not even light, can escape it.    ASTRO

1960s       In the United States a creationist movement begins to gain strength advocating the belief that God literally created all life forms as described in the biblical book of Genesis and demanding that this doctrine be taught in the public schools. Creationists have never accepted the theory of evolution as expressed by Charles Darwin in 1859.    BIO

1960s       Ecology becomes identified for the first time with environmental concerns like pesticides, pollutants, and preservation. Prior to this time, ecology was linked mostly to agriculture and related economic issues.    BIO

1960s       American biologist Daniel Mazia uncouples centrosomal and nuclear replication in the fertilized eggs of sand dollars and sea urchins. With his assistants he proves that centrosome replication can occur in the absence of nuclear replication.    BIO

1960s       In Czechoslovakia, new macromolecular substances are synthesized, leading to the development of hydrophilic gels used in soft contact lenses and soft surgical implants.    CHEM

1960s       By mid-decade some scientists observe what they consider evidence of global warming due to the greenhouse effect. *See* 1863, EARTH.    EARTH

1960s       Biofeedback becomes popular as a short-term therapy to help people learn healthy responses to stress. Biofeedback enables one to learn, with the aid of a machine, when one is controlling one's response (slowing the heart rate and breathing, and relaxing the muscles). Eventually, one learns to control one's responses without the machine. Biofeedback will come to be used to treat migraine headaches, anxiety and panic attacks, and hypertension.          PSYCH

1960s     Encounter groups, meant to help emotionally well people achieve high-level mental health, begin to reach their peak in American society. By discussing the meaning of life on a profound level, these groups try to cultivate self-actualization.          PSYCH

1960s     Family therapy, the treatment of entire families, develops as a result of theories claiming that many mental illnesses are caused by abnormal family communication patterns.          PSYCH

1960s     Stanford psychiatrist and scientist Kenneth Colby develops a computer program simulating a neurotic individual. Colby's artificial neurotic will be followed by a more sophisticated artificial paranoid Colby will name Perry. Both programs are attempts to use computers to study the structure of mental illness and create a dynamic model that trainee analysts can practice on.          PSYCH

1960s     American neurophysiologist David H. Hubel and Swedish neurobiologist Torsten Wiesel find in experiments that a patch placed over one eye of a kitten during its critical period of neural growth leads to permanent blindness in that eye. The blindness is not reversible because, while the one eye is patched, inputs arriving from the unpatched eye take over the visual cortex's allotment for the patched eye. The two scientists will share the 1981 Nobel Prize for physiology or medicine with American neuropsychologist Roger W. Sperry (*see* 1967, PSYCH).          PSYCH

1960s     Many political scientists react against behaviorism (*see* c. 1945, SOC), which they view as having placed an excessive emphasis on methodology and the preservation of the status quo. They call instead for more emphasis on contemporary problems and human values.          SOC

1960s     The emerging field of cognitive anthropology seeks to understand the structure of cultures as systems of knowledge.          SOC

1960s     By the early part of this decade, radar is in use for such civilian purposes as air traffic control, weather forecasting, and police procedures.          TECH

1960s     Japanese scientists working for Sony perfect the Chromotron color television tube and develop the tunnel, or Esaki, diode, which is able to conduct electrons faster than an ordinary transistor. *See also* 1973, PHYS.          TECH

1960s     Late in the decade, Nils Nilsson, Bertram Raphael, and their colleagues at the Stanford Research Institute develop a robot they name Shakey that is able to distinguish boxes from pyramids and follow simple instructions.          TECH

1960      At L'Anse aux Meadows in north Newfoundland, Helge Ingstad and George Decker rediscover a Viking settlement dating to the 11th century, indicating that Vikings settled in North America several centuries before Columbus.          ARCH

1960        American astronomer Frank Drake organizes Project Ozma, a 400-hour radio
            search for extraterrestrial life that yields negative results.                    ASTRO

1960        American astronomer Allen Sandage identifies several starlike objects emitting
            radio waves, the first being 3C48. Maarten Schmidt will show in 1963 that
            these objects are quasars.                                                        ASTRO

1960        American zoologists Kenneth Norris and John Prescott determine that marine
            mammals (in this case dolphins) use echolocation to find the range and direc-
            tion of objects in the water.                                                     BIO

1960        On May 16, University of California–Berkeley biochemists A. Tsugita and
            Heinz Fraenkel-Conrat describe their discovery of the first definite link
            between a mutation, or change in inheritance code, and an alteration in the
            molecule manufactured according to that code.                                     BIO

1960        *Tiros I*, launched by the United States, is the first weather satellite. It is
            equipped to take thousands of photographs of Earth and its cloud cover and
            transmit them back.                                                               EARTH

1960        American geologist Harry H. Hess proposes the concept of sea-floor spreading,
            a key idea in plate tectonics. Hess suggests that new crust forms at rifts, espe-
            cially in the sea floor, where lithospheric plates move apart. *See also* 1912, EARTH
            EARTH

## Chaos Theory

In 1961, Massachusetts Institute of Technology meteorologist Edward N. Lorenz was
forced to pause while running a lengthy computer calculation of weather patterns. He
didn't want to start over from scratch, so he saved some of his intermediate results and,
when he came back, had the computer begin from that new starting point. He then discovered that the final results were quite different from those he had gotten earlier by
running the same calculation uninterrupted.

Searching for the source of the discrepancy, Lorenz found that the computer had
rounded off the figures slightly differently when saving them than when using them
continuously. Although this discrepancy affected only the eighth decimal place in the
original numbers, it was enough to cause enormous differences in the final results.
Lorenz had found that weather systems are highly sensitive to initial conditions. They
are, in short, chaotic systems. The weather in New York on the third Sunday of next
December cannot be predicted, because it depends on the initial conditions around the
globe today—and those conditions cannot be known with complete accuracy.

Since Lorenz's discovery, chaos theory—the study of chaotic systems, using nonlinear equations that involve several variables—has been applied not only to weather but
to turbulent flow, planetary dynamics, electrical oscillations, and many other areas.

| c. 1960 | The "new math," an educational system that constructs mathematical relationships from set theory, is introduced in American public schools.   MATH |

| 1960 | The General Conference of Weights and Measures sets a new standard for the meter: 1,650,763.73 wavelengths of the spectral line of a certain isotope of krypton.   MISC |

| c. 1960 | Fossil bones of primitive Triassic saurischians found in the Ischigualasto beds of Argentina appear to be those of the oldest dinosaurs known at the time.   PALEO |

| 1960 | An international team of geologists discovers dinosaur footprints on the Arctic island of West Spitzbergen (Svalbard). The tracks will be identified in 1961 as those of the early Cretaceous dinosaur the *Iguanodon*.   PALEO |

| 1960 | Scientists use the Mössbauer effect (*see* 1958, PHYS) to test the general theory of relativity (*see* 1916, PHYS). Monochromatic gamma rays fired from the top of a building at a crystal at its base prove to increase in wavelength, owing to the stronger gravitational field at the bottom, an effect predicted by the theory of general relativity. German physicist Rudolf L. Mössbauer for whom the effect was named, will share next year's Nobel Prize for physics with American physicist Robert Hofstadter.   PHYS |

| 1960 | American physicist Luis W. Alvarez, who will win the 1968 Nobel Prize for physics, discovers resonance particles, which exist for so short a time ($10^{-24}$ second) that they can be regarded as the excited state of more stable particles.   PHYS |

| 1960 | Polish-American chemist Leo Sternback discovers a drug that will be marketed as Librium, a benzodiazepine for the treatment of anxiety and tension.   PSYCH |

| 1960 | American physicist Theodore Harold Maiman invents the laser (light amplification by stimulated emission of radiation), a device that produces an intense beam of coherent light. The laser will have many applications in physics research, industry, electronics, and surgery.   TECH |

| 1960 | The first electronic wristwatch, the Accutron, is developed by Bulova. It operates with a tuning fork that vibrates 360 times per second.   TECH |

| 1960 | Aluminum cans come into use in the United States as containers for soft drinks and food products.   TECH |

| 1961 | Riding aboard the *Vostok 1*, Soviet cosmonaut Yuri A. Gagarin becomes on April 12 the first human to reach outer space and the first to orbit Earth. ASTRO |

| 1961 | American astronaut Alan B. Shepard Jr. becomes the first American in space. His *Freedom 7* space capsule makes a 15-minute suborbital flight on May 5.   ASTRO |

1961    U.S. president John F. Kennedy promises on May 21 to send a man to the moon and back by the end of the decade. *See* 1969, ASTRO.          ASTRO

1961    Virgil I. Grissom becomes the second American in space, during a suborbital flight aboard the *Liberty Bell 7* on July 21.          ASTRO

1961    Soviet cosmonaut Gherman S. Titov, aboard the *Vostok 2*, becomes the second Soviet in space on August 7 and the first human to spend more than a day in space, completing 17 Earth orbits during his 25½-hour flight.          ASTRO

1961    Biochemists at the Oak Ridge National Laboratory in Tennessee observe in a test tube for the first time the genetic process by which proteins are synthesized.          BIO

1961    American physicist Albert Ghiorso and his colleagues discover the element lawrencium.          CHEM

1961    American meteorologist Edward N. Lorenz begins to develop the mathematics that will become chaos theory.          MATH

1961    Chinese mathematician Hua Luogeng applies operations research to wheat-harvesting problems.          MATH

1961    The rubella (German measles) virus is identified and isolated. A live vaccine for long-lasting immunity from it will become available within the next decade.          MED

1961    American physicist Murray Gell-Mann develops a classification system for the elementary particles called hadrons (*see* 1962, PHYS), categorizing them in families according to properties that vary regularly in value. He calls his system the Eightfold Way. Israeli physicist Yuval Ne'emen independently develops a similar system around the same time.          PHYS

1961    British physicist Jeffrey Goldstone formulates what will become known as Goldstone's theorem, which predicts the existence of a spin-zero massless particle called a Goldstone boson in certain situations of symmetry.          PHYS

1961    Soviet military scientists set a record, which will remain unbroken, for the largest nuclear explosion, testing a 58-megaton weapon.          TECH

1962    The unmanned U.S. spacecraft *Mariner 2* completes the first flyby of Venus, transmitting pictures back to Earth.          ASTRO

1962    Circling Earth three times on February 20 aboard the *Friendship 7*, John Glenn becomes the first American to orbit the planet.          ASTRO

1962    American author and scientist Rachel Carson publishes *Silent Spring*, an alarming and revealing glimpse into how chemicals in the environment damage ecosystems.          BIO

1962        American chemist Linus C. Pauling and Austrian-born French biochemist Émile Zuckerkandl suggest that changes in genetic material can be used as a kind of biological clock to date the time one species separated from another.
                                                                                    **BIO**

1962        British-born Canadian chemist Neil Bartlett combines the noble gas xenon with platinum fluoride to produce xenon fluoroplatinate in the first known case of a noble gas bonding with another element to form a compound. **CHEM**

1962        Ukrainian academician Victor M. Glushkov becomes director of the Institute of Cybernetics of the Ukrainian Academy of Sciences. He and the institute are important in developing the theory and application of computers and informatics in the Soviet Union.                                       **MATH**

1962        Japanese physicians introduce the first flexible fiber-optic endoscope, a device consisting of a tube and an optical system for seeing inside a hollow organ or body cavity.                                                          **MED**

1962        In Britain the first beta-adrenergic blocking agent (a drug used mostly to treat heart disorders) is developed.                                          **MED**

1962        American chemist Linus C. Pauling, winner of the 1954 Nobel Prize for chemistry, is awarded the 1962 Nobel Peace Prize for his crusade against nuclear weapons. The 1963 nuclear test ban treaty will be based in part on a draft written by Pauling. In subsequent years Pauling will become known for his staunch advocacy of megadoses of vitamin C.                        **MISC**

1962        In South Africa, paleontologists A. W. Crompton and Alan J. Charig report the discovery of the oldest known ornithischian dinosaurs, dating from the late Triassic.                                                               **PALEO**

1962        British physicist Brian D. Josephson predicts what will become known as the Josephson effects, a group of electrical results that occur at low temperatures when two superconducting materials are separated by a thin layer of insulation. He will share the 1973 Nobel Prize for physics with Japanese physicist Leo Esaki (*see* 1973, PHYS) and Norwegian-American physicist Ivar Giaever.                    **PHYS**

1962        German-born British physicist Heinz London develops a technique for inducing very low temperatures with a mixture of helium-3 and helium-4. With this and other methods, temperatures of a millionth of a degree above absolute zero will eventually be obtained.                                      **PHYS**

1962        American physicists Leon Max Lederman, Melvin Schwartz, and Jack Steinberger confirm the existence of two types of neutrinos, one associated with the muon, one with the electron. Scientists will later infer the existence of a third type, associated with the tauon (*see* c. 1974, PHYS). For their discovery, Lederman, Schwartz, and Steinberger will share the 1988 Nobel Prize for physics.                                                            **PHYS**

1962      L. B. Okun coins the term *hadron* to describe the class of particles including protons and neutrons that are affected by the strong force. *See also* 1954, PHYS.

PHYS

1962      German psychiatrist Karl Leonard uses for the first time the term *bipolar disorder* to describe manic-depressive psychosis. *See also* 1854, PSYCH.      PSYCH

1962      American behaviorist Abraham Maslow publishes *Toward a Psychology of Being*. In this work and his 1954 *Motivation and Personality*, Maslow describes two basic types of human motivation: deficiency motivation (the need for shelter, food, and water) and growth motivation (the striving for knowledge and self-actualization).      PSYCH

1962      British anthropologist Victor Turner publishes his *Forest of Symbols*, a seminal work on African ritual and symbolism.      SOC

1962      The United States launches *Telstar I*, the first commercial communications satellite, which provides television and voice communications between the United States and Europe. *See also* 1945, ASTRO.      TECH

1962      Diet-Rite Cola becomes the first low-calorie soda with a sugar substitute to be sold nationally. This beverage is sweetened with cyclamate.      TECH

1962      The Aluminum Corp. of America helps to develop a can with discardable pull tabs, an innovation test marketed in Virginia with Iron City Beer.      TECH

1963      The radio telescope at Arecibo, Puerto Rico, is completed, with a dish 305 meters (1,000 feet) in diameter.      ASTRO

1963      Dutch-American astronomer Maarten Schmidt discovers the first quasar when he identifies the large red shift of object 3C273, a very distant extragalactic radio source receding at great speed. Hong-Yee Chiu will coin the term *quasar* to describe it the following year.      ASTRO

1963      The American satellite *Syncom 2* is the first satellite to be launched into a geosynchronous orbit, stationary above a given longitude on Earth.      ASTRO

1963      Scientists detect hydroxyl groups (combinations of one hydrogen and one oxygen atom) in space, providing the first evidence that interstellar space contains matter in forms other than individual atoms.      ASTRO

1963      Astronaut L. Gordon Cooper becomes the first American to spend more than a day in space, orbiting Earth 22 times on May 15–16 in the last flight of the Project Mercury.      ASTRO

1963      Riding aboard the *Vostok 6*, Soviet cosmonaut Valentina V. Tereshkova becomes on June 16 the first woman in space. On June 19 she completes the last of 48 Earth orbits achieved during a 71-hour flight.      ASTRO

1963            In September, Columbia University geneticist Ruth Sager reports finding the genetic system called nonchromosomal inheritance. This system involves genes but follows different rules from the chromosomal system, including the fact that nonchromosomal genes do not seem to mutate as chromosomal ones do, that they appear to be transmitted to the offspring by the female only, that the two systems have different sorting times, and that the systems produce different numbers of possible kinds of progeny.                    BIO

1963            Geologists discover the phenomenon of periodic magnetic reversal in the earth's crust, evident from the pattern of alternating magnetic polarity in the ocean floor near mid-ocean rifts. The discovery lends support to the theory of sea-floor spreading and plate tectonics (*see* 1960, EARTH).                    EARTH

1963            American mathematician Paul J. Cohen shows that German mathematician Georg Cantor's continuum hypothesis concerning transfinite numbers (*see* c. 1895, MATH) is neither consistent nor inconsistent with the axioms of set theory. MATH

1963            Valium (diazepam), marketed now by Roche Laboratory, will quickly become the most widely used tranquilizer in the world.                    PSYCH

1963            The word *psychedelic* is first used. It originally means mind manifesting but will soon become associated with drug intoxication and visual hallucinations.    PSYCH

1963            Researchers A. Carlson and M. Lindquist are the first to propose the dopamine hypothesis, that the neurotransmitter dopamine is linked to schizophrenia. It will become one of the most researched biochemical theories of schizophrenia since the illness was identified.                    PSYCH

1963            After studying child development with Jean Piaget in Geneva, artificial intelligence (AI) specialist Seymour Papert begins working with Marvin Minsky at the Massachusetts Institute of Technology (MIT). Papert will develop the AI program LOGO as a language for helping children develop their problem-solving skills. He will also hypothesize that many steps in mental growth are based less on the acquisition of new skills than on gaining new ways to administer already established abilities, a concept that becomes known as Papert's principle.                    PSYCH

1963            MIT computer scientist Joseph Weizenbaum writes the AI program Eliza, to parody a Rogerian psychoanalyst's noncommittal questioning style. Weizenbaum then is appalled at how attached some of Eliza's "psych patients" become to their mentor. In his 1976 book *Computer Power and Human Reason*, he denounces artificial intelligence and questions why people become willing to accept machines as all-knowing and all-powerful.                    PSYCH

1963            The father of artificial intelligence, John McCarthy, leaves MIT, where he had developed Lisp, the most popular AI language in the United States, to establish another major AI laboratory at Stanford University.                    PSYCH

| | |
|---|---|
| 1963 | The first commercial nuclear reactor, Jersey Central Power's Oyster Creek facility, is opened. **TECH** |
| 1963 | The electronic transistorized telephone service called Touch-Tone is marketed in Pennsylvania by AT&T. **TECH** |
| 1964–1974 | At Liujiaxia, China, archaeologists discover the stone Buddhist sculptures of the Bingling Temple Grottoes and several prehistoric sites, including Jijiachuan and Qinweijia. **ARCH** |
| 1964 | The U.S. *Ranger 7* spacecraft transmits more than 4,000 photographs of the moon's surface before crashing. **ASTRO** |
| 1964 | The U.S. unmanned spacecraft *Mariner 4* completes the first flyby of Mars. **ASTRO** |
| 1964 | In May, German-American physicist Arno A. Penzias and American astronomer Robert W. Wilson (both of whom who will share the 1978 Nobel Prize for physics with Soviet physicist Pyotr Kapitza) detect radio-wave background radiation indicating an average temperature of the universe of 3 K (degrees Kelvin) and corroborating the Big Bang theory (*see* 1949, **ASTRO**). **ASTRO** |
| 1964 | Three Soviets aboard the *Voskhod 1* are the first humans to ride as a team aboard a single capsule. **ASTRO** |
| 1964 | Egyptian-born British biologist and geneticist William Donald Hamilton writes on the genetic evolution of social behavior, claiming that the traits of social species like ants and bees can be explained as a mechanism designed to transmit genes. This concept will later develop into the field of sociobiology. **BIO** |
| 1964 | American microbiologists Keith Porter and Thomas F. Roth discover, embedded in the cell membrane of an egg cell, the first cell receptors. **BIO** |
| 1964 | Australian geneticist Pamela Abel, working with German geneticist T. A. Trautner at the University of Cologne, reports on June 14 that evidence has been found showing the genetic code of life to be universally the same in all living things. Abel and Trautner report taking genes from one organism and making them work in the environment of another, completely alien, organism. **BIO** |
| c. 1964 | American biochemist R. Bruce Merrifield invents a simplified technique for synthesizing proteins and peptides. Later automated, this method will become useful in gene synthesis in the 1980s. **CHEM** |
| 1964 | An international research program is developed to take advantage of a period of minimal solar activity known as "the year of the quiet sun." **EARTH** |
| c. 1964 | American physician Stanley Dudrick introduces total parenteral nutrition (TPN), an intravenous feeding system that meets the total caloric needs of a patient unable to eat or drink normally. **MED** |

1964            British anthropologist Louis Leakey and his colleagues announce the discovery of fossils of *Homo habilis* in the Olduvai Gorge of what will become Tanzania. The earliest known member of the genus *Homo*, this first hominid species to make stone tools is believed to have lived as early as 2.5 million years ago.                                                                PALEO

1964            American physicist Murray Gell-Mann further develops his Eightfold Way of classifying hadrons (*see* 1961, PHYS) by reference to more fundamental particles he calls quarks, an allusion to James Joyce's novel *Finnegans Wake*. A few kinds of quarks and their oppositely charged counterparts, antiquarks, interact to form all the many varieties of hadrons, a class of particles that includes protons and neutrons. Quarks have fractional rather than whole electric charges.                                                                PHYS

1964            A particle with a strangeness number of –2 is discovered. Its properties correspond precisely to those that Gell-Mann had predicted for an empty spot in his classification system called the Eightfold Way. This discovery lends credibility to Gell-Mann's quark theory.                                                          PHYS

1964            American physicists Val L. Fitch and James W. Cronin, who will share the 1980 Nobel Prize for physics, disprove the accepted belief that CP (charge conjugation and parity, two properties of particles) is always conserved when they discover that neutral kaons occasionally violate CP conservation. The CPT theorem, which will become generally accepted, adds a third characteristic, time (T), to the symmetry.                                                          PHYS

1964            American physicists Sheldon L. Glashow and James D. Bjorken propose the existence of quarks possessing a property they call charm. The hypothesis will be confirmed by fellow American physicists Samuel C. C. Ting and Burton Richter in 1974.                                                                    PHYS

1964            British physicist Peter Higgs predicts the existence of a spin-zero particle with a nonzero mass, later called the Higgs boson.                                    PHYS

1964            In Japan, a high-speed passenger train service, Shinkansen, is inaugurated in the Tokyo-Osaka corridor. Efforts to develop a "maglev," or "magnetic levitation," railroad system begin at about the same time.                                TECH

1965            Astronomers Herbert Friedman, Edward Byram, and Talbot Chubb discover an intense X-ray source in the constellation Cygnus, which becomes known as Cygnus X-1. *See* 1970, ASTRO.                                                          ASTRO

1965            Cosmic masers are discovered. These are interstellar gas clouds whose intense radio emission lines indicate that their molecules are being pumped to more highly excited levels by the radiation of nearby stars.                          ASTRO

1965            Astronomers at the Arecibo Observatory in Puerto Rico discover the retrograde rotation of Venus and the rotation of Mercury.                                    ASTRO

1965        On March 18, Soviet cosmonaut Alexei A. Leonov becomes the first human to "walk" in space, leaving his spacecraft, the *Voshhod 2*, while in orbit.        ASTRO

1965        The first American two-person crew in space, Virgil I. Grissom and John W. Young, carries out the first in-orbit maneuvers of a manned spacecraft, the *Gemini 3*, during a June 3–7 mission.        ASTRO

1965        Edward H. White II becomes the first American to walk in space and the first human to use a personal propulsion pack during a space walk. His 36-minute EVA (extravehicular activity) takes place outside his *Gemini 4* spacecraft during a June 3–7 mission.        ASTRO

1965        Coming within one foot of each other on December 15, the *Gemini 6A* and the *Gemini 7* become the first manned spacecraft to rendezvous in space.        ASTRO

1965        In April, the United States launches SNAP-10A (systems for nuclear auxiliary power), the first and only American nuclear reactor to be placed in orbit. In contrast, the Soviet Union will launch 33 reactors into space from 1968 to 1988, most of them to power spy satellites. *See* 1978, ASTRO.        ASTRO

1965        Chinese biochemists synthesize biologically active insulin.        BIO

1965        It is discovered that algae chloroplasts have their own DNA.        BIO

1965        French biochemist Jacques Monod wins the Nobel Prize for physiology or medicine for his work with François Jacob and André Lwoff on the regulatory activities of genes. Monod is considered the discoverer of the operon system that controls bacteria gene action.        BIO

1965        Scientists now suspect the existence of "hotspots," junctures at tectonic plates through which heat leaks up into the ocean, and hotspots will in fact be detected in the early 1970s.        EARTH

1965        Chinese mathematician Feng Kang publishes a general convergence theory justifying the finite element method. His publication remains largely unknown in the West, where the same type of theory is independently developed.        MATH

1965        Using computer power, Hugh C. Williams and his colleagues discover the first complete solution to the "cattle of the sun" problem posed by Archimedes in the third century B.C. The solution has in excess of 200,000 digits.        MATH

1965        American paleontologist Elso Sterrenberg Barghoorn discovers the first microfossils, the fossilized remains of ancient single-celled organisms dating as far back as 3.5 billion years.        PALEO

1965        Japanese physicist Yoichiro Nambu and his colleagues develop the concept of color charge, a property of quarks.        PHYS

1965        Psychologists Ronald Melzack and Patrick Ward develop the gate control the-
            ory of pain, which holds that selective brain processes increase or decrease sen-
            sitivity to pain.                                                           PSYCH

1965        Austrian-born psychoanalyst Anna Freud, the daughter of Sigmund Freud,
            publishes *Normality and Pathology in Childhood*, a cumulation of her theories
            on child psychotherapy and the prevention of mental illness.               PSYCH

1965        American pediatrician and psychoanalyst Donald Winnicott emphasizes that
            infancy is a critical time in human development that is relevant to later
            psychopathology.                                                           PSYCH

1965        American consumer advocate Ralph Nader exposes the safety defects of
            American automobiles, particularly the Chevrolet Corvair, in his book *Unsafe
            at Any Speed*.                                                             TECH

1965        On November 9, one of the biggest electrical blackouts in history occurs when
            a faulty relay in a Canadian power plant leads to a loss of electricity in New
            York City and much of the northeastern United States and southern Canada. TECH

1966        The U.S. *ESSA I* satellite becomes the first weather satellite capable of viewing
            the whole Earth.                                                          ASTRO

1966        On February 3 the Soviet unmanned spacecraft *Luna 9* makes the first soft
            landing on the moon. *See also* 1969, ASTRO.                              ASTRO

1966        The Soviet unmanned spacecraft *Luna 10* completes the first orbit of the moon.
            *See also* 1968, ASTRO.                                                   ASTRO

1966        The *Gemini 8* docks with an unmanned target vehicle on March 16–17 in the
            first space docking.                                                      ASTRO

1966        On June 2 the unmanned spacecraft *Surveyor 1* becomes the first U.S. vehicle
            to land on the moon. It transmits photographs of the lunar surface for six
            weeks.                                                                    ASTRO

1966        Austrian ethologist Konrad Lorenz publishes his controversial book *On
            Aggression*. Lorenz, who pioneered the study of animal behavior (ethology),
            argues that animals—including humans—inherit many of their behavioral pat-
            terns such as aggression and maternal bonding. In this book he argues that the
            impersonal weapons of war have allowed humans to develop an unnatural
            level of aggression. He will share the 1973 Nobel Prize for physiology or med-
            icine with Austrian-born German zoologist Karl von Frisch (*see* 1920, BIO) and
            Dutch-born British zoologist Niko Tinbergen.                               BIO

1966        American writer Robert Ardrey publishes *The Territorial Imperative*, which
            argues that humans, like other animals, are driven by territoriality.      BIO

1966      Chinese seismologists succeed in forecasting the Xingtai earthquake this year, based on study of tilt and crustal deformation. In 1975, the Haicheng earthquake will also be predicted.      EARTH

1966      Chinese mathematician Chen Jingrun makes progress on proving the Goldbach conjecture when he proves that every sufficiently large even number can be expressed as the sum of a prime and a second number which is either a prime or a product of two primes.      MATH

1966      In the United States, the first antiviral drug to block influenza infections, amantadine hydrochloride, is licensed.      MED

1966–1969      Chinese leader Mao Zedong leads the Cultural Revolution, a violent movement to purge the nation of bourgeois influences and restore Communist purity. Universities are closed and scientific societies and journals suspended. The study of pure science and mathematics in China is discouraged in favor of research considered to yield immediate practical benefits.      MISC

1966      British psychiatrist Gordon Allen German sets up the first east-central African academic psychiatry unit, in Uganda. He will show that mental disorders are as prevalent in developing nations as in industrialized ones.      PSYCH

1966      British-American chemist and psychologist Raymond Cattell publishes his *Handbook of Multivariate Experimental Psychology*. This research is the practical application of the Cattell 16PF personality inventory, one of the major personality tests used in North America.      PSYCH

1966      Americans William Howell Masters, a gynecologist, and Virginia Johnson, a psychologist, publish the first of their reports on the psychology, physiology, and anatomy of human sexual activity, *Human Sexual Response*. They will devise methods of sex therapy after investigating human sexuality via the electroencephalograph (EEG), the electrocardiogram (ECG), and motion picture cameras in a laboratory setting.      PSYCH

1966      A partial meltdown occurs on October 6 at the Fermi nuclear reactor near Detroit, when a metal plate comes loose and blocks the cooling water.      TECH

1967      British astronomer Jocelyn Bell discovers the first pulsar, in the constellation Vulpecula. This object emitting intense, regular radio-wave pulses will turn out to be a rapidly rotating neutron star.      ASTRO

1967      The unmanned Soviet *Venera 4* spacecraft is the first to enter the atmosphere of Venus.      ASTRO

1967      The *Venera 4* parachutes a probe into the atmosphere of Venus, which is discovered to be composed mostly of carbon dioxide. The U.S. spacecraft the *Mariner 5* flies by Venus on the following day.      ASTRO

1967        The first deaths of U.S. astronauts in the line of duty take place on January 27, when a flash fire in the *Apollo 1* space capsule during a test at Cape Kennedy, Florida, kills Virgil I. Grissom, Edward H. White II, and Roger Chaffee.    ASTRO

1967        The first human death during a space mission occurs when the Soviet *Soyuz 1* spacecraft crashes on April 24 during reentry, killing cosmonaut Vladimir M. Komarov.    ASTRO

1967        American entomologist and sociobiologist Edward O. Wilson and his colleague R. H. MacArthur publish *The Theory of Island Biogeography*, marking the beginning of a school of ecology that focuses on biogeographical equilibrium, or balanced and stable ecosystems.    BIO

1967        Austrian-born British molecular biologist Max F. Perutz, who shared the 1962 Nobel Prize for chemisty with British biochemist John C. Kendrew (*see* 1957, BIO) and his colleague Hilary Muirhead build the first high-resolution model of the atomic structure of oxyhemoglobin.    BIO

1967        British biologist John B. Gurden is the first to successfully clone a vertebrate, in this instance a South African clawed frog, using the technique of nuclear transplantation.    BIO

1967        American geneticist Sewall Wright receives the National Medal of Science for his work in genetic studies and evolution research. He originated the mathematical theory of evolution, which argues that mathematical chance, as well as mutation and natural selection, affect evolutionary change.    BIO

1967        American biochemists at the Roswell Park Memorial Institute in Buffalo, New York, discover the complex structure of a protein enzyme called ribonuclease that breaks down ribonucleic acid (RNA). Since ribonuclease exerts control over cell growth, the discovery of its structure is thought to help explain why cancer cells spread.    BIO

1967        American molecular biologists Walter Gilbert, Benno Müller-Hill, and Mark Ptashne isolate and identify for the first time two of the cell substances believed to control the process of making genes either operational or dormant.    BIO

1967        On March 18, the U.S. oil tanker *Torrey Canyon* is grounded off the coast of Cornwall, England, creating an oil spill that damages 120 miles of British and French coastline.    EARTH

1967        American cardiovascular surgeon René Favaloro develops the coronary artery bypass operation to graft on additional blood vessels in the heart to get around narrowed or obstructed arteries. It will come into general use as a treatment for coronary artery disease.    MED

1967        Marburg virus is discovered in Germany in a shipment of African monkeys. The microbe is a filovirus, a member of a hitherto unknown family of filament-like viruses that can readily elude the immune system. *See also* 1976, MED.

MED

1967    American biochemist Maurice Hilleman develops a live-virus vaccine against mumps.

MED

1967    By now fluoridation—the addition of fluoride to the water supply with the aim of combating tooth decay—has been widely adopted throughout the United States. *See also* 1945, MED.

MED

1967    On December 3, South African cardiovascular surgeon Christiaan Barnard performs the first human heart transplant, in Cape Town. His patient, Louis Washkansky, lives for 18 days before succumbing to postoperative pneumonia.

MED

1967    The United Nations World Health Organization (WHO) begins a worldwide vaccination campaign to eradicate smallpox. In 1980, after several years with no known cases, the disease will be declared vanquished. Smallpox will be the first disease to be eliminated by vaccination.

MED

1967    American anthropologist Elwyn Simons discovers the skull of the primate *Aegyptopithecus*, which at 30 million years of age is the oldest known ancestor in the line leading to humans.

PALEO

1967    American psychologist Aaron T. Beck designs the Beck Depression Inventory (BDI), a test to measure the depth of a person's depression.

PSYCH

1967    American neuropsychologist Roger W. Sperry, at the California Institute of Technology (Caltech), reports on his research concerning the split brain. This type of radical surgery on patients with severe seizures involves severing the corpus callosum, a network of fibers that connects the brain's two hemispheres. As a result, each hemisphere operates in isolation. This research will lead to further studies of how each hemisphere specializes in processing information. Sperry will share the 1981 Nobel Prize for physiology or medicine with American neurophysiologist David H. Hubel and Swedish neurobiologist Torsten Wiesel (*see* 1960s, PSYCH).

PSYCH

1967    The Amana Refrigeration Co. introduces the first small microwave oven in the United States for home use.

TECH

1968    The Soviet unmanned spacecraft *Zond 5* becomes the first to return to Earth after orbiting the moon.

ASTRO

1968    Astronomers at Green Bank, West Virginia, discover a pulsar or neutron star in the Crab Nebula, thus corroborating the Baade-Zwicky theory (*see* 1934, ASTRO) that neutron stars form in the aftermath of supernovae. *See also* 1054, ASTRO.    ASTRO

1968    Using the U.S. Third Orbiting Solar Observatory, astronomers discover gamma radiation emanating from the center of the Milky Way.

ASTRO

1968        Scientists detect water and ammonia molecules in interstellar clouds, thereby
            showing that complex compounds can form in space.                              ASTRO

1968        The U.S. spacecraft *Pioneer 9*, launched on November 8, will achieve an orbit
            around the sun and return solar-radiation data to Earth.                       ASTRO

1968        During the *Apollo 8* mission (December 21–27), American astronauts Frank
            Borman, James A. Lovell Jr., and William A. Anders are the first humans to
            orbit the moon and the first to see its dark side, the one never visible from
            Earth.                                                                         ASTRO

1968        The U.S. House of Representatives declares Lake Erie a "dead" lake, due to its
            pollution levels.                                                              BIO

1968        In August, University of Illinois chemical geneticist Sol Speigelman announces
            at the 12th International Congress of Genetics that he has developed the first
            method of observing evolution in a test tube. It will allow scientists not only
            to observe but to manipulate molecular events associated with evolutionary
            change under controlled laboratory conditions.                                 BIO

1968        The U.S. government declares Bikini Island, the former site of nuclear bomb
            tests, to be "safe" and its displaced inhabitants return. Ten years later, howev-
            er, the medical hazards of nuclear fallout are reassessed and the inhabitants
            again removed.                                                                 MED

1968        American cardiovascular surgeons Charles Dotter and Melvin Judkins intro-
            duce angioplasty, the widening or unblocking of a blood vessel or heart valve
            by using a balloon catheter.                                                   MED

1968        In Japan, the first research institute opens in the Tsukuba Science City, a
            planned center for scientific research.                                        MISC

1968        Iranian scholar Seyyed Hossein Nasr publishes *Science and Civilization in Islam*.
            In Muslim countries, the book stimulates a renewed interest in science as a
            component of Islamic civilization.                                             MISC

1968        American paleontologist Robert Bakker proposes that dinosaurs were warm-
            blooded and highly active, not cold-blooded and sluggish, as had been previ-
            ously believed. Bakker's ideas will gradually gain adherents as well as
            detractors.                                                                    PALEO

1968        American physicists trap neutrinos emanating from the sun in an under-
            ground tank in South Dakota, but the quantities collected are only one-third
            that predicted by solar theory.                                                PHYS

1968        American physicists Steven Weinberg and Sheldon L. Glashow, with Pakistani
            physicist Abdus Salam, propose the electroweak theory, which gives a unified
            description of the electromagnetic and weak interactions. The three scientists
            will share the 1979 Nobel Prize for physics.                                   PHYS

1968        American behaviorist B. F. Skinner writes about his technique of programmed
            instruction in *The Technology of Teaching*, which presents ordered information
            to students, each bit of which must be understood before the student can pro-
            ceed. Many teaching machines will then be designed to incorporate Skinner's
            ideas.                                                                    PSYCH

1968        The three-digit emergency telephone number 911 is first used, in New York
            State. Over the next two decades it will become widely instituted across the
            country. *See also* 1937, TECH.                                          TECH

1968        The Jacuzzi whirling bath is demonstrated in California by Jacuzzi Bros., a
            farm pump manufacturer.                                                  TECH

1969        Astronomers Thomas Gold and Franco Pacini develop their theory that pulsars
            are neutron stars rotating at a rapid rate.                              ASTRO

1969        Japanese geologists discover meteorites on the Antarctic ice cap.        ASTRO

1969        On January 14–15, the Soviet *Soyuz 4* and *Soyuz 5* spacecraft are the first
            manned vehicles to dock in space.                                        ASTRO

1969        In a dress rehearsal (May 18–26) for the moon landing, the American *Apollo 10*
            lunar lander descends to within 50,000 feet of the moon's surface.       ASTRO

1969        At 10:56 P.M. E.D.T. on July 20 in the course of the July 16–24 *Apollo 11* mis-
            sion, American astronaut Neil A. Armstrong becomes the first human to set
            foot on the moon. Edwin E. "Buzz" Aldrin Jr. follows him onto the surface
            while Michael Collins orbits in the command module.                      ASTRO

1969        On October 11–13, the Soviet spacecraft *Soyuz 6, 7,* and *8* orbit simultaneous-
            ly, in the first triple launch of manned spacecraft.                     ASTRO

1969        During the November 14–24 mission of the U.S. spacecraft *Apollo 12* com-
            pletes the second manned lunar landing, as Charles Conrad Jr. and Alan L.
            Bean become the third and fourth men on the moon.                        ASTRO

1969        The modern five-kingdom classification of living things is by now firmly estab-
            lished. Organisms are grouped into the kingdoms of Monera or Prokaryotae
            (bacteria), Protista or Protoctista (algae, protozoans, slime molds), Fungi,
            Plantae, and Animalia.                                                   BIO

1969        Exploratory researchers at Merck Laboratories in Rahway, New Jersey, and
            Rockefeller University in New York City announce independently that they
            have synthesized the enzyme ribonuclease for the first time.             BIO

1969        Harvard University research scientists report the isolation of a gene from
            an organism.                                                            BIO

1969        American physicist Albert Ghiorso and his colleagues discover element 104, rutherfordium.                                                    CHEM

1969        The American underwater laboratory Tektite I houses scientists studying the physiological and psychological reactions of humans to a hostile, isolated environment.                                                              EARTH

1969        The U.S. oil tanker *Manhattan* becomes the first commercial ship to navigate the Northwest Passage.                                           EARTH

1969–1970   In a voyage of about two months, Norwegian explorer Thor Heyerdahl sails across the Atlantic from Morocco to Barbados in the *Ra II*, a reed boat made in the ancient Egyptian fashion to demonstrate that such a voyage was possible in antiquity.                                                        EARTH

1969        American experimental psychologist John Bowlby extends his earlier studies of childhood attachment and loss to the hospital setting. He identifies the separation stages of protest, despair, and detachment that occur when a child is hospitalized. His findings will lead to related research on the human grief process in death and dying.                                                        PSYCH

1969        The Internet has its origins in the Advanced Research Project Agency Net (ARPANET), a computer communications network founded by the U.S. Department of Defense.                                                       TECH

1969        Chemical and biological warfare materials are banned from production in the United States by President Richard Nixon.                          TECH

1970s       A series of Soviet *Venera* spacecraft—*Venera 8* (1972), *9* and *10* (1975), and *11* and *12* (1978)—study the surface and atmosphere of the planet Venus.   ASTRO

1970s       American astronomer John A. Eddy, following up on the work of British astronomer Edward W. Maunder (*see* 1893, ASTRO), discovers that there have been several periods of very low sunspot activity throughout history, called Maunder minima. One such period occurred in 1645–1715, another in 1400–1510; both were also periods of extreme cold.                    ASTRO

1970s       The cloning of plants from protoplasts becomes an active area of research. BIO

1970s       Superovulation (hormone-induced excess) and embryo transfer become routine in the U.S. cattle industry. These steps increase cattle production by thousands of calves per year.                                             BIO

1970s       At General Electric, biochemist Ananda Chakrabarty develops oil-eating bacteria. This new strain will be the subject of a 1980 U.S. Supreme Court ruling stating that "a live, human-made microorganism is patentable subject matter."  BIO

1970s       American scientists at the Brookhaven National Laboratory on Long Island build a cell that contains both plant and animal cells.                 BIO

1970s        Between now and the 1980s, biochemical systematics advance to reveal that a range of common animals once thought to be a single species is in fact complexes of several different species, based on DNA or protein differences.    BIO

1970s–1980s   American biochemists Sidney Altman and Thomas R. Cech discover independently that RNA is not just a passive carrier of genetic information but can process such information, actively promote chemical reactions, and even reproduce itself. In 1989 Altman and Cech will share the Nobel Prize for chemistry    BIO

1970s        American researchers succeed in teaching sign language to two primates, Washoe the chimpanzee and Koko the gorilla. Scientists disagree whether the results prove that such animals are capable of authentic language or have merely undergone conditioning.    BIO

1970s        Several new schools of tectonics emerge in China, including those of Chen Guoda, Huang Jijing, and Zhang Wenyou.    EARTH

1970s        In Czechoslovakia, mathematicians resolve theoretical questions related to fourth-generation computers and develop algorithms for dynamic and final optimization.    MATH

1970s        Interest in the ancient Chinese medical practice of acupuncture grows in the West. In China, research focuses on the use of acupuncture as anesthesia.    MED

1970s        Australian immunologist Peter C. Doherty and Swiss immunologist Rolf M. Zinkernagel discover how the immune system identifies cells infected with viruses. They will share the 1996 Nobel Prize for physiology or medicine for their research.    MED

1970s        American cardiologists Meyer Friedman and Ray Rosenman are the first to identify a behavior pattern known as Type A, characterized by impatience, a rapid pace, and trying to do too many things at one time. It is suspected at the time that Type A behavior leads to cardiac risk and mental stress.    PSYCH

1970s        Assertiveness training in group settings is used to enhance individual social skills and self-concept. Its basis is the belief that when people react passively to others it can make them feel mistreated and used.    PSYCH

1970s–1990s   Psychological self-help groups become widespread. The common therapeutic factors of these groups are helping others, discussing a shared experience, forming support networks, sharing information, gaining feedback, and learning special methods of coping.    PSYCH

1970s        The basic action of the neurotransmitter called GABA (gamma-aminobutyric acid) is worked out. Unlike most other transmitters, GABA works as an inhibitor and, along with serotonin (the transmitter involved in sleep and sensory perception), helps keep the mind from running amok.    PSYCH

1970s        British psychologist Lawrence Weiskrantz studies blindsight, a visual phenom-
             enon occurring in people who develop visual-field gaps following brain
             injuries. Blindsight allows people to identify objects in their blind areas with-
             out their being aware of it, because they have vision they do not know they
             have. Weiskrantz's experiments suggest that different aspects of vision are sep-
             arately processed and that vision itself is processed separately from awareness. PSYCH

1970         American archaeologist J. M. Adovasio claims to have discovered human
             remains at Meadowcroft, Pennsylvania, dating from 19,000 years ago, some
             7,500 years earlier than previously known sites (*see* 1952, ARCH). Claims of sim-
             ilar antiquity will be made for other sites in the Americas but their dates will
             be disputed, leading in the 1990s to an unresolved controversy about the date
             of the first human migration to the Americas. Some will place it at about
             12,000 years ago, others as early as 35,000 years ago. *See also* 1997, ARCH. ARCH

1970         A satellite observatory is launched to locate and study celestial X-ray sources.
             Information from the satellite provides evidence that Cygnus X-1, discovered in
             1965, is the first known black hole.                                    ASTRO

1970         The Soviet *Venera 7* spacecraft becomes the first to land on the surface of
             Venus.                                                                   ASTRO

1970         British physicist Stephen Hawking suggests that black holes may evaporate
             over long periods as they gradually release subatomic particles.          ASTRO

1970         Large reflecting telescopes are completed at Kitt Peak, Arizona, and Mauna
             Kea, Hawaii. A 100-meter (328-foot) radio telescope is completed at Bonn,
             Germany.                                                                 ASTRO

1970         The Chinese and Japanese launch their first artificial satellites.        ASTRO

1970         Sri Lanka–born American biochemist Cyril Ponnamperuma discovers several
             kinds of amino acids in a meteorite, showing that amino acids have been
             formed beyond Earth.                                                     ASTRO

1970         The U.S. *Apollo 13* lunar mission (April 11–17) is aborted when an oxygen tank
             malfunctions. After several tense days, the astronauts return safely to Earth.
                                                                                      ASTRO

1970         The so-called telomere hypothesis is proposed as an explanation for certain
             characteristics of cellular aging. This argument states that a small amount of
             DNA from telomeres (chromosomal ends) is lost each time DNA replicates
             itself, and these accumulated deletions eventually result in cellular senescence. BIO

1970         Molecular biologists at the University of California–Berkeley report fusing
             together two separate genes inside bacteria to form a single enzyme-producing
             gene that performs the functions of both genes. They claim this is an impor-
             tant clue as to how evolution occurred at the most basic level of molecular
             activity.                                                                BIO

1970        The human growth hormone is synthesized.                                    CHEM

1970        American physicist Albert Ghiorso and his colleagues discover element 105, hahnium.                                                                      CHEM

1970        The first Earth Day is celebrated in San Francisco, California, on March 21, the first day of spring (vernal equinox). Elsewhere the event is celebrated on April 22, 1970. In subsequent years, international Earth Day celebrations will be variously held on both March 21 and April 22, in conjunction with a world-wide "green movement" expressing concern about environmental damage.
EARTH

1970        Cyclones originating in the Bay of Bengal hit the low-lying islands and coasts of Bangladesh on November 13, killing an estimated 200,000 to 500,000 people.                                                                            EARTH

1970        Following the Cultural Revolution (*see* 1966–1969, MISC), medical schools begin to reopen in China, but physicians are now expected to be trained in three rather than five or six years.                                        MED

1970        French anthropologist Louis Dumont publishes *Homo Hierarchicus*, an influential study of the caste system in India.                                SOC

c. 1970     Researchers develop techniques to use fine glass fibers to conduct light, which can be modulated to carry pulses of information. This technology, called fiber optics, will revolutionize communication in coming decades, replacing copper wires with cheaper, less bulky glass fibers.                          TECH

1970        British-born American physicist Albert Victor Crewe invents the scanning electron microscope.                                                           TECH

1970        The Concorde supersonic jet airplane reaches speeds of two times the speed of sound.                                                                       TECH

1970        American scientist Ted Hoff, working for Intel, invents the microprocessor, a silicon chip containing the central processor of a computer. The versatile chip will lead to the proliferation of small, inexpensive computers for home and business use. Intel microprocessors will be marketed commercially for the first time in 1971.                                                              TECH

1971        Italian astronomer Paolo Maffei discovers Maffei One and Maffei Two, member galaxies of the Local Group.                                           ASTRO

1971        Orbiting Mars, the U.S. unmanned *Mariner 9* spacecraft becomes the first such vehicle to orbit another planet.                                          ASTRO

1971        American physicist Irwin Ira Shapiro discovers what are now known as super-liminal sources—components of quasars that appear to be moving away from each other faster than the speed of light.                              ASTRO

1971        British physicist Stephen Hawking proposes the existence of mini black holes formed when the universe was created. He suggests that these might be detected by a final explosive evaporation that would take place only now, after 15 billion years of slow evaporation.                                                    ASTRO

1971        *Apollo 14*, launched on January 31 and returning to Earth on February 9, completes the third successful manned lunar landing, the first after the near-disaster of *Apollo 13* in 1970.                                                    ASTRO

1971        On April 19, the Soviet Union launches the *Salyut 1*, the first Earth-orbiting space station. *Soyuz 11* cosmonauts occupy the station for 23 days (June 7–29) but are killed on June 30 by loss of air while returning to Earth.                                                    ASTRO

1971        U.S. spacecraft *Apollo 15*, launched on July 26 and returning to Earth on August 7, completes the fourth successful manned lunar landing. Its Lunar Rover becomes the first wheeled vehicle to ride on the moon's surface.                                                    ASTRO

1971        American entomologist and sociobiologist Edward O. Wilson publishes his fundamental work *The Insect Societies*.                                                    BIO

1971        British ethologist Jane Goodall publishes *In the Shadow of Man*, an account of her years observing the chimpanzees of Gombe Stream National Park, Tanzania.                                                    BIO

1971        Chinese plant breeders are the first to apply the haploid breeding procedure to wheat. The methodology will be used widely for many kinds of crops, including rice, maize, and eggplant.                                                    BIO

1971        The use of the insecticide DDT (dichloro-diphenyltrichloroethane) is banned in the United States for all but essential uses because of its being linked with severe bird and animal birth anomalies.                                                    BIO

1971        Thousands of gallons of dioxin waste from an herbicide factory spill on the roads of Times Beach, Missouri, causing a major toxic chemical disaster. By 1983 the town's entire population will have been evacuated.                                                    EARTH

1971        British paleontologist Harry Whittington begins a major reexamination of Charles Walcott's interpretation of the Cambrian fossils of the Burgess Shale (*see* 1909, PALEO). Over the next two decades, Whittington and colleagues Simon Conway Morris and Derek Briggs will propose that the Burgess Shale fauna include many phyla (basic body plans) that are now extinct, in contradiction to Walcott's view that these fauna were early examples of present-day phyla. *See also* 1989, BIO.                                                    PALEO

1971        American behaviorist B. F. Skinner publishes *Beyond Freedom and Dignity*, repeating his belief that behavioral free will is an illusion. In it he argues that when humans are behaving freely they are free only from negative reinforcement; their behavior is still dependent on positive reinforcement from their past and is always shaped by the expected consequences.                                                    PSYCH

1971    The National Railroad Passenger Corp., also known as Amtrak, is appointed by Congress to assume all U.S. passenger train business to stem the decline of private passenger rail service as railroads face increased competition from planes, buses, and cars.    TECH

1971    Chicago's Union Stockyards close, ending the city's century-old role in meat production.    TECH

1972    The U.S. spacecraft *Apollo 16*, launched on April 16, completes the fifth manned lunar landing, returning to Earth on April 27.    ASTRO

1972    The U.S. spacecraft *Apollo 17*, launched on December 7, completes the sixth manned lunar landing, returning to Earth on December 9.    ASTRO

1972    The Marine Protection, Research, and Sanctuaries Act, or more familiarly the Ocean Dumping Act, gives the U.S. Environmental Protection Agency and the National Oceanic and Atmospheric Administration important control over waste dumping. It is an early effort to protect delicate oceanic ecosystems.    BIO

1972    The first of a series of U.S. *Landsat* satellites is launched to study the earth, including its mineral and agricultural resources.    EARTH

1972    An earthquake on December 23 in Managua, Nicaragua, destroys much of the center of the capital, killing at least 5,000 people. The Somoza government fails to distribute internationally provided relief supplies.    EARTH

1972    The lumpectomy procedure is introduced for the treatment of breast cancer. Instead of removing the entire breast, as during a mastectomy, this operation removes only the cancerous tissue and leaves the remaining part of the breast intact.    MED

1972    The first computerized axial tomography (CAT or CT) scanner goes into operation. Combining computer and X-ray technology, the device provides physicians with clearer, more detailed information about tissue than X rays alone can. It will be used to study soft organs such as the brain and will have applications outside medicine (*see* 1991, EARTH). The CAT scanner's inventors are British engineer Godfrey Hounsfield and American physicist Allan Cormack, who will share the 1979 Nobel Prize for physiology or medicine.    MED

1972    American paleontologists Niles Eldredge and Stephen Jay Gould publish their theory of punctuated equilibrium, which holds that evolution proceeds through relatively short bursts of rapid change followed by long periods of stasis.    PALEO

1972    American physicist Murray Gell-Mann, winner of the 1969 Nobel Prize for physics, establishes quantum chromodynamics (QCD), a theory describing how quarks combine to form hadrons in terms of a characteristic called color charge.    PHYS

| | |
|---|---|
| 1972 | In October, using laser beams, American physicist Kenneth M. Evenson and his colleagues obtain a new level of precision in measuring the speed of light: 186,282.3959 miles per second.                                                    PHYS |
| 1972 | In his doctoral work at the Massachusetts Institute of Technology, artificial intelligence researcher Terry Winograd develops one of the best-known computer language programs, which he calls SHRDLU. With it he tries to simplify words and make them less likely to be misinterpreted.                  PSYCH |
| 1972 | ARPANET developers (*see* 1969, TECH) introduce electronic mail, or e-mail. TECH |
| 1973 | On May 14 the U.S. launches *Skylab*, its first orbiting space station.        ASTRO |
| 1973–1974 | Astronauts aboard the U.S. space station *Skylab* set a new space endurance record of 84 days.                                                                  ASTRO |
| 1973 | The unmanned U.S. spacecraft *Mariner 10* becomes the first to visit the two planets Venus and Mercury, transmitting the first television pictures of Venus and completing the first flyby of Mercury.                                        ASTRO |
| 1973 | The Soviet unmanned spacecraft *Mars 2* and *3* are the first to enter the atmosphere of Mars. The capsules land but stop transmitting shortly after that. *See* 1976, ASTRO.                                                                       ASTRO |
| 1973 | American physicist Edward P. Tryon proposes his theory that the universe originated as a random quantum fluctuation in a vacuum, given the prediction of quantum mechanics that particles can appear and disappear in a vacuum.       ASTRO |
| 1973 | After passing through the asteroid belt, the U.S. unmanned spacecraft *Pioneer 10*, launched March 3, 1972, becomes the first probe to fly past Jupiter, on December 3. In 1986, it will also become the first spacecraft to leave the solar system.                                                                                       ASTRO |
| 1973 | For the first time, a calf is created from a frozen embryo.                    BIO |
| 1973 | Genetic engineering begins when American biochemists Stanley Cohen and Herbert Boyer show that if DNA is broken into fragments and combined with new genes, these genes can be inserted into bacterial cells, where they will reproduce whenever the cells divide in two.                            BIO |
| 1973–1974 | Many industrialized countries receive a rude awakening about energy dependence when an oil embargo from October 1973 to March 1974 by the Organization of Petroleum Exporting Countries (OPEC) touches off energy shortages.                                                                                 EARTH |
| 1973–1974 | Chinese mathematicians Yang Le and Zhang Guanghou contribute to value distribution theory when they establish that the number of deficient values of a function $f$ equals the number of Borel directions for $f$.              MATH |

| | |
|---|---|
| 1973 | American paleontologist John Ostrom argues that birds are direct descendants of dinosaurs, a proposition that will inspire heated debate. **PALEO** |
| 1973 | Pakistani physicist Abdus Salam suggests that a grand unified theory (GUT) attempting to combine the strong, weak, and electromagnetic interactions would imply that protons are slightly unstable and will occasionally decay to positrons and neutrons. The first GUT will be presented in 1974. **PHYS** |
| 1973 | Japanese physicist Leo Esaki, a researcher into crystal rectifiers, shares the Nobel Prize for physics for showing that resistance sometimes decreases with increasing current, a phenomenon he attributes to quantum mechanical tunnelling. He shares the prize with British physicist Brian D. Josephson (*see* 1962, PHYS) and Norwegian-American physicist Ivar Giaever. **PHYS** |
| 1973 | Physicists predict the existence of solitons, which are stable, particle-like, solitary wave states. **PHYS** |
| 1973 | Physicist Paul Musset and his colleagues discover neutral currents in neutrino reactions, a discovery that tends to confirm the electroweak theory. **PHYS** |
| 1973 | Physicist David Politzer theorizes that quarks exhibit asymptotic freedom in that the forces between them become weaker as the distance between them grows shorter, then vanishes entirely when the distance reaches zero. **PHYS** |
| 1973 | Scottish scientists at Aberdeen University isolate endorphins, which, with a structure and action similar to morphine, act as the brain's own opiate. Endorphins will come to be considered natural painkillers. **PSYCH** |
| 1973–1976 | The first examples of neurotransmitters are isolated. The discovery of these molecular substances, which serve as information processors in the brain, will provide neuroscientists with the material to describe and evaluate fundamental aspects of cognition. **PSYCH** |
| 1973 | American anthropologist Clifford Geertz publishes *The Interpretation of Culture*, outlining an influential approach to the study of cultural symbols. **SOC** |
| 1973 | American researchers introduce Ethernet technology, which will become the dominant computer network technology. **TECH** |
| 1973 | American physicists build a continuous-wave laser that can be tuned or modulated. **TECH** |
| 1973 | The Universal Product Code (UPC) system is promoted by U.S. supermarket owners and food producers to speed the process of checking by electronically scanning the price of products. **TECH** |
| 1974 | The U.S. unmanned spacecraft *Pioneer 11* transmits the first close-up color photographs of Jupiter, then travels on to Saturn (*see* 1979, ASTRO). **ASTRO** |

1974    American astronomer Charles Kowal discovers Leda, the 13th known satellite of Jupiter. In 1975, he will find a 14th satellite.                    ASTRO

1974    The Anglo-Australian telescope, jointly owned by Britain and Australia, is commissioned. Operating on Siding Spring Mountain in Australia, it will be vital in studying the eruption of Supernova 1987A (*see* 1987, ASTRO).                    ASTRO

1974    The skeleton structure of a cell, its cytoskeleton, is revealed for the first time by using monoclonal antibodies and fluorescence.                    BIO

1974    Scientists in the United States and the Soviet Union discover element 106, unnilhexium.                    CHEM

1974    American scientists F. Sherwood Rowland and Mario Molina show that chlorofluorocarbons, or CFCs, such as Freon (*see* 1930, CHEM) released from spray cans and refrigeration units can erode the ozone layer in the upper atmosphere, permitting more ultraviolet (UV) radiation to reach the earth's surface. Such an increase in UV rays could raise the incidence of skin cancer and eye cataracts and also disrupt ecosystems by destroying ocean plankton and soil bacteria. Concerned about protecting the ozone layer, the United States will ban the use of CFCs in spray cans. In 1995 Rowland and Molina will share the Nobel Prize for chemistry with Dutch chemist Paul Crutzen for their contributions to the study of the ozone layer.                    EARTH

1974    Chinese seismologists analyze records of earthquakes in China dating back 3,000 years to compile data that can be used in forecasting earthquakes.    EARTH

1974    Belgian mathematician Pierre Deligne proves the last of French mathematician André Weil's conjectures concerning algebraic topology (*see* 1946, MATH). Weil's conjecture is a generalized version of the Riemann hypothesis (*see* 1857, MATH), which remains unconfirmed.                    MATH

1974    Studying Markov processes, Chinese mathematician Hou Zhending finds necessary and sufficient conditions that a given density matrix $Q$ arises from a unique transition matrix $P$ of probabilities.                    MATH

1974    At the CASTAFRICA conference, ministers of 32 African states frame recommendations to stimulate research in science and technology. A second CASTAFRICA conference will follow in 1987.                    MISC

1974    In East Africa, American paleontologist Donald Johanson discovers the partial skeleton of Lucy, an australopithecine dating back more than 3 million years. Lucy would have been three and a half feet tall and walked erect. Her kind is given the species name *Australopithecus afarensis*. This skeleton is the most complete ever found for a hominid of this period.                    PALEO

1974    American physicists Samuel C. C. Ting and Burton Richter independently discover a new subatomic unit called the J-Psi particle that provides evidence for the existence of "charmed" quarks (*see* 1964, PHYS). Two years later Ting and Richter will share the Nobel Prize for physics.                    PHYS

c. 1974    American physicist Martin L. Perl discovers the tau particle or tauon, a type of lepton. In 1995 he and American physicist Frederick Reines (*see* 1956, PHYS) will share the Nobel Prize for physics.                    PHYS

1974    American physicists Sheldon L. Glashow and Howard Georgi set forth the first grand unified theory (GUT), unifying the strong, weak, and electromagnetic forces.                    PHYS

1974    A silicon photovoltaic cell for harnessing solar power is developed by engineer Joseph Lindmayer, the head of Solarex, Inc.                    TECH

1974    A text-editing computer with a cathode-ray tube video screen and its own printer is put on the American market by Vydek.                    TECH

1974    American computer scientists Robert E. Kahn and V. G. Cerf publish their design of a protocol to meet the needs of an open-architecture network environment. Eventually called the Transmission Control Protocol/Internet Protocol (TCP/IP), it will become the standard for the Internet.                    TECH

1974    Japan experiments with nuclear-powered sailing on the nuclear ship *Mutsu*. However, the ship's reactor springs a leak, rousing public protest and putting the project on hold for more than a decade.                    TECH

1974    In Japan, the worldwide oil crisis spurs introduction of the Sunshine program, devoted to developing new forms of energy technology, including solar, geothermal, and hydrogen. A related program, Moonlight, will be launched in 1978 to study energy conservation methods.                    TECH

1975    Farmers in Shensi Province, China, discover the tomb of Chinese emperor Ch'in Shih Huang Ti (*d.* 210 B.C.), which proves notable for 7,500 life-sized terra-cotta human statues placed there as guards for the deceased emperor.    ARCH

1975    Astronomer Alan E. E. Rogers rediscovers the concept of very long baseline interferometry for improving the resolution of radio telescopes, an idea originally proposed, but not implemented, by Roger Jennison in 1953.                    ASTRO

1975    The first docking in space between U.S. and Soviet spacecraft occurs on July 17, as the *Soyuz 19* performs a rendezvous with an *Apollo* spacecraft.                    ASTRO

1975    The Japanese space agency NASDA introduces its first launch vehicle, the N-I rocket, and its first engineering test satellites. During the coming decades, Japan will grow in its spacefaring capacity.                    ASTRO

1975    American gene researchers David Baltimore, Howard Temin, and Renato Dulbecco share the Nobel Prize for physiology or medicine for their work on interactions between tumor viruses and the genetic material of cells.                    BIO

1975        Argentinean-born British geneticist César Milstein announces the use of genet-
            ic cloning to create monoclonal antibodies (MABs). This cloning process
            allows antibodies to be custom-made to neutralize one specific antigen.
            Milstein will share the 1984 Nobel Prize for physiology or medicine with
            British immunologist Niels K. Jerne and German immunologist Georges
            Köhler.                                                                    BIO

1975        American mathematician Benoit Mandelbrot coins the term *fractals* to describe
            irregular mathematical patterns and structures generated by a process that
            involves successive subdivision.                                          MATH

1975        Mitchell J. Feigenbaum discovers what will become known as Feigenbaum's
            number (approximately 4.6692), the ratio that the consecutive differences of
            iterated functions tend to approach.                                      MATH

1975        In Hungary, research on the fusion of hot plasma begins. Hungarian
            researchers contribute to the development of plasma-diagnostic devices and
            the study of surface waves and instabilities in plasma.                   PHYS

# Personal Computers

The first personal computer on the market had no keyboard, no monitor, and no software. It was simply a set of parts that the user assembled and programmed by flicking
the little switches on its front panel. However inauspicious, the Altair computer, first
marketed in 1975, marked the beginning of the personal computer industry.

The Altair was developed by Ed Roberts, owner of MITS, a struggling calculator company in Albuquerque, New Mexico. He knew that in 1971 scientists at Intel, a young
company in northern California's Silicon Valley, had introduced something called a
microprocessor, a silicon chip that contained the central processor of a computer.
Because of its small size and versatility, it could potentially lead to a new generation of
small, inexpensive computers, but the leading computer manufacturers saw no market
for such things.

In the midst of a calculator price war, Roberts was desperate for a new product, so
he built a small computer based on Intel's 8080 microprocessor. Called the Altair and
sold at the ridiculously low price of $500, the computer first appeared on the cover of
*Popular Electronics* magazine in January 1975. Within weeks, Roberts's company could
barely keep up with demand. Altair enthusiasts formed clubs to discuss the product,
write programs, and design add-on devices. As one member of Silicon Valley's
Homebrew Club said, "You read about technological revolutions, the Industrial
Revolution, and here was one of those sort of things happening and I was a part of it."

That member was Stephen Wozniak, who, with Steve Jobs, invented a computer
inspired by the Altair. In 1977 the second version of that machine, the Apple II, became
the first computer on the market to be accessible not just to hobbyists but to the general public. From then on, computers increasingly became a visible part of everyday life.

1975    The Betamax, a home video recorder, is introduced to the market by the Sony Corp. of Japan.      TECH

1975    The first personal computer, the Altair, is put on the market by American inventor Ed Roberts.      TECH

1975    Americans William Henry Gates III and Paul Gardner Allen found Microsoft, which will become the world's most successful manufacturer of computer software.      TECH

1976    The 6-meter (236-inch) reflecting telescope on Mount Semirodriki in the U.S.S.R. becomes the world's largest but remains inoperative, due to technical problems.      ASTRO

1976    The U.S. unmanned spacecraft *Viking 1* and *2* are the first to complete successful landings on Mars (*see* 1973, ASTRO), transmitting back pictures of the planet's surface. The *Viking 1* lands on July 20, *Viking 2* on September 3. Both craft continue to operate for several years, going silent by 1982.      ASTRO

1976    American astronomers discover rings around the planet Uranus.      ASTRO

1976    Astronomers discover a covering of frozen methane on Pluto.      ASTRO

1976    Astronomer Tom Kibble predicts the existence of cosmic strings—very thin, massive objects millions of light-years in length formed by ripples in the universe following the Big Bang. These strings would account for the observed large-scale structure of the universe, including cosmic voids (regions of apparently empty space) and galactic superclusters.      ASTRO

1976    The assertion that four colors are needed to color any map is verified computationally.      MATH

1976    Chinese mathematician Su Buqing, known for his studies in projective differential geometry, applies his research to solving problems in shipbuilding. MATH

1976    Cimetidine (Tagamet) becomes available for the treatment of peptic ulcers. It is the first of many drugs made to block the action of histamine and inhibit gastric secretions. By 1990 it will be the most prescribed drug in the United States.      MED

1976    Chinese researchers perform clinical studies on traditional herbal pharmacology, according to a report by a delegation from the U.S. National Academy of Sciences. Among the drugs under study is trichosanthin, used as an abortifacient since at least the Ming dynasty in the 1500s.      MED

1976    In Cameroon, a national commission is formed to study the role of traditional African medicine in modern health care. Interest in integrating traditional healers and remedies with contemporary medicine is widespread in Africa.      MED

1976            In Zaire and the Sudan, the filovirus Ebola (named for Zaire's Ebola River) emerges for the first time, killing hundreds of people. Ebola infection is characterized by acute hemorrhaging, with mortality rates of 50 percent to 90 percent.                                                                                                MED

1976            American health authorities investigate a severe form of pneumonia after an outbreak of the disease kills 29 American Legion convention attendees in Philadelphia. This so-called Legionnaires' disease is found to be caused by the *Legionella pneumophilia* bacteria, which thrives in a variety of moist conditions.                                                                                            MED

1976            American physician and medical researcher Baruch S. Blumberg is awarded the Nobel Prize for physiology or medicine for his discovery of the hepatitis B virus. He shares the prize with American pediatrician and virologist D. Carleton Gajdusek (*see* 1995, MED).                                            MED

1976            Chinese gynecologists develop the technique of chorionic villus sampling to aid in the early diagnosis of congenital birth defects. This test, involving a tissue sample from the placenta, provides results earlier than an amniocentesis.     MED

1976            The Islamic Solidarity Conference on Science and Technology in Riyadh, Saudi Arabia, urges a revival of Islamic science.                                             MISC

1976            American psychologist Herbert Benson publishes *The Relaxation Response*, concerning the physiological effect at work during therapeutic techniques used to control panic and general anxiety symptoms.                                      PSYCH

1976            Austrian psychologist Bruno Bettelheim publishes *The Uses of Enchantment*, in which he argues that the "evil" in fairy tales is valuable to children in that it can help them recognize and assimilate "good" and "bad" parts of their own psychological makeup.                                                                        PSYCH

1976            American economist Milton Friedman wins the Nobel Memorial Prize in economic sciences. As a leader of the monetarist school, a branch of neoclassical economics that opposes Keynesian government intervention, he supports laissez-faire policies and argues that monetary policy (control of the money supply) is the most important factor in stabilizing the economy.                            SOC

1976            Americans Stephen Wozniak and Steve Jobs design a prototype for a computer that will be the first product of Apple Computer.                             TECH

1976            Facsimile, or fax, machines, which transmit type or images via telephone lines, gain in popularity for office use.                                                TECH

1976            On January 21, Air France and British Airways begin the first regularly scheduled commercial flights of supersonic transports (SSTs). Air France flies from Paris to Rio de Janeiro, British Airways from London to Bahrain.                  TECH

1977    American physicist Alan Guth postulates an inflationary universe, one that underwent exponential expansion after the Big Bang.                    ASTRO

1977    Beginning on December 10, Soviet cosmonauts set a new space endurance record of 96 days aboard the space station *Salyut 6*.                    ASTRO

1977    Examining ancient records of Halley's comet and also of sunspot activity dating back thousands of years, Chinese astronomers provide data on the durability of comets and of the 11-year sunspot cycle.                    ASTRO

1977    Scientists aboard the submersible *Alvin* discover deep ocean vents near the Galapagos Islands, where hot, mineral-laden water spews into the sea. The vents sustain an ecology of sulfur-eating bacteria and other life forms, including large clams and tube worms.                    EARTH

1977    Amendments to the Clean Water Act give the U.S. Environmental Protection Agency more authority to regulate waste discharges into rivers, lakes, and coastal waters as awareness about pollution continues to grow.                    EARTH

1977    American physicist Leon Max Lederman discovers the upsilon particle, which supports the quark theory of baryons.                    PHYS

1977    At a Yale University conference (February 4–6) on behavioral medicine a new branch of medicine formally comes into existence. As an extension of psychomatic medicine, behavioral medicine investigates the psychosocial factors in illness and health. A well-known type of behavioral medicine is biofeedback (*see* 1960s, PSYCH).                    PSYCH

1977    The Apple II computer is marketed by American inventors Stephen Wozniak and Steve Jobs. It is the first personal computer to be accessible not just to hobbyists but to the public at large. *See also* 1975 and 1976, TECH.                    TECH

1977    The first successful human-powered aircraft, the *Gossamer Condor*, invented by American Paul MacCready, is flown three miles on August 23 by Bryan Allen.                    TECH

1978    Chinese archaeologists publish reports of their discoveries of the Dawenkou Culture, a Neolithic culture showing signs of incipient statehood, and of the Erlitou Culture, which may have been the first true state-type culture known to Chinese archaeology.                    ARCH

1978    The Soviet satellite *Cosmos-954*, containing a nuclear reactor, falls to Earth, showering Canada with radioactive debris.                    ASTRO

1978    Vladimir Remek of Czechoslovakia, aboard the Soviet *Soyuz 28*, becomes the first person in space who is not from the United States or the Soviet Union.    ASTRO

1978    American astronomer James Christy discovers Charon, Pluto's only satellite.                    ASTRO

| | |
|---|---|
| 1978 | The first known satellite of an asteroid is discovered orbiting the asteroid Herculina. ASTRO |
| 1978 | The U.S. robotic spacecraft *Pioneer 12* is launched on May 20 on what is originally intended to be a one-year mission to study Venus. Far exceeding expectations, it will send more than 100 gigabits of data, including radar pictures of most of the planet's surface, back to Earth over the next 14 years. *See* 1992, ASTRO. ASTRO |
| 1978 | The *Pioneer 13* U.S. robotic spacecraft, launched in August, carries four probes into the atmosphere of Venus on December 9 while its companion, *Pioneer 12*, relays data back to Earth. ASTRO |
| 1978 | American molecular geneticists Daniel Nathans and Hamilton O. Smith along with Swiss geneticist Werner Arber win the Nobel Prize for physiology or medicine for their discovery of restriction enzymes and the enzymes' application to problems in molecular genetics. BIO |
| 1978 | The genome of the virus SV40 is determined, the first step in working out the human genome, which is a complete single set of chromosomes with its associated genes. BIO |
| 1978 | In England, the world's first successful human pregnancy by in vitro (test tube) fertilization comes to term as Louise Brown is delivered by doctors Patrick Steptoe and Robert Edwards on July 25. BIO |
| 1978 | The United States launches the satellite *Sensat I* to study the earth's oceans. EARTH |
| 1978 | British scientists introduce the Laparoscope, a type of endoscope used to examine the fallopian tubes, appendix, gallbladder, and liver for disease and obstruction. MED |
| 1978 | A National Science Conference marks the beginning of a resurgence of scientific research in China. At the conference, China introduces a "four modernizations" policy, including modernization in science and technology. MISC |
| 1978 | In Montana, American paleontologists John R. Horner and Bob Makela discover the first known nest of baby dinosaurs, indicating that dinosaur babies were cared for by adults. The new species is called the *Maiasaura* ("good mother lizard"). *See also* 1979, PALEO. PALEO |
| 1978 | Using deuterium as fuel, the Princeton Large Torus nuclear fusion reactor attains a temperature of 60 million °F, for 1/20 of a second. In so doing it reaches nearer to the temperature of the sun (100 million °F) than any other reactor, though it is still not a practical source of energy. TECH |
| 1979 | Aboard the *Salyut 6*, Soviet cosmonauts begin a new space endurance record of 175 days. ASTRO |

1979    The U.S. unmanned spacecraft *Voyager 1* and *Voyager 2*, launched in 1977, reach the planet Jupiter, transmitting back to Earth spectacular images and abundant information. Among the discoveries are a ring around the planet, two new moons, and details of the surfaces of Io, Ganymede, Europa, and Callisto, the four moons first observed by Galileo in 1610. Both spacecraft will go on to Saturn and Uranus, in 1980–1981 and 1986. Then *Voyager 2* will go on to Neptune in 1989.                                                      ASTRO

1979    The U.S. unmanned spacecraft *Pioneer 11* becomes the first probe to reach Saturn, where it discovers several new moons and the planet's magnetic field. ASTRO

1979    Japan launches Hakucho, the first X-ray-observing astronomical satellite. ASTRO

1979    British physician Dick Rees, using the nine-banded armadillo as a source of the vaccine organism, discovers the first leprosy vaccine.                          MED

1979    From excess traces of iridium in late Cretaceous rocks, American physicist Luis W. Alvarez theorizes that a large comet or asteroid struck Earth 65 million years ago, raising clouds of dust that reduced the amount of solar radiation penetrating the atmosphere and triggering the mass extinctions of that period. Among the victims of this hypothetical meteorite were the dinosaurs. *See also* 1991, EARTH.                                                          PALEO

1979    From now into the 1980s, American paleontologist John R. Horner, digging in Montana, will discover evidence of *Maiasaura* (*see* 1978, PALEO) colonial nesting grounds and herding behavior. This herd is believed to have comprised 10,000 dinosaurs. Horner will also discover egg clutches laid by hypsilophodontid dinosaurs. His findings will provide new insight into the social behavior of dinosaurs.                                          PALEO

1979    Scientists discover evidence of gluons, the exchange particles that bind quarks together.                                                              PHYS

1979    The World Health Organization publishes its *International Classification of Diseases* (ICD-9-CM), which covers mental diseases and will prove compatible with the U.S. classification system DSM-III. *See* 1952, PSYCH.         PSYCH

1979    The American Psychological Association gives its Distinguished Scientific Award to South African–born psychiatrist Joseph Wolpe, for his work in the systematic desensitization of military patients traumatized by combat.   PSYCH

1979    A partial meltdown occurs on March 28 in Unit 2 of the Three Mile Island nuclear reactor near Harrisburg, Pennsylvania, as a result of design flaws, operator mistakes, and mechanical failure. Some radioactive material is released and nearly 150,000 people are evacuated from the surrounding area.      TECH

1980s   Unmanned flights to Venus in this decade include the Soviet spacecraft *Venera 15* and *16* (1983). In 1986 the American spacecraft *Vega 1* and *2* drop probes on Venus while en route to Halley's comet. *See* 1986, ASTRO.           ASTRO

| | |
|---|---|
| 1980s | Homeobox genes are discovered. Found in all animals, these genes are important in directing the development of an embryo. **BIO** |
| 1980s | In the Sudan and Sahel regions of Africa, new high-yield, drought-resistant, hybrid sorghum seeds are tested. Other plant-breeding experiments aimed at remedying Africa's persistent food shortages involve white corn, peanuts, and cassava. **BIO** |
| 1980s | American animal researchers begin using computer monitoring in animal studies to evaluate physiological and behavioral reactions to different experiences. **BIO** |
| 1980s | Late this decade, researchers at the Cetus Corp. of California develop the polymerase chain reaction (PRC), a genetic engineering technique that uses the enzyme DNA polymerase to make thousands of copies of small samples of genetic material. **BIO** |
| 1980s | The problem of acid rain begins to gain international attention. Acid rain, atmospheric water contaminated with industrial pollutants, causes long-term devastation to the environment. **EARTH** |
| 1980s | In terms of global averages, the 1980s go on record as the warmest decade since recording began in the 19th century. **EARTH** |
| 1980s | New discoveries are made about four-dimensional spaces, by such mathematicians as Mike Freedman, Simon Donaldson, and Clifford Taubes. Freedman finds a way of classifying some of these spaces, Donaldson shows that some are not smooth, and Taubes demonstrates that the infinity of non-smooth four-dimensional spaces is uncountable. **MATH** |
| 1980s | Some employers provide access to mental health programs for their employees on a low-cost or even no-cost basis. These Employee Assistance Programs (EAPs) provide confidential counseling on a short-term basis. After sufficient time has elapsed, it will be shown that EAPs reduce absenteeism and resignations, thus providing a more stable work force. **PSYCH** |
| 1980s | American neurophysiologist Michael M. Merzenich and American psychophysiologist Jon Kaas develop the idea that a "hard-wired" brain circuit—one with fixed neural connections—does not adequately explain phantom pain, the perception of pain from a missing limb. Experiments on a monkey with an amputated finger indicate that a remapping of the cortex takes place in such situations. **PSYCH** |
| 1980s | In African countries such as Ethiopia, Mauritius, and Zimbabwe, biomass—fuel obtained from forestry and crop residues and animal wastes—is increasingly studied as an energy source. **TECH** |
| 1980 | The Very Large Array (VLA) radio telescope in Socorro, New Mexico, opens for business. Its resolution is equivalent to that of a single 17-mile dish. **ASTRO** |

| | |
|---|---|
| 1980 | The quasar 3C273 is observed emitting gamma rays. Scientists then discover a nebulous region around it, suggesting that it may be in the center of a galaxy. ASTRO |
| 1980 | Uwe Fink and others discover a thin atmosphere on Pluto. ASTRO |
| 1980 | Aboard the *Salyut 6*, Soviet cosmonauts begin a new space endurance record of 184 days. ASTRO |
| 1980 | The U.S. spacecraft *Voyager 1* flies by Saturn. With *Voyager 2*, which will arrive there on August 27, 1981, this probe sends back copious information to Earth, discovering two new Saturnian moons in addition to the 12 known ones. ASTRO |
| 1980 | Lymphocyte adhesion molecules are identified, leading to a burst of information on adhesion molecules' structure, expression, and function. This knowledge will prove vital to understanding intercellular interactions as they affect the human immune system and disease processes like AIDS. BIO |
| 1980 | British biochemist Frederick Sanger wins the Nobel Prize for chemistry for his development of methods of determining nucleotide sequences of the nucleic acids DNA and RNA. He shares half the prize with American molecular biologist Walter Gilbert; their sequencing techniques will have widespread application in the field of genetic engineering. The other half goes to American biochemist Paul Berg for his studies of the biochemistry of nucleic acids. BIO |
| 1980 | The U.S. *Magsat* satellite maps Earth's magnetic field. EARTH |
| 1980 | The volcano Mount St. Helens in Washington State erupts on May 18, killing dozens of people. EARTH |
| 1980 | American mathematician Robert Griess Jr. and his colleagues finish a comprehensive classification of finite simple groups, the building blocks of modern algebra. MATH |
| 1980 | Mathematicians Leonard Adleman and Robert Rumely develop a new test for prime numbers. MATH |
| c. 1980 | Apheresis, a new technique of giving blood, is introduced. It allows only a single component like plasma, platelets, or white cells to be taken from circulation, reducing the chances of hepatitis transmission and rejection reactions. MED |
| 1980 | Physicians in Europe and the United States start removing bone marrow in patients getting large doses of radiation during therapy. This marrow is frozen and saved for later reimplantation. MED |
| 1980 | American and Soviet scientists suggest that neutrinos, previously believed to be massless, do have mass, possibly $1/13,000$ that of electrons. PHYS |

| | |
|---|---|
| 1980 | At Stanford University, the undulator, a device to increase the power of synchrotron radiation, is invented. **PHYS** |
| 1980 | German physicist Klaus von Klitzing, who will win the 1985 Nobel Prize for physics, discovers the quantum Hall effect, an observable example of quantum behavior. Named after American physicist Edwin H. Hall, who first observed it in 1880, this effect involves discrete, not continuous, changes in resistance in a plate in a magnetic field at very low temperatures. **PHYS** |
| 1980 | British research psychiatrist T. J. Crow publishes his hypothesis that schizophrenia is a "two-syndrome" disease process, and names two schizophrenia subtypes as Type I and Type II. The first condition has a sudden onset and responds well to antipsychotic medication; the second develops slowly and responds poorly to such drugs. **PSYCH** |
| 1980 | Scientists develop the scanning tunneling microscope, which can produce images of individual atoms on the surface of a material. **TECH** |
| 1980 | American businessman Ted Turner establishes the Cable News Network (CNN). Over the next few years, cable television stations of all kinds will proliferate. **TECH** |
| 1980 | Rollerblades, bootlike skates that each have a row of four wheels, are patented by Canadian hockey player Scott Olsen. **TECH** |
| 1980 | In the first long-distance solar-powered flight, American Janice Brown flies six miles on December 3 in the aircraft *Solar Challenger*. **TECH** |
| 1981 | Stephen Boughn discovers variations of 0.3 percent in directions 90° apart in the cosmic background radiation. **ASTRO** |
| 1981 | Joseph P. Cassinelli discovers R136a, the most massive star yet known, 2,500 times more massive than the sun. **ASTRO** |
| 1981 | William B. Hubbard theorizes that there is a partial ring around Neptune. *See* 1984, **ASTRO**. **ASTRO** |
| 1981 | Hyron Spinrad and John Stauffer discover the most distant galaxies yet known, about 10 billion light-years away. **ASTRO** |
| 1981 | John Stocke discovers narrow-line quasars, which have spectra consisting of narrow emission lines. **ASTRO** |
| 1981 | A. D. Linde and, independently, Andreas Albrecht and Paul Steinhardt develop the theory of the new inflationary universe, building on the ideas of Alan Guth. *See* 1977, **ASTRO**. **ASTRO** |
| 1981 | The U.S. space shuttle *Columbia*, the first spacecraft designed for regular reuse, is launched on April 12 on its first voyage around Earth, with Robert L. |

Crippen and John W. Young as crew. This shuttle is also known as the Space Transportation System (STS).                                                              ASTRO

1981        The U.S. spacecraft *Voyager 2* flies by Saturn. *See* 1980, ASTRO.                    ASTRO

1981        Geneticists in China are the first to successfully clone a fish, the golden carp.     BIO

1981        The entire sequence of nucleotides in the DNA of a mitochondrion, the cell's energy producer, is determined.                                                    BIO

1981        Soviet scientists discover element 107, unnilseptium.                                 CHEM

1981        Scientists develop a technique for producing "glassy," extraordinarily light and strong metal alloys from rapidly cooled molten metal.                            CHEM

1981        The first experimental work in Ocean Acoustic Tomography is conducted by Robert C. Spindel and Peter F. Worcester studying such below-the-surface features of oceans as temperatures and currents.                                            EARTH

1981        British scientists introduce the nuclear magnetic resonator, a diagnostic tool used extensively to study the brain and spinal cord, heart, major blood vessels, joints, eyes, and ears. This technique makes use of magnetic resonance imaging (MRI) to provide images of the body's organs and structures without surgery or radiation.                                                                      MED

1981        A vaccine against serum hepatitis is approved in the United States. It will be in heavy demand well into the next decade.                                           MED

1981        In June, the Centers for Disease Control in Atlanta, Georgia, report unusual cases of pneumocystis pneumonia among homosexual men. The cases will lead to the diagnosis of the new and deadly ailment AIDS (Acquired Immune Deficiency Syndrome). *See* 1982, MED.                                              MED

---

# Super Vision

The world's most powerful microscope is the scanning tunneling microscope (STM) invented at the IBM Zürich laboratory in Switzerland in 1981. With a magnification factor of 100 million, it can resolve to 100th the diameter of a single atom. This device works by holding a fine conducting probe to the surface of a sample. The probe's tip tapers down to a single atom. As electrons tunnel between the sample and the probe, the probe's movement yields a contour map of the surface.

In 1990, IBM scientists in California used an STM to reposition individual xenon atoms on a nickel surface. In the process they succeeded in producing the world's smallest graffiti—the initials IBM spelled out in atoms.

| | |
|---|---|
| 1981 | Archaeologists in northern Spain discover the remains of a Neanderthal religious sanctuary. Its limestone altar and remnants of burnt offerings indicate that Neanderthals practiced religious rituals. PALEO |
| 1981 | In the Awash River Valley of Ethiopia hominid fossil bones are discovered dating from 4 million years ago. PALEO |
| 1981 | American psychologist Eleanor Rosch expands her theory of prototypes and basic level categories, challenging Aristotle's classical theory of categorization and establishing categorization as a subfield of cognitive psychology. PSYCH |
| 1981 | American economist James Tobin wins a Nobel Prize for his studies concerning the impact of financial markets on spending and investment. SOC |
| 1981 | Adam Heller, Barry Miller, and Ferdinand Thiel develop a liquid junction cell that converts up to 11.5 percent of solar energy to electric power. TECH |
| 1981 | The first IBM personal computer, employing the Microsoft operating system MS-DOS, is introduced on August 12. TECH |
| 1982 | The *Mary Rose*, the flagship of King Henry VIII, is raised from the bottom of England's Portsmouth Harbor, where French warships sank it on July 19, 1545. Artifacts found inside include musical instruments, board games, boots, and jerkins. ARCH |
| 1982 | The unmanned Soviet spacecraft *Venera 13* and *14* complete the first successful soft landings on Venus. ASTRO |
| 1982 | Soviet cosmonauts aboard the space station *Salyut 7*, launched on April 19, set a new space endurance record of 211 days. ASTRO |
| 1982 | On the fifth flight of the U.S. space shuttle *Columbia*, its first operational mission, the crew successfully deploys a satellite. ASTRO |
| 1982 | Chinese scientists achieve total synthesis of yeast alanine tRNA, a nucleic acid. BIO |

## The Heaviest Element

The element with the highest atomic number and heaviest atomic mass is provisionally known as unnilennium (Une). Produced by West German scientists on August 29, 1982, its atomic number is 109 (i.e., it has 109 protons in its nucleus) and its atomic mass 266.

Unnilennium is not, however, the most recent element to be discovered. Element 108, provisionally called unniloctium (Uno), was identified in West Germany in 1984 from observations of three atoms. Soviet scientists made a less well substantiated claim to have discovered this element later that year.

1982        Genetically identical twin calves Chris and Becky are born. Their embryo was
            split by Colorado State University researcher Tim Williams when it was about
            a week old; then the two halves were implanted in the uteri of separate cows.
                                                                                    BIO

1982        West German scientists discover element 109, unnilennium.              CHEM

1982        Mathematician Ronald Bracewell introduces a quicker version of the Hartley
            technique to replace the Fourier transform. It becomes known as the Hartley-
            Bracewell algorithm or the Hartley transform.                         MATH

1982        British physician Michael Epstein identifies a herpeslike virus found in a type
            of lymphoma and associated with infectious mononucleosis, the Epstein–Barr
            virus.                                                                  MED

1982        The American pharmaceutical company Eli Lilly markets the first genetically
            engineered human insulin.                                               MED

1982        In August, a fatal immune system disorder transmitted sexually or through
            contaminated blood is termed AIDS (Acquired Immune Deficiency
            Syndrome). At the time of discovery, the highest-risk groups are homosexual
            men and intravenous drug users, but it will spread to other groups. Eroding
            the body's ability to fight disease, AIDS manifests itself through such ailments
            as pneumonia and a form of cancer known as Kaposi's sarcoma (*see* 1872,
            MED). Over the next few years it will become a worldwide epidemic. *See also*
            1984, MED.                                                               MED

1982        On December 2, the first artificial heart is implanted, in the chest of a 62-
            year-old Utah man suffering from heart disease. The operation is a success
            and the patient's discharge is planned, but 92 days after the surgery the
            patient develops a flulike illness and soon dies.                        MED

1982        Physicist Blas Cabrera reports the discovery of a magnetic monopole, a parti-
            cle with a single magnetic pole, as predicted by the grand unified theory.
            However, this discovery is not confirmed by further experiments.        PHYS

1982        Roger Schank of Yale University publishes *Dynamic Memory: A Theory of
            Reminding and Learning in Computers and People*, in which he describes his
            attempts to develop and write an artificial intelligence program capable of
            understanding what it reads and of drawing upon its memory to come to con-
            clusions and answers as humans do.                                     PSYCH

1983        Scientists develop a chemical method of dating objects based on changes
            observable in obsidian.                                                ARCH

1983        The satellite known as IRAS is launched on a 10-month mission to search for
            the infrared radiation that would indicate planet formation around stars
            beyond the sun. IRAS does discover such evidence, around the star Beta

Pictoris, 56 light-years from Earth. Astronomers will come to believe that a disk of gas and dust surrounds this star. *See also* 1985 and 1991, ASTRO.    ASTRO

1983    Sally Ride becomes the first American woman in space, during the second flight of the second space shuttle *Challenger*, launched on June 18. On this mission a satellite is deployed and retrieved.    ASTRO

1983    On its third mission, the space shuttle *Challenger*, launched on August 30, carries the first African-American in space, Guion Bluford Jr.    ASTRO

1983    American geneticist Barbara McClintock is awarded the Nobel Prize for physiology or medicine for her discovery of mobile genes in the chromosomes of plants.    BIO

1983    The world's first artificially made chromosome is created at Harvard University.    BIO

1983    The process of group transfer polymerization (GTP) is introduced.    CHEM

1983    American scientist Carl Sagan and others theorize that a nuclear war could trigger a nuclear winter, in which fusion explosions raise clouds of dust that reduce sunlight enough to cause mass starvation and extinctions. This theory is inspired by American physicist Luis W. Alvarez's hypothesis (*see* 1979, PALEO) that a meteorite collision indirectly exterminated the dinosaurs.    EARTH

1983    American physician John E. Buster reports on an artificial insemination procedure in which a female donor receives the sperm of the prospective father, the sperm fertilizes the donor's ovum, and then the egg is gently washed out of the donor's uterus and implanted in the uterus of the infertile patient. Buster claims a 40 percent success rate for this technique.    MED

1983    In Kenya, paleontologists find a jawbone of *Sivapithecus*, a primate 16 million to 18 million years old.    PALEO

1983    Italian physicist Carlo Rubbia and Dutch physicist Simon van der Meer discover three exchange particles with the mass predicted by the electroweak theory: the two W bosons, one positive ($W^+$) and one negative ($W^-$), and the neutral Z boson ($Z^0$). In the following year they will share the Nobel Prize for physics.    PHYS

1983    The Center for the Study of Language and Information is established at Stanford University, combining resources from such language-related fields as psychology, philosophy, linguistics, and computer science.    PSYCH

1983    French neurologist A. Roch-Lecours discovers that humans are born with two language areas in the brain. The right hemisphere has a certain potential that is lost in adult life as the left begins to dominate early in life, probably within the first year.    PSYCH

1983        Cellular telephones, made by Motorola, are first test-marketed in Chicago. These phones use computers and multiple transmitters to receive and transfer calls.                                                                TECH

1983        Aspartame-based Nutrasweet is used for the first time to sweeten beverages. In the next decade it will be used in a variety of foods, including salad dressings and desserts. *See also* 1958, TECH.                                        TECH

1984        Archaeologists discover Altit-Yam, an underwater site off the coast of Israel that preserves the remains of an 8,000-year-old settlement.                      ARCH

1984        Lindow man is discovered in a peat bog in Germany. This 2,200-year-old preserved body is believed to be that of a Druid victim of human sacrifice.     ARCH

1984        Scientists at the European Southern Observatory, near Santiago, Chile, confirm the existence of a partial ring around Neptune, as suggested by William B. Hubbard in 1981.                                                     ASTRO

1984        Indian astronomer J. C. Bhattacharyya discovers two more rings of Saturn.  ASTRO

1984        On the fourth *Challenger* mission (February 3–7), two American astronauts use jet-propelled backpacks in the first untethered space walks.                   ASTRO

1984        On February 8, Soviet cosmonauts Leonid Kizim, Vladimir Solovyov, and Oleg Atkov begin setting a new space endurance record, spending 237 days (until October 2, 1985) aboard the *Salyut 7*.                                     ASTRO

1984        American astronauts on the fifth *Challenger* mission, launched on April 7, deploy the Long Duration Exposure Facility, an orbiting platform designed for long-range space experiments. On the same mission a disabled satellite is captured, repaired, and redeployed for the first time.                                ASTRO

1984        The *Discovery*, the third space shuttle, is launched August 30 on its first voyage.                                                                      ASTRO

1984        Large-scale biological research begins in private industry with the establishment of the Monsanto Life Sciences Research Center in Missouri built to create new drugs, crop plants, and microbial pesticides.                      BIO

1984        American scientist Allan Wilson at the University of California–Berkeley clones a pair of gene fragments from a preserved pelt of an animal that has been extinct for hundreds of years, the South African quagga, related to the zebra.                                                                BIO

1984        Steen A. Willadsen clones sheep from embryo cells.                          BIO

1984        American geneticists analyzing DNA find that chimpanzees are more closely related to humans than either are to gorillas or other apes; the genetic difference is 1 percent. From this evidence it is deduced that humans and chimpanzees diverged from a common ancestor some 5 million to 6 million years ago.                                                                                    BIO

1984–1985    Severe droughts lead to famine in Ethiopia–Eritrea. The famine is accompanied by epidemic disease and complicated by the refusal of warring factions to allow free passage of aid shipments. More than 1 million people die.      BIO

1984    West German scientists discover element 108, unniloctium.      CHEM

1984    An electrically conductive polymer called MEEP is introduced; it will be applied to the manufacture of lightweight batteries.      CHEM

1984    In Bhopal, India, a leak on December 3 of lethal methyl isocyanate gas from a Union Carbide plant creates a toxic cloud over the city, killing more than 3,500 people and injuring at least 200,000 more.      EARTH

1984    In April, French scientist Luc Montagnier of the Pasteur Institute in Paris and Robert Gallo, a physician with the U.S. National Cancer Institute, announce their discovery of a virus believed to cause AIDS. They call it HTLV-III or HIV, for human immunodeficiency virus. A dispute will arise over who initially discovered the virus, with the French scientist being officially credited in 1987.      MED

1984    Near Lake Turkana, Kenya, British paleontologist Richard Leakey discovers the nearly complete skeleton of Turkana boy, who lived some 1.6 million years ago. He is seen as one of the earliest confirmed specimens of *Homo erectus*, though some classify him as *Homo ergaster*, a closely related species.      PALEO

1984    A fossilized jawbone found by Andrew Hill in Kenya is believed to come from a 5-million-year-old specimen of *Australopithecus afarensis*, among the earliest known hominids.      PALEO

1984    American scientist D. Schechtman and others discover the first quasi crystal, a crystal-like substance that violates the rules for crystal patterns.      PHYS

1984    The first one-megabit random access memory (RAM) chip is developed in the United States by Bell Laboratories. It stores four times as much data as any chip produced to date.      TECH

1985    The luxury ocean liner *Titanic* is located on the ocean bottom in the North Atlantic. Using a remote-controlled camera, French and American oceanographers study this ship, which sank in 1912 after hitting an iceberg.      ARCH

1985    Construction begins on the Keck telescope, the world's largest, on Mauna Kea in Hawaii. The reflecting telescope will be completed in 1990, with a 387-inch (9.82 meter) primary mirror consisting of 36 separate segments.      ASTRO

1985    James R. Houck discovers eight infrared galaxies, located by the IRAS satellite. *See* 1983, ASTRO.      ASTRO

1985        Mark Morris discovers string-shaped radio sources, possibly low-energy cosmic strings (*see* 1976, ASTRO) at the center of the Milky Way.                    ASTRO

1985        Observing an eclipse of Pluto by its satellite Charon, Edward F. Tedesco determines that Pluto's diameter is less than 1,900 miles.                    ASTRO

1985        Neil Turok theorizes that cosmic strings are responsible for the formation of the groups of galaxies called Abell clusters.                    ASTRO

1985        The *Atlantis,* the fourth space shuttle, makes its first flight.                    ASTRO

c. 1985      The U.S. Department of Agriculture announces the success of the first bioinsecticide.                    BIO

1985        Indian-born American botanist Subhash Minocha succeeds in producing clones of a Venus flytrap.                    BIO

1985        American primatologist Dian Fossey is murdered, probably by enemies made while protecting the mountain gorillas of Rwanda's Virunga Mountains from poachers. Her years observing the gorillas, beginning in 1967, were described in her book *Gorillas in the Mist* (1983).                    BIO

1985        American chemists Richard Smalley and Robert F. Curl Jr. and British chemist Harry W. Kroto discover buckminsterfullerene, a third form of pure carbon (in addition to graphite and diamond) composed of hollow, geodesic, spherical molecules of 60 atoms each. Chemists foresee a wide range of applications in industry and medicine for these "buckyballs" and for related forms of pure carbon, collectively known as fullerenes. The molecules are named in honor of American architect R. Buckminster Fuller, designer of the geodesic dome, which was a popular structure, especially in industry and the military, in the 1950s. In 1996 Smalley, Curl, and Kroto will share the Nobel Prize for chemistry after their buckyball theory will have been proved by two other scientists.                    CHEM

1985        American researchers report the discovery of lanxides, substances with characteristics of both metal and ceramics.                    CHEM

---

## Elephant Communication

In 1985, while studying a herd of elephants, Cornell University animal researcher Katherine Payne detected a spasmodic throbbing in the air that seemed to correspond with movements of the elephants' foreheads. Further study with ultrasonic recording equipment showed that the sound and the movements were not coincidental. Elephants in fact communicate by using low-frequency sounds. They can locate each other with this technique even when initially separated by great distances, and male elephants use it to find ready-to-mate females that are miles out of smelling range.

1985    The British Antarctic Expedition detects a "hole" that forms annually in the ozone layer above Antarctica. The opening represents a substantial reduction in the naturally occurring concentration of ozone.                    EARTH

1985    Positron emission tomography (PET) scans are developed to reveal the metabolic activity levels of the brain and heart. These scans show the rate at which abnormal and healthy tissues consume glucose and other biochemicals.    MED

1985    In August, scientists at the National Institutes of Health suggest that other viruses, in addition to HTLV-III (HIV), may cause AIDS. *See* 1984, MED.    MED

1985    In December, researchers at the University of California–San Francisco announce that passage of the AIDS virus is blocked by using condoms.    MED

1985    By now, the Traditional Medicine Research Unit in Tanzania, formed in 1974, has interviewed more than 1,000 traditional African healers and identified 2,300 plant specimens used in treating more than 1,000 diseases.    MED

1985    In an underground test, a nuclear X-ray laser produces X rays 1 million times brighter than previously obtained.    PHYS

1985    In Japan, the JT-60 tokamak, an experimental nuclear fusion reactor, begins operation.    PHYS

1985    Psychiatrist Leopold Bellak proposes that many supposed cases of schizophrenia are misdiagnoses, claiming that as many as 10 percent of the schizophrenia cases diagnosed are really examples of what will be called attention-deficit disorder (ADD).    PSYCH

1985    American neurologist and mathematician Marvin Minsky, one of the founders of artificial intelligence, publishes *The Society of Mind*, in which he argues that intelligence arises from the interaction of a network of simpler processes or agencies that are not in themselves intelligent.    PSYCH

1985    The National Science Foundation forms the NSFNET to allow scientists to connect electronically with NSF supercomputers. It is one of several academic computer networks that sprout into being in the 1970s and 1980s; collectively, they are known as the Internet, the global network of all computers able to share information electronically. By the 1990s, the Internet will include commercial and recreational users as well, with millions of people online.    TECH

1986    Archaeologists in Mexico discover a stone stele from about A.D. 1, inscribed with writing in an undecipherable language.    ARCH

1986    Harold L. Dibble develops a computerized surveying system, based on an electronic theodolite (a surveying device for measuring horizontal and vertical angles), for use in archaeological digs.    ARCH

1986      At Tell Leilan, in northern Iraq (formerly Mesopotamia), archaeologist Harvey Weiss discovers the largest cache of cuneiform tablets since the 1930s. These 1,100 tablets are inscribed in the extinct Semitic language Akkadian and date from the 18th century B.C.      ARCH

1986      Astronomers discover that the Milky Way and its neighboring galaxies in the local supercluster are moving toward a hypothetical Great Attractor, a point in the direction of the Southern Cross. *See* 1990, ASTRO.      ASTRO

1986      The U.S. spacecraft *Voyager 2* flies by Uranus, sending pictures and information back to Earth. Ten new moons of Uranus are discovered.      ASTRO

1986      The U.S. unmanned spacecraft *Pioneer 10*, launched 14 years ago, in 1972, becomes the first spacecraft to leave the region of the solar system where the planets orbit. Along the way it became the first craft to fly past Jupiter, in 1973.      ASTRO

1986      In the worst space flight disaster in history, the space shuttle *Challenger* explodes 73 seconds after takeoff on January 28, killing its crew of seven, including teacher Christa McAuliffe.      ASTRO

1986      On February 20 the Soviet Union launches the *Mir* space station, in which cosmonauts will set new records for continuous habitation in space. *See also* 1997, ASTRO.      ASTRO

1986      Several spacecraft make close approaches to Halley's comet, including the Japanese probes *Suisei* (March 8) and *Sakigake* (March 10) and the European Space Agency probe *Giotto* (March 13).      ASTRO

1986      The U.S. Department of Agriculture grants the first license to market a living organism produced by genetic engineering. Registered to the Biologics Corp. of Omaha, it is a virus to vaccinate against a herpes disease in swine.      BIO

1986      Brooklyn-born American physicist Arthur Ashkin discovers a new method for observing and manipulating biological particles by applying radiation pressure with a laser.      BIO

1986      While researching cancer growth patterns, American scientists discover the first gene known to inhibit cell growth.      BIO

1986      On August 21, in Cameroon, a cloud of toxic gas rises from Lake Nios, which has a volcanic crater. The cloud kills approximately 1,700 people and injures more than 500 more.      EARTH

1986      Mathematician Ramachandran Balasubramanian and his colleagues prove a conjecture by Edward Waring dating from 1770, that every natural number is the sum of, at most, 19 fourth powers.      MATH

1986    Using the radiation pressure of a laser, Arthur Ashkin and associates trap indi-
        vidual living organisms and individual atoms, permitting new methods of
        observation and handling.                                              MISC

1986    American anthropologists Tim White and Donald Johanson discover fossil
        remains, including the first known limb bones, of OH62, a female *Homo
        habilis* dating back 1.8 million years.                                PALEO

1986    A complete frog, dating back 35 million to 40 million years, is found fossilized
        in amber in the Dominican Republic.                                    PALEO

1986    American linguist Joseph H. Greenberg uses the technique of mass compari-
        son of Native American languages to classify them into three groups repre-
        senting successive waves of migration from northeast Asia.             PALEO

1986    Swiss physicist Karl Alexander Müller and German physicist Johannes Georg
        Bednorz discover superconductivity in certain ceramics at temperatures of 30
        K (degrees Kelvin), which is very cold, but warmer than any results obtained
        so far. This discovery leads to further experiments yielding superconductivity
        at still higher temperatures, making practical applications possible. In 1987
        they will share the Nobel Prize for physics.                           PHYS

1986    Working independently, American and German researchers observe individual
        quantum jumps in single atoms for the first time.                      PHYS

1986    Ephraim Fishbach claims to have discovered a fifth fundamental force referred
        to as the hypercharge, detectable in certain subatomic interactions.   PHYS

1986    American biophysicist Michael Phelps and pediatric neurologist Harry
        Chugani use positron emission tomography (PET) scans (*see* 1985, MED) to dis-
        cover that the primary brain metabolic activity in infants five weeks or younger
        occurs in areas of the brain that control the primitive sensory and motor activ-
        ities. The results suggest that newborns have a limited capacity for high-order
        functioning, including thought, a capacity that increases slowly as they grow.
                                                                               PSYCH

1986    Nintendo video games are introduced in the United States by the Japanese
        game-manufacturing company of the same name, founded in 1898. By the
        start of the next decade, U.S. sales of the games will top $3 billion.  TECH

1986    In the worst nuclear accident in history, the Chernobyl 4 reactor in the Soviet
        republic of Ukraine undergoes a meltdown on April 26. Thirty-one firemen
        and plant workers are killed and radioactive fallout covers a wide region. TECH

1986    Americans Dick Rutan and Jeana Yeager complete the first nonstop around-the-
        world flight without refueling. Their airplane, *Voyager*, departs on December 14
        from Edwards Air Force Base in California, to which it returns on December 23.
                                                                               TECH

1987            Further remains of Clovis people named for the New Mexico town near where
               they were first found are identified in Washington State. These prehistoric
               ancestors of Native Americans lived about 11,500 years ago. *See also* 1997, ARCH.

                                                                                          ARCH

1987            American astronomer R. Brent Tully reports the discovery of the Pisces-Cetus
               supercluster complex, the largest known structure in the universe.          ASTRO

1987            American astronomers Benjamin Zuckerman and Eric E. Becklin report the
               discovery of a brown dwarf star in orbit around the white dwarf star Giclas
               29-38.                                                                        ASTRO

1987            On February 24, Canadian astronomer Ian Shelter discovers a supernova in
               the Large Magellanic Cloud, a galaxy near the Milky Way. This Supernova
               1987A is the closest supernova since the one observed by Kepler in 1604.

                                                                                          ASTRO

1987            Aboard the space station *Mir*, Soviet cosmonaut Yuri V. Romanenko sets a new
               record for a single endurance flight—326.5 days, beginning February 8 and
               ending December 29.                                                          ASTRO

1987            The U.S. Supreme Court strikes down a Louisiana law requiring the teaching
               of "creation science" in public schools whenever evolution is taught. This rul-
               ing follows a string of lawsuits pitting evolution against creationism in public
               schools.                                                                      BIO

1987            Plant pathologists Steve Lindow of the United States and Nickolas Panopoulos
               of Greece develop a mutant of a common parasite, *Pseudomonas syringae*. This
               genetically altered bacterium is designed to retard frost formation on plants,
               giving them the ability to withstand some subzero temperatures. The mutant
               bacteria is applied on a field on April 24, marking the first time scientists are
               allowed to release artificially created microbes into the environment.        BIO

1987            In Czechoslovakia, metal-oxide ceramics are developed that are superconduc-
               tive at temperatures as high as 170 K (degrees Kelvin).                      CHEM

1987            Chemists H. Naarmann and N. Theophilou develop a polyacetylene-iodine
               compound that serves as an efficient conductor of electricity.              CHEM

1987            In Montreal, an international agreement is signed in September restricting the
               release of ozone-destroying chlorofluorocarbons (CFCs) into the atmosphere.

                                                                                          EARTH

1987            The Antarctic ozone hole (*see* 1985, EARTH), which appears in late autumn peri-
               odically, is larger in magnitude and duration than in previous years. The layer
               of ozone over Antarctica is less than 50 percent of its 1979 value.        EARTH

1987            Tretinoin (Retin-A), a prescription product used to treat acne for more than a
               decade, is shown to improve the skin's quality and diminish wrinkles.

However, the potential side effects of its use, such as skin irritation, will deter some from using this "youth potion."                                                  MED

1987            Mexican surgeon Ignacio Navarro develops a surgical procedure to treat severe cases of Parkinson's disease. The procedure involves implanting the patient's adrenal tissue into the brain, causing the brain to produce dopamine, the substance deficient in Parkinson's disease.                                           MED

1987            The anti-AIDS drug AZT (azidothymidine, or zidovudine) receives approval from the U.S. Food and Drug Administration.                                        MED

1987            Near the Milk River, in Alberta, Canada, Kevin Aulenback discovers dinosaur eggs containing fossilized unhatched dinosaurs, in only the second such find.  PALEO

1987            Scientists at IBM produce a standing wave called a dark pulse soliton, which propagates through an optical fiber without spreading.                            PHYS

1987            Physicist Michael K. Moe and others determine that selenium-82 has the longest half-life ever recorded for a radioactive substance.                             PHYS

1987            Researchers in Mexico, Europe, and the United States perform the first fetal brain tissue transplant into patients suffering from Parkinson's disease.  PSYCH

1987            The American drug company Eli Lilly introduces the antidepressant Prozac (fluoxetine), which initially appears to have few side effects and will become the most widely prescribed antidepressant in the country. Later, however, it will become the center of a controversy over charges that taking it can lead to suicide, murder, and self-mutilation. Despite these charges, the Food and Drug Administration will not require warning labels on it and Prozac will continue to be prescribed worldwide.                                               PSYCH

1988            The Shroud of Turin, a linen cloth marked with an image of a bearded man and believed by many to be the burial shroud of Jesus, is dated with carbon-14 at about 1300, some 1,300 years too late to have covered Jesus. The evidence now shows that the cloth was woven soon before it was first displayed in France in the 1350s.                                                           ARCH

1988            On Rome's Palatine Hill a wall dating from the seventh century B.C. is discovered, which supports the legendary date of 735 B.C. for the founding of Rome.                                                                             ARCH

1988            Simon J. Lilly identifies a galaxy that is 12 billion light-years away, indicating that it formed 12 billion years ago, early in the universe's history.       ASTRO

1988            In July, the Soviet Union launches *Phobos 1* and *2*, unmanned spacecraft designed to study the Martian moon Phobos, but loses contact with both spacecraft before their missions can be completed.                            ASTRO

1988    The space shuttle *Discovery* is launched on September 29 in the first U.S. manned space mission since the *Challenger* disaster of January 28, 1986.    ASTRO

1988    The development of a method to identify a person from the DNA in a single hair is announced.    BIO

1988    American biochemist Sidney Fox makes proteinlike substances called proteinoids that self-organize in water to form cell-like units known as microspheres that closely resemble ancient protocells.    BIO

1988    Scottish researchers report that some cancers involve the loss of a specific piece of genetic material, bolstering the theory that cancer development is caused by activation of cancer-causing genes and the loss of cancer-controlling genes in the body.    BIO

1988    It is estimated that some 10 million chemical compounds are known to science, with the number growing by approximately 400,000 every year.    CHEM

1988    Government scientist James E. Hansen testifies before the U.S. Senate that he is "99 percent sure that accumulation of greenhouse gases is responsible for global warming trends." His remarks contribute to a growing public sense of urgency on this issue.    EARTH

1988    Scientists will determine that 1988 was the warmest year on record for average temperatures worldwide as thousands of heat-related deaths occur internationally. Links to the greenhouse effect are suspected but cannot be confirmed.    EARTH

1988    A Joint Symposium on Ozone Depletion, Greenhouse Gases, and Climate Change is held at the U.S. National Academy of Sciences.    EARTH

1988    Mathematician Silvio Micali and his colleagues report the development of a method for generating purely random numbers, based on the problem of factoring large numbers that are the products of two large primes.    MATH

1988    In Israel, fossils of early humans are found that bear many characteristics of modern *Homo sapiens*. Dating back 90,000 to 100,000 years, they are more than twice as old as the previously known specimens of modern humans.    PALEO

c. 1988    American psychologist David Sack and colleagues at the National Institute of Mental Health find that partial sleep deprivation can reverse some of the effects of severe depression.    PSYCH

1988    Canadian psychiatrist Colin Ross conducts a major study and discovers that out of 236 persons diagnosed as having multiple personality disorder, 41 percent had previously been diagnosed as schizophrenics. This result represents a trend toward more careful examination of persons presenting symptoms of probable mental illness and it also illustrates growth in the recognition of multiple personality disorder.                    PSYCH

1988    University of California–Irvine psychiatrist Richard Haier reports that high intelligence may be the result of an efficiently organized brain. His experiments with positron emission tomography (PET) scans (*see* 1985, MED) of people taking a series of visual tests show that those who performed best used less energy in the cortical areas of the brain, where abstract reasoning takes place, than those who performed poorly. Haier claims that his experiments show that the high scorers used their brains more efficiently.                    PSYCH

1988    The positron transmission microscope is invented.                    TECH

1989    American archaeologists excavate the Babylonian city of Mashkan-shapir. Built about 1840 B.C., it is one of the oldest known cities.                    ARCH

1989    Astronomers report evidence that a pulsar—an extremely dense, rapidly spinning star—has formed in the debris left over from Supernova 1987A.                    ASTRO

1989    Astronomers theorize that the Andromeda and M32 galaxies have black holes at their centers.                    ASTRO

1989    A whirlpool of rotating gas expelled from the core of the Milky Way, possibly caused by a black hole at the galaxy's center, is discovered.                    ASTRO

1989    While mapping the location of galaxies, astrophysicists discover the largest structure yet known in the universe, a sheet of galaxies that comes to be called the Great Wall.                    ASTRO

1989    Studies of the Saturnian moon Titan indicate that it is not covered by a global ocean, so that in all the solar system Earth remains the only known body with liquid on its surface.                    ASTRO

1989    The U.S. unmanned spacecraft *Magellan* is launched on March 4 from the shuttle *Discovery* in the first instance of a space probe being launched from a shuttle. *See also* 1990, ASTRO.                    ASTRO

1989    In August, the U.S. spacecraft *Voyager 2* becomes the first probe to fly past Neptune. It discovers three new moons, detects volcanic activity on the Neptunian moon Triton, and confirms the existence of partial rings around the planet.                    ASTRO

1989    On October 18 the United States launches the *Galileo* spacecraft, which is scheduled to reach Jupiter by 1995 to study the planet's atmosphere and satellite system. *See also* 1990, ASTRO.                    ASTRO

1989    Col. Vladimir Titov and Musa Manarov set a team endurance flight record aboard the Soviet space station *Mir* on December 21 after 366 days beginning December 21, 1988.                    ASTRO

1989        In his book *Wonderful Life*, Harvard biologist Stephen Jay Gould argues that the "explosion" of life forms (570 million years ago) produced many more basic body plans than exist today. This controversial theory, based on recent reinterpretations of Canada's Burgess Shale fossils (*see* 1971, PALEO), suggests that only a fraction of the phyla then living survived the mass extinction at the end of the Cambrian period.                                                           BIO

1989        The Texas State Board of Education formally votes, for the first time, to require the teaching of evolution in all biology textbooks.                                          BIO

1989        American geneticists Steven Rosenberg, R. Michael Blaese, and W. French Anderson discover human gene transfer.                                                     BIO

1989        The first robot honeybee able to use the waggle dance to communicate with other bees is developed by a team of engineers and entomologists.                     BIO

1989        Scientists determine that dogs have some color vision.                                  BIO

1989        In May, for the first time, American geneticists at the National Institutes of Health inject genetically engineered nonhuman cells into a human patient. These cells will be used to mark and trace other cells in an experimental therapy for skin cancer.                                                                          BIO

1989        Under millions of atmospheres of pressure, hydrogen is converted into a metal-like phase that may be superconducting.                                             CHEM

1989        Scientists from the United States and India find evidence that tectonic plates clashed some 2.5 billion years ago in what is now India's Kolar schist belt, indicating that tectonic processes were under way early in the earth's history.                                                                                      EARTH

1989        Using a computer model, William F. Ruddman and John E. Ketzbach demonstrate that a tectonic uplifting of the Tibetan plateau and the Rocky Mountains caused global cooling and weather patterns that may have set off the recent ice ages.                                                                                          EARTH

1989        American scientists complete an ocean-mapping project that reveals that the Mid-Atlantic Ridge is comprised of a string of 16 spreading centers.              EARTH

1989        The oldest known rock is discovered, dated at 3.96 billion years, soon after the earth's formation 4.6 billion years ago.                                               EARTH

1989        More than 80 nations, including the United States and the 12 nations of the European Community, agree to plans to phase out ozone-destroying chlorofluorocarbons (CFCs) by the year 2000.                                                       EARTH

1989        On March 24, the *Exxon Valdez* oil tanker is the source of a massive oil spill of about 250,000 barrels in Prince William Sound off the Alaskan coast. The spill causes a series of related ecological disasters, including 350,000 to 2.4 million seabirds killed by the spilled oil, according to an estimate in a 1991 study.          EARTH

1989      Using a new algorithm to compute pi, mathematicians extend the calculation of pi to 1 billion digits.          MATH

1989      Researchers Francis Collins of the United States and La-Chee Tsui of Canada discover the gene that causes cystic fibrosis, the most common deadly genetic disease in North America.          MED

1989      Researchers announce that they have pinpointed a set of genes that seem to make some families more susceptible than others to the debilitating nerve disease multiple sclerosis (MS).          MED

1989      American scientists David Goeddel, William Korh, Diane Pennica, and Gordon Behar are named inventors of the year for their invention of t-PA, a drug used to dissolve blood clots in heart attack patients.          MED

1989      Research determines that azidothymidine (AZT) can slow the progression of AIDS in HIV-infected patients who present no symptoms of the disease. However, studies also show that some patients develop viruses resistant to AZT. *See* 1987, MED.          MED

1989      A strain of Ebola virus infects a population of macaque monkeys at a quarantine facility in Reston, Virginia, but does not cause disease in the laboratory workers it infects.          MED

1989      In Culpeper, Virginia, quarry workers uncover the largest set of dinosaur tracks in North America: about 1,000 well-preserved footprints dating from 210 million years ago.          PALEO

1989      Based on evidence from fossil skulls, scientists argue that Neanderthals and anatomically modern humans evolved near each other, possibly interbreeding, in the Near East as long ago as 145,000 years.          PALEO

1989      Researchers at the University of Utah claim to have discovered cold fusion of atomic nuclei, with a resulting release of energy, at room temperature. The experiment is never successfully reproduced and the claim is generally discredited.          PHYS

1989      Japanese physicists report the first experimental confirmation that the sun generates neutrinos.          PHYS

1989      Funding for the construction of the Superconducting Super Collider is approved by the U.S. Congress, which will vote to terminate the unfinished project in 1993.          PHYS

1989      American researchers James W. Tetrud and J. William Langstrom report that they have developed Deprenyl, the first drug shown to delay symptoms of neurological disease. It prevents brain-cell death and slows the progress of Parkinson's disease.                                                                                    PSYCH

1989        The University of Minnesota Press releases a new version of the world's most widely used psychological profile test, the MMPI (*see* 1940s, PSYCH). Criticized for an alleged sexist bias, the test was reviewed and redeveloped to be gender neutral. It also includes more recent psychological disorders such as those relating to drug abuse, eating disorders, and Type A personality.                    PSYCH

1989        University of Texas researchers present the first evidence that stuttering and spasmodic dysphonia (difficulty in speaking) are caused not by emotional disturbance but by biochemical abnormalities in the brain.                                       PSYCH

1989        In Japan, 48 companies co-found the Laboratory for International Fuzzy Engineering Research (LIFE), aimed at researching applications for "fuzzy logic" computer theory.                                                                          TECH

1990s       The United Nations estimates that the population of the lesser-developed and developing countries will rise by almost 3 billion by the year 2020.              BIO

1990        Microscopic study of horses' teeth removed from a site in Ukraine shows that horses were ridden about 6,000 years ago, considerably earlier than according to previous estimates.                                                                     ARCH

1990        Scientists confirm the existence of a concentration of mass called the Great Attractor, which changes the rate at which the Milky Way and nearby galaxies spread apart as the universe expands. *See also* 1986, ASTRO.                       ASTRO

1990        By launching the *Muses-A* satellite into lunar orbit, Japan becomes the third country, after the United States and the Soviet Union, to send a spacecraft to the moon. The flight demonstrates the fuel-saving efficacy of a gravity-assisted "slingshot" approach in reaching the moon.                                             ASTRO

1990        The U.S. unmanned spacecraft *Galileo*, launched October 18, 1989, flies by Venus in February, gaining a gravity assist on its way to Jupiter. *See also* 1991, ASTRO.
                                                                                                  ASTRO

1990        The Hubble Space Telescope, the first telescope intended for permanent Earth orbit, is launched on April 25 aboard the space shuttle *Discovery*. A joint project of the National Aeronautics and Space Administration (NASA) and the European Space Agency, this telescope proves to have obscured vision resulting from technical flaws that another shuttle crew will repair in December 1993.                                                                                        ASTRO

1990        The U.S. unmanned spacecraft *Magellan*, launched May 4, 1989, reaches Venus on August 10. In the next two years it will use radar to penetrate the thick Venusian cloud cover and transmit back detailed maps of most of the planet's surface.                                                                                            ASTRO

| | |
|---|---|
| 1990 | The United States launches the unmanned spacecraft *Ulysses* on October 6 to study the poles of the sun and the interstellar space above and below those poles. The craft is scheduled to approach the sun first in 1994, then again in 1995. *See also* 1992, ASTRO. <div align="right">ASTRO</div> |
| 1990 | Tigers conceived through in vitro fertilization are born, in Omaha, Nebraska. BIO |
| 1990 | Researchers expand the genetic alphabet of four nucleotides by adding two artificial nucleic acids that can be recognized and built into new DNA and RNA molecules by cellular biochemical machinery. <div align="right">BIO</div> |
| 1990 | New evidence indicates that humans are destroying tropical rain forests at a rate faster than previously believed. <div align="right">BIO</div> |
| 1990 | American physicist Donald Huffman and German colleague Wolfgang Krätschmer patent a method for producing buckminsterfullerene molecules (buckyballs) in larger quantities than previously possible. *See* 1985, CHEM. CHEM |
| 1990 | Inexpensive synthetic diamonds are developed that conduct heat 50 percent better than natural ones and withstand 10 times as much laser power. The synthetic diamonds are made from isotopically purified carbon. <div align="right">CHEM</div> |
| 1990 | Scientists make aerogels, very low density solid materials, out of silica and other substances. <div align="right">CHEM</div> |
| 1990 | Scientists from the United States and the Soviet Union discover the first known fresh-water geothermal vents, in the floor of Lake Baikal, Russia. These vents confirm that the lake is a spreading area, a place where new crust is forming. Should this spreading continue, over the course of several hundred million years Asia could split apart and the lake become an ocean between two tectonic plates. <div align="right">EARTH</div> |
| 1990 | Undersea core samples drilled in the Ontong–Java Plateau provide evidence for a superplume, a giant mass of hot material that bursts through to the earth's crust in a relatively short time, in the Cretaceous period. According to current theory, superplumes affect the shape of the crust, tectonic movement, geomagnetism, climate, and the course of evolution. <div align="right">EARTH</div> |
| 1990 | Hundreds of climate experts sign a statement predicting global warming unless the nations of the world act to stop the increase of greenhouse gases in the atmosphere. However, the United States refuses to set specific targets for limiting carbon dioxide emissions. <div align="right">EARTH</div> |
| 1990 | Meteorologists develop new techniques to determine the probability that atmospheric chaos will disrupt any given long-term forecast. <div align="right">EARTH</div> |
| 1990 | In the San Francisco Bay Area, an earthquake measuring 7.1 on the Richter scale kills 67 people and injures more than 3,000 on October 17. <div align="right">EARTH</div> |

1990        A 155-digit Fermat number is factored by two computer scientists, breaking
            existing records.                                                              MATH

1990        The *Journal of the American Medical Association* publishes the results of a study
            linking dopamine receptor genes to alcoholism.                                 MED

1990        The *New England Journal of Medicine* publishes a study supporting the idea that
            genetic factors influence weight gain.                                         MED

1990        Daniell Rudman of the Medical College of Wisconsin (Milwaukee) and col-
            leagues report that treatment with the human growth hormone can reverse
            the physical effects of aging. Mass production of this hormone has been
            made possible by the advent of genetic engineering.                            MED

1990        British physicians Norman Winston and Alan Handyside are the first to
            implant embryos screened in a test tube for genetic defects. Handyside says
            that refining the technique will allow physicians to screen for any genetic dis-
            order. Previously, genetic screening was not possible until the 10th week of
            pregnancy, when an amniocentesis could be performed.                            MED

1990        American rheumatologist Lawrence E. Shellman discovers the gene that caus-
            es osteoarthritis, the most common form of arthritis.                          MED

1990        Several researchers present evidence of a link between very low frequency elec-
            tromagnetic fields and human cancer.                                           MED

1990        Researchers discover that the number of years a person smokes cigarettes is the
            most important risk factor in carotid artery disease.                          MED

1990        Researchers find that a high-fiber diet can help protect against colon cancer.  MED

1990        Secondhand tobacco smoke—smoke inhaled by nonsmokers—is called a
            "known human carcinogen" in a draft Environmental Protection Agency
            report.                                                                        MED

1990        On September 14, the first United States government-approved infusion of geneti-
            cally engineered cells into a human for therapeutic purposes is successfully per-
            formed when a four-year-old girl with an inherited immune disorder (adenosine
            deaminase deficiency) begins receiving monthly injections of genetically engi-
            neered white-blood cells.                                                      MED

1990        In December, the U.S. government approves the first significantly new contra-
            ceptive in 25 years, Norplant. This hormone-releasing system, to be implanted
            in a woman's arm for long-term protection, is introduced as the most effective
            contraceptive on the market. Like any hormone-based contraceptive, it may
            produce side effects.                                                          MED

1990        Scientists present evidence that humans settled Australia as early as 50,000
            years ago.                                                                     PALEO

1990        Scientists analyze DNA from a magnolia leaf that is 20 million years old, the
            oldest genetic material ever tested. It provides insight into the evolution of
            plants.                                                                    PALEO

1990        In Egypt, paleontologists discover the fossil remains of a 40-million-year-old
            whale with feet, providing a clue to the evolution of cetaceans.           PALEO

1990        Scientists report discovering the fossil of *Sinornis*, a 135-million-year-old bird
            from China. More recent and less primitive in appearance than *Archaeopteryx*
            (*see* 1861, PALEO), it is the oldest bird with modern flight features.     PALEO

1990        Using a scanning tunneling microscope, researchers at IBM are able to move
            individual atoms on a surface for the first time.                          PHYS

1990        Physicists use new data on Z particles to develop strict limits concerning the
            number of particle families and refine their estimates of the top quark
            mass.                                                                      PHYS

1990        American neuroscientist Solomon Snyder grows a human brain cell in a lab at
            the Johns Hopkins University School of Medicine. This cell, a neuron from the
            most highly evolved portion of the brain, will allow for more detailed studies
            of the brain.                                                              PSYCH

1990        Psychologists discover that faces usually found to be attractive have features
            that approximate the mathematical average of all faces in the area's popula-
            tion.                                                                      PSYCH

1990        Two separate studies of twins, one on men and another on women, suggest
            that genes may have an important influence on the development of one's sex-
            ual orientation.                                                           PSYCH

1990        "Smart" materials and structures are developed that sense such conditions as
            pressure and temperature, then respond with changes in their properties, such
            as their conductivity or shape.                                            TECH

1991        The 5,000-year-old body of a man preserved in ice is discovered in the Alps
            between Austria and Italy. Dubbed the "iceman," he carries tools that provide
            clues to life in Europe about 3000 B.C.                                    ARCH

1991        American astronomers discover a quasar 12 billion light-years away, the most
            distant object ever identified.                                           ASTRO

1991        The *Magellan* space probe completes its first radar survey of the planet Venus,
            mapping more than 90 percent of its surface, revealing thousands of cracks
            and craters, enormous lava flows, and features indicating quakes and volcanic
            activity.                                                                 ASTRO

1991        British astronomers report indirect evidence for a planet orbiting a distant pul-
            sar or neutron star.                                                      ASTRO

1991    Using the Hubble Space Telescope's spectrograph, astronomers analyze the disk of gas and dust orbiting the star Beta Pictoris (*see* 1983, ASTRO). By late 1992 the Hubble telescope will uncover evidence of similar protoplanetary disks around 15 new stars in the Orion nebula.                    ASTRO

1991    The U.S. Gamma Ray Observatory is launched on April 7 into Earth orbit to study celestial gamma-ray sources, particularly supernovae, quasars, neutron stars, and black holes.                    ASTRO

1991    On October 29 the U.S. unmanned spacecraft *Galileo* takes the first close-up photograph of an asteroid in space when it captures the image of 951 Gaspra from a distance of 10,000 miles.                    ASTRO

1991    The World Resources Institute estimates that the world's forests are being destroyed by deforestation at a rate of 80 acres per minute, or some 40 to 50 million acres a year.                    BIO

1991    At a meeting of the International Union of Biological Sciences in Amsterdam, British mycologist David Hawksworth estimates that the total of fungi types worldwide could be as high as 1.6 million. At the same conference on biodiversity, some biologists claim that humans may share the world with 100 million other species but are rapidly causing hundreds of extinctions via global warming, habitat destruction, and the introduction of species foreign to a region.                    BIO

1991–1992    Under a scanning electron microscope, American geologist John Watterson studies gold grains (placer gold) from Lillian Creek, Alaska, and discovers that the grains are attached to bacteria. Among the possible explanations are that the gold is a chemical residue left after bacterial breakdown in the humic acids of Alaskan soil, or that the gold comes from extracellular enzyme activity. Watterson estimates that a 0.1 millimeter gold grain takes at least a year to grow.                    BIO

1991    The British journal *Nature* publishes a study by British scientists claiming to have discovered the gene on the Y chromosome that determines maleness in mice. When the researchers injected the gene into female mouse embryos, some of the embryos became male.                    BIO

1991    Contrary to the theories of Gregor Mendel (*see* 1866, BIO), geneticists find that genes may behave differently, depending on which parent they were inherited from.                    BIO

1991    In the United States, on September 26, the privately financed *Biosphere 2* project begins. Eight men and women are locked in a sealed structure containing five sample earth environments. Over the next two years they will study the feasibility of sustaining a closed ecology. The project draws much media attention, though its scientific standards are criticized in many quarters.                    BIO

1991        American organic chemist Joel Hawkins uses X-ray diffraction to generate the
            first image of a buckyball molecule (*see* 1985, CHEM), corroborating the exis-
            tence of this form of pure carbon, also called buckminsterfullerene. American
            physicist Arthur Hebard demonstrates that buckyballs doped with potassium
            or rubidium are superconductive.                                            CHEM

1991        The newly revised U.S. Clean Air Act identifies 189 chemicals commonly
            found in the air as toxic. Environmental Defense Fund senior scientist Michael
            Oppenheimer says that more than 150 million Americans live in areas where
            air pollution levels still violate federal health standards.                EARTH

1991        The Environmental Protection Agency estimates that more than 1 million of
            America's 5 million underground storage tanks, most of which hold petrole-
            um products, are currently leaking. The EPA asserts that the damage done to
            groundwater supplies and ecosystems is only beginning to be evaluated. EARTH

1991        Computerized axial tomography (CAT or CT scanning) is applied to studies of
            the earth's interior. On that basis, the International Association of Seismology
            and Physics of the Earth's Interior develops new seismic-wave travel timetables
            that supersede the 1940 Jeffreys and Bullen tables.                         EARTH

1991        American scientists R. C. Capo and D. J. DePaolo report that the ratio of two iso-
            topes of strontium can be used to determine past climates.                  EARTH

1991        The eruption of Mount Pinatubo on Luzon Island in the Philippines is the
            largest volcanic eruption in the 20th century. Within 10 days its cloud spreads
            7,000 miles to reach from Indonesia to Central Africa. The quantities of sulfur
            dioxide, ash, and aerosol material issued by it lead to lower global tempera-
            tures and accelerate the erosion of the ozone layer.                        EARTH

1991        American geologists Eldridge M. Moore and Ian W. D. Dalziel propose that
            Antarctica and the western coast of North America were originally linked 500
            million years ago.                                                          EARTH

1991        American geologists led by Haraldur Sigardsson analyze glass fragments from
            Haiti that confirm the theory (*see* 1979, PALEO) that a large object from outer
            space rammed into Earth 65 million years ago.                               EARTH

1991        The United Nations Intergovernmental Negotiating Committee on Climate
            Change meets in Chantilly, Virginia. Its delegates, from 130 countries, structure
            a treaty to "curb the threat of global warming."                            EARTH

1991        Retreating Iraqi troops set hundreds of Kuwaiti oil wells on fire during the
            Persian Gulf War through February and March. Smoke from the burning wells
            blankets the area, and soot is detected well north of Turkey, as far west as
            Egypt, the Sudan, and Ethiopia, and as far east as India and China. The smoke
            also contributes to a cyclone in Bangladesh that takes more than 100,000
            lives.                                                                      EARTH

1991     Twenty-four countries with interests in Antarctica sign a treaty banning oil
         exploration there for the next 50 years.                                    EARTH

1991     As of this year there have been 13 epidemiological studies linking secondhand
         cigarette smoke and diseases in nonsmoking people exposed to smoke.   MED

1991     British geneticists claim to have located a gene that, when mutated, causes a
         hereditary type of Alzheimer's disease.                                    MED

1991     A characteristic gene mutation is found in people prone to developing colon
         cancer.                                                                     MED

1991     The *New England Journal of Medicine* reports on the longest and largest estro-
         gen-after-menopause investigation, which concludes that the benefits of estro-
         gen replacement, evaluated on an individual basis, outweigh the risk of
         cancer.                                                                     MED

1991     Researchers at Israel's Weizmann Institute of Science say that it may be possi-
         ble to vaccinate against the insulin-dependent form of diabetes after manipu-
         lating T-cells, which has allowed researchers to keep mice from developing dia-
         betes.                                                                      MED

1991     The first woman named to the National Inventors Hall of Fame in Akron,
         Ohio, is Gertrude B. Elion, an American biochemist who helped develop drugs
         to fight leukemia, septic shock, and tissue rejection in patients having kidney
         transplants. In 1988 she was awarded the Nobel Prize for physiology or med-
         icine.                                                                      MED

1991     The American Heart Association recommends limiting meat in the diet, claim-
         ing it is not a necessary food.                                            MED

1991     By now, more than 366,000 cases of AIDS have been reported in 162 countries
         and 10 million adults are believed to be infected with the HIV virus that causes
         AIDS. The World Health Organization estimates that by the year 2000 some 40
         million people will be infected.                                           MED

1991     In Argentina, paleontologist Rodolfa Coria discovers fossil remains of the 100-
         ton sauropod *Argentinosaurus*, the biggest known land animal in history.
                                                                                    PALEO

1991     French diver Henri Cosquer discovers paintings and engravings in an undersea
         cave in the Calanque region of France. The artwork, which includes paintings
         of marine birds called auks, dates from at least two different periods—one
         27,000 years ago, the other 19,000 years ago.                              PALEO

1991     Physicists in England achieve controlled nuclear fusion with a mixture of tri-
         tium and deuterium (two forms of hydrogen) producing almost 2 megawatts
         of power, a new record for experimental fusion reactors. Though this fuel is

more efficient than the pure deutrium previously used, fusion has still not become a commercially viable energy source.          PHYS

1991          The journal *Cell* reports that geneticists have located the gene that causes the most common form of inherited mental retardation, the "Fragile X" syndrome.          PSYCH

1991          American neurobiologist Simon LeVay announces that the brains of homosexual men are structurally different from those of heterosexual men. The affected brain area, a segment of the hypothalamus, is believed to influence male sexual behavior.          PSYCH

1991          American cognitive scientist Daniel C. Dennett publishes *Consciousness Explained*, in which he argues that there is no central, conscious "audience" in the brain, nor a single, unified stream of consciousness. Rather, consciousness consists of multiple drafts composed by neural processes of "content fixation" playing semi-independent roles and generating the illusion of a single, conscious self.          PSYCH

1991          American psychologist Jan Belsky argues that girls who grow up in dangerous environments have a tendency to experience the onset of puberty earlier than other girls. Her controversial explanation is that children in such conditions are encouraged by evolution to have offspring early and often, to increase the chances that some will survive.          SOC

1991          CERN, the European Organization for Nuclear Research, releases the World Wide Web. Developed by Tim Berners-Lee, the Web begins as a networked information project and will develop into the most widely used form of accessing the Internet.          TECH

1992          A 5,000-year-old city, possibly the legendary trade city Ubar, is discovered in the Arabian Desert.          ARCH

1992          European and American astronomers announce the discovery of two black holes, V404 Cygni and Nova Muscae, each orbited rapidly by a star.          ASTRO

1992          European astronomer Mart de Groot demonstrates that the blue supergiant star P Cygni has brightened steadily over the past 300 years, thus providing evidence of stellar evolution.          ASTRO

1992          The Hubble Space Telescope detects the hottest star yet known, the white dwarf NGC 2440. Its surface temperature is 360,000° C, more than 30 times hotter than the sun's.          ASTRO

1992          A cometlike object titled QB1 is found at a distance of 39 to 45 astronomical units (the distance between Earth and the sun), at least as far out as Pluto. It may be the first known comet in the hypothetical Kuiper belt. *See* 1950, ASTRO.          ASTRO

1992    Studies indicate that the Milky Way galaxy, as well as thousands of others, is moving across the sky at 375 kilometers per second.                                        ASTRO

1992    The unmanned U.S. spacecraft *Ulysses* (*see* 1990, ASTRO) flies by Jupiter on February 8, using the planet's gravity to speed it on its way toward the sun. The spacecraft studies Jupiter's magnetic field in the process.                          ASTRO

1992    The Cosmic Background Explorer (COBE) science team reports on April 23 that the U.S. *COBE* satellite, launched in 1989, has discovered small temperature variations or "ripples" in the universe's microwave background radiation. These variations support the Big Bang theory by indicating that there were fluctuations in the density of gas in the early universe soon after the Big Bang. Then, as the universe expanded, density variations like these led to the formation of galaxies, galaxy clusters, and other large-scale structures.          ASTRO

1992    The fifth U.S. space shuttle, *Endeavour*, is launched on May 7 on its first mission, during which astronauts retrieve a satellite stranded in a useless orbit and launch it into the correct one.                                           ASTRO

1992    The U.S. space shuttle *Endeavour*, launched on a subsequent mission on September 12, carries the first female African-American astronaut, Mae C. Jemison, and the first married couple in space, Mark C. Lee and N. Jan Davis.                                                           ASTRO

1992    The *Mars Observer* spacecraft, launched on September 25, is the first U.S. probe sent to Mars since the *Viking* in 1975. The rendezvous with Mars, planned for August 1993, will fail to take place due to loss of radio contact with the probe.                                                           ASTRO

1992    On the 500th anniversary of Columbus's discovery of America, U.S. astronomers begin a planned 10-year search of the sky for radio signals indicating extraterrestrial intelligence.                                          ASTRO

1992    National Aeronautics and Space Administration scientists lose radio contact on October 8 with the *Pioneer 12*, which is presumed to have broken up as it plunged through the upper atmosphere of Venus. This robotic spacecraft has remained operational for 14 years, far exceeding initial expectations. *See* 1978, ASTRO.                                          ASTRO

1992    Teams of scientists working in the United States, France, and elsewhere, complete the first comprehensive maps of two human chromosomes, the Y chromosome and chromosome 21. These maps are an important step forward for the Human Genome Project, a multiyear effort to determine the entire human genome.                                          BIO

1992    American biologists Joseph Manson and Richard Wrangham report that their recent studies show that all primate aggression, including that of humans, is more complex than just a stress reaction or an atavistic animal instinct. They claim that primate aggression is an ancient evolutionary strategy linked more to coalition building and maintaining harmony than to wanton murder and violence. BIO

1992     Scientists report locating what they think to be the oldest and largest living organism on earth, a giant mold called *Armillaria bulbosa*, found growing beneath a Crystal Falls, Michigan, forest. BIO

1992     An 11-day United Nations Conference on Environment and Development, known as the Earth Summit, is held in Rio de Janeiro, Brazil. Some of the issues discussed are a funding agreement to provide environmental aid to developing countries, a so-called World Statement of Principles for the sustainable management and conservation of forests, and various other conventions on climate change and biodiversity. BIO

1992     Researchers from Yale University and New York City's American Museum of Natural History report extracting DNA fragments from an extinct termite embedded in amber for 30 million years, proving that genetic molecules can survive far longer than had been thought possible. BIO

1992     The National Institutes of Health withdraws funds from an academic conference searching for a genetic basis for criminal behavior, objecting to the notion that violence and crime might have genetic causes. BIO

1992     Scientists isolate a cluster of proteins that imitates the process by which baker's yeast copies its genes. This experiment proves to be a major step toward understanding how higher organisms copy their genes. BIO

1992     A panel of experts finds that DNA fingerprinting is useful in identifying criminals, but insists that standards for its use must be developed. BIO

1992     While studying specimens of the rock called shungite, Russian-American mineralogist Semeon Tsipursky discovers naturally occurring buckyballs, or buckminsterfullerene molecules, which had previously been found only in the laboratory (*see* 1985, CHEM). CHEM

1992     American scientists succeed in forming the first buckyball polymers and in making diamonds from buckyballs at lower temperatures and pressures than are needed to transform graphite into diamonds. Like diamonds and graphite, buckyballs are a form of pure carbon. CHEM

1992     Scientists for the NEC Corp. in Japan synthesize buckytubes, hollow cylinders of carbon atoms that may be useful for their great strength and electric conductivity. CHEM

1992     The ozone hole over Antarctica grows to its largest size ever. EARTH

1992     Scientists studying the climate of the last ice age by analyzing samples of ancient ice from Greenland determine that the climate changed significantly in that period over spans as short as one or two years. EARTH

1992        Scientists find new laboratory evidence that the atmospheres of Mars, Venus,
            and Earth were formed partly from noble gases trapped in the icy nuclei of
            comets.                                                                    EARTH

1992        On June 28, the largest earthquake in California in 40 years (*see* 1952, EARTH)
            strikes near the town of Landers in southern California. The quake registers 7.4
            on the Richter scale, but casualties are light due to the sparseness of the pop-
            ulation in the area.                                                       EARTH

1992        Using a Cray supercomputer, British scientist David Slowinski discovers the
            largest prime number to date, called the 30-second Mersenne prime, which is
            227,832 digits long.                                                       MATH

1992        Harvard University Medical School's *Harvard Health Letter* lists what it con-
            siders the seven best screening tests in preventive medicine, based on health
            professionals' guidelines: blood-pressure screening for hypertension, serum
            cholesterol for coronary artery disease, a stool smear for occult or hidden
            blood, the sigmoidoscopy for colon and rectal cancer, a clinical exam for
            breast cancer, mammography for breast cancer, and the Pap smear for cer-
            vical cancer.                                                              MED

1992        Researchers find evidence that Alzheimer's disease may develop as the result of
            an imbalance in two biochemical pathways that break down a precursor of
            beta amyloid, an ingredient of the plaques that attack the brain in this disease.
                                                                                       MED

1992        Researchers are divided on the benefits of the drug azidothymidine (AZT) in
            prolonging the lives of AIDS patients.                                     MED

1992        Three studies find that the symptoms of Parkinson's disease can be improved
            through transplanting fetal brain tissue.                                  MED

1992        American paleontologist Paul Sereno identifies the 230-million-year-old
            *Herrerasaurus* as the earliest carnivore in the dinosaur family. This 400-pound,
            15-foot-long reptile was probably an ancestor of *Tyrannosaurus rex*.      PALEO

1992        The "Eve" hypothesis is undermined when flaws are revealed in molecular
            research that had led scientists in the 1980s to trace the human lineage back
            to a common female ancestor in Africa some 200,000 years ago.             PALEO

1992        Li Tianyuan of China and Dennis Etler of the United States announce the dis-
            covery of two Chinese skulls dating back an estimated 350,000 years. The fos-
            sils may be the oldest complete human skulls yet found in eastern Asia. Li and
            Etler claim that these specimens are intermediate between *Homo erectus* and
            modern humans.                                                             PALEO

1992        In the Yucatán Peninsula of Mexico, geologists find evidence of a crater from
            the crash of a meteorite or comet 65 million years ago, an event hypothesized
            by American physicist Luis W. Alvarez in 1979. Paleontologists disagree on

whether this collision was responsible for the extinction of the dinosaurs.

PALEO

1992    The U.S. federal government, the Cheyenne River Sioux tribe, and the private Black Hills Institute of Geological Research become embroiled in a dispute over ownership of a *Tyrannosaurus rex* skeleton referred to as Sue.          PALEO

1992    In February, American paleoanthropologists Andrew Hill and Steven Ward announce their discovery of the oldest *Homo* specimen yet known. This skull fragment, from Kenya's Lake Baringo Basin, is 2.4 million years old, half a million years older than previously known specimens. Its discovery corroborates the theory that stone-tool-making hominids first emerged about 2.5 million years ago.          PALEO

1992    Researchers observe the unusual behavior of electrons confined to spaces small enough for quantum effects to become significant.          PHYS

1992    Researchers make the best measurement to date of the neutrino's mass.          PHYS

1993    American archaeologists John S. Justeson and Terrence Kaufman report their decipherment of an epi-Olmec stone stela, or monument, from Mexico dating from A.D. 159.          ARCH

1993    Archaeologists discover evidence of a human campsite in northern Alaska dating from 11,700 years ago, the first solid evidence of human activity in the northern part of the migration route from Asia to the Americas believed to have been traveled by ancestors of the Native Americans. *See* 10,000 B.C., EARTH.          ARCH

1993    American astronomer Douglas Lin presents evidence that the Milky Way is much larger than formerly believed and is surrounded by a halo of "dark matter" invisible to telescopes. These findings support the view that most of the universe is composed of such dark matter.          ASTRO

1993    American scientists report that *Voyagers* 1 and 2, now far beyond Pluto, have detected intense, low-frequency radio emissions from the heliopause, the outer boundary of the solar system. This point, marking the juncture where interstellar gases interact with the solar wind, is believed to be 8 billion to 12 billion miles from the sun. *Voyager 1* is expected to reach it in about 15 years. ASTRO

1993    Evidence from the *COBE* satellite (*see* 1992, ASTRO) indicates that 99.97 percent of the radiant energy of the universe was released within a year of the Big Bang. This discovery lends support to the Big Bang theory.          ASTRO

1993    On March 27, Supernova SN1993J appears in Galaxy M-81, about 12 million light-years from Earth. This supernova's X-ray emissions are the first to be analyzed by an X-ray camera orbiting Earth.          ASTRO

1993    In August, the National Aeronautics and Space Administration loses radio contact with the *Mars Observer* spacecraft (*see* 1992, ASTRO), bringing the mission to an end before the planned rendezvous with Mars.    ASTRO

1993    In December, space shuttle astronauts succeed in repairing the optical problems of the Hubble Space Telescope. *See* 1990, ASTRO.    ASTRO

1993    The British journal *Nature* reports that British geneticists have identified the gene whose mutation leads to amyotrophic lateral sclerosis (ALS, or Lou Gehrig's disease), an incurable muscle-wasting neural disease.    BIO

1993    American researchers at the National Cancer Institute claim to have linked a genetic marker on the X chromosome to homosexual orientation.    BIO

1993    Researchers clone human embryos for the first time.    BIO

1993    American chemists synthesize the largest molecule yet created in a laboratory from carbon and hydrogen alone. This lumpy, ball-shaped molecule consists of 1,134 carbon atoms and 1,146 hydrogen atoms.    CHEM

1993    A devastating flood, perhaps the worst in U.S. history, hits the Midwest. The Mississippi and Missouri Rivers and their tributaries overflow their banks, submerging towns and farmland in several states and wreaking havoc in such cities as Des Moines, Iowa; Kansas City, Kansas; and St. Louis, Missouri.    EARTH

1993    In September, a 6.1 magnitude earthquake near Latur, India, destroys more than 20 villages, killing about 10,000 people.    EARTH

1993    British mathematician Andrew Wiles proves Fermat's Last Theorem (*see* 1637, MATH), perhaps the most famous unsolved mathematical problem.    MATH

1993    U.S. health officials investigate a mysterious, deadly disease afflicting Navajos in New Mexico and Arizona. The disease is determined to be linked to a virus found in rodent droppings.    MED

1993    American molecular biologist Raúl J. Cano extracts DNA from a weevil fossilized in amber for 120 million to 135 million years.    PALEO

1993    American paleontologist John R. Horner reports the discovery of red blood cells in the fossilized leg bone of a 65-million-year-old *Tyrannosaurus rex* found in Montana. Horner hopes that DNA can be extracted from the dinosaur's cells.    PALEO

1993    American and Mongolian paleontologists announce the discovery of a 75-million-year-old fossil animal they believe to be a flightless bird transitional between dinosaurs and modern birds. Other scientists disagree, saying that the creature, *Mononychus*, was a dinosaur not ancestral to modern birds.    PALEO

1993            Spanish paleontologists find evidence that hominids in Spain were beginning
                to evolve Neanderthal features as early as 300,000 years ago, much earlier than
                previously thought.                                                      PALEO

1993            American physicists run a supercomputer calculation that appears to confirm
                the theory of quantum chromodynamics. *See* 1972, PHYS.                      PHYS

1993            The United States Congress votes to terminate funding for the unfinished
                Superconducting Super Collider. *See* 1989, PHYS.                            PHYS

1993            Princeton University researchers set a new record for energy production from
                controlled nuclear fusion, generating more than three megawatts, and later
                five megawatts, of power.                                                PHYS

1993            Researchers develop nanowires, virtual strings of atoms of a few nanometers
                in thickness that may be useful in magnetic recording technology.        TECH

1993            The movement toward digital libraries, accessible via the Internet, attains two
                notable landmarks: the Library of Congress begins to put its holdings online
                and the British Library begins digitizing the 11th-century manuscript of
                *Beowulf.*                                                               TECH

1994            American archaeologists report evidence of local production of tin in Turkey
                as early as 3,000 B.C. Previously, it was believed that tin in the Bronze Age
                Middle East was imported from such distant areas as Afghanistan.         ARCH

1994            American scientists report the discovery of genetic traces of tuberculosis infec-
                tion in a 1,000-year-old pre-Columbian mummy from Peru, proving that
                Europeans did not introduce the disease to the Americas.                 ARCH

1994            The unmanned U.S. spacecraft *Clementine* maps the moon, 22 years after the
                last American lunar mission in 1972.                                     ASTRO

1994            Fragments of Comet Shoemaker-Levy 9 smash into Jupiter, providing
                astronomers with a flood of data about the planet's atmosphere and the effects
                of cataclysmic impacts.                                                  ASTRO

1994            American astronomer Alexander Wolszczan discovers indirect evidence of at
                least three planets orbiting a pulsar in Virgo, 3,000 light-years away.  ASTRO

1994            British astronomers report the discovery of the closest known galaxy, a dwarf
                galaxy in the constellation Sagittarius, 50,000 light-years from the Milky Way's
                center.                                                                  ASTRO

1994            The U.S. spacecraft *Galileo* transmits the first complete image of a moon orbit-
                ing an asteroid, 243 Ida.                                                ASTRO

1994            British researchers discover a "deep biosphere" of bacteria living at depths of
                1,700 feet or more below the ocean floor.                                BIO

1994        A small herd of Przewalski's horses, bred in captivity, is successfully returned
            to the wild in Mongolia, where the species had been extinct.                    BIO

1994        The first live specimen of *Pseudoryx nghetinhensis,* or *sao la,* a wild Vietnamese
            cow relative, is captured. The animal had been unknown until 1993, when the
            first evidence of its existence began surfacing.                                 BIO

1994        American volcanologists Tobias Fischer and Stanley Williams discover a pat-
            tern of geological signs that can help predict when a volcano is about to erupt.
            The research results from a 1993 eruption in Colombia that severely injured
            Williams and killed nine others.                                                 EARTH

1994        On January 17, Los Angeles suffers a magnitude 6.7 earthquake centered in
            Northridge that kills 61 people and costs an estimated $15 billion in damages.
                                                                                             EARTH

1994        An American medical team reports the first account of successful gene therapy
            to be published in a scientific journal. The team used a receptor gene that low-
            ers cholesterol to treat a patient suffering from an inherited high-cholesterol
            disorder.                                                                        MED

1994        American researchers report the discovery of a gene mutation that may lead to
            many forms of cancer.                                                            MED

1994        In Britain, 11 people are killed by a rare but grisly Group A streptococcus infec-
            tion known as necrotizing fasciitis, or "flesh-eating disease."                  MED

1994        In Pakistan, paleontologist J. G. M. Thewissen and his colleagues discover the
            most solid fossil evidence to date of an ancestral whale intermediate between
            land and sea. The 50-million-year-old whale, *Ambulocetus natans,* had large
            hind legs that were functional on land and in water.                             PALEO

1994        In Ethiopia, researchers from Ethiopia, Japan, and the U.S. discover fossil
            remains of *Australopithecus ramidus* (4.4 million years old).                   PALEO

1994        American anthropologist Donald Johanson and his colleagues piece together
            the first nearly complete skull of the human ancestor *Australopithecus afarensis*
            from fragments found in 1992 at Hadar, Ethiopia. *See also* 1974, PALEO.         PALEO

1994        American scientists report the discovery of the earliest known land life, tubu-
            lar microorganisms from Arizona dating from 1.2 billion years ago.               PALEO

1994        New Zealand physicist Daniel F. Walls and his colleagues show theoretically
            that the relationship between the complementarity and uncertainty principles
            must always hold in quantum experiments employing closely spaced double
            slits.                                                                           PHYS

1994        An international team of physicists at Fermilab in Illinois discovers evidence
            for the top quark, a fundamental particle sought for nearly two decades.   PHYS

1994    A comprehensive survey, the results of which are published in the *Archives of General Psychiatry*, indicates that nearly half of all American adults experience a mental disorder at some point in their lives.                    PSYCH

1994    Studying the 19th-century brain damage case of Phineas Gage, American brain researchers Hanna and Antonio Damasio discover evidence that the ventro-medial region of the brain governs social behavior.                    PSYCH

1994    Canadian researcher Jeff Dahn and his colleagues develop a lithium battery with a water-based electrolyte that might prove safe, cheap, and powerful enough for use in electric cars.                    TECH

1995    In Peru, American archaeologist Johan Reinhard discovers the 500-year-old frozen body of an Inca girl, killed as a human sacrifice on the summit of Mt. Ampato at the age of 12-14.                    ARCH

1995    Egyptologist Kent Weeks at the American University in Cairo discovers a 3,000-year-old tomb that may be the burial place of the sons of Ramses II.                    ARCH

1995    The U.S. spacecraft *Galileo* reaches Jupiter and launches a probe into the giant planet's atmosphere.                    ASTRO

1995    The Solar and Heliospheric Observatory (SOHO) is launched by the European Space Agency and National Aeronautics and Space Administration to study the sun from a point 1 million miles from Earth.                    ASTRO

1995    The ultraviolet telescope system Astro-2, launched into orbit, yields the best views ever of UV radiation from celestial objects.                    ASTRO

1995    A planet is found orbiting the star 51 Pegasus. The Jupiter-sized world is the first extrasolar planet to be found around a star similar in mass and age to the sun. By the end of 1996, about a dozen planets will have been found outside the solar system.                    ASTRO

1995    American researchers led by Craig Venter succeed for the first time in deci-phering the complete genome of an organism: the bacterium *Haemophilus influenzae*.                    BIO

1995    American scientists discover a protein called leptin that regulates body weight in mice.                    BIO

1995    Wolves are reintroduced into Yellowstone National Park after an absence of 60 years.                    BIO

1995    On January 17, a major earthquake strikes Kobe, Japan, killing 5,100 people and injuring 26,800. The quake registers 7.2 on the Richter scale.                    EARTH

1995          The concentration of ozone-destroying chlorofluorocarbons in the atmos-
              phere begins to decrease after more than 20 years of steady increase, probably
              due to international efforts to restrict release of the compounds. *See* 1987,
              EARTH.                                                                          EARTH

1995          Using computer models of the earth's climate, American and British scientists
              discover new evidence of global warming as a result of human activity.    EARTH

1995          German geochemist Kaj Hoernle theorizes that a single "hot sheet" of subter-
              ranean rock feeds all the volcanoes of a region stretching from the Canary
              Islands to Germany.                                                         EARTH

1995          Japanese geologist Kenji Satake discovers that an earthquake that rocked North
              America's Pacific Northwest in 1700 may have had a magnitude of 9, making
              it one of the largest in history.                                           EARTH

1995          In Kikwit, Zaire, an outbreak of Ebola virus infects 315 people, killing 244—a
              mortality rate of 77 percent.                                                 MED

1995          An epidemic of dengue fever in Latin America and the Caribbean sickens over
              140,000 people and kills 38.                                                  MED

1995          American medical researchers Byron Caughey and Peter Lansbury present evi-
              dence that spongiform encephalopathies, including scrapie and mad cow dis-
              ease, are caused by protein crystals called prions. The evidence supports the
              theories of American pediatrician and virologist D. Carleton Gajdusek, who
              shared the Nobel Prize for physiology or medicine in 1976 with American
              physician and medical researcher Baruch S. Blumberg (*see* 1976, MED).    MED

1995          Paleontologists in Argentina discover a carnivorous dinosaur called
              *Giganotosaurus* that was at least as large as *Tyrannosaurus rex*, previously
              believed to be the largest land predator in earth's history.              PALEO

1995          Biologists in Arizona discover fossil remains of the earliest known frog,
              *Prosalirus bitis*, dating to 190 million years ago.                      PALEO

1995          The most complete hominid foot yet found is reconstructed when South
              African paleoanthropologist Ronald Clarke pieces together the foot bones of a
              3-million- to 3.5-million-year-old australopithecine.                     PALEO

1995          American physicists Carl Weiman and Eric Cornell succeed in producing a
              Bose-Einstein condensate at 20 nanokelvins above absolute zero. In describing
              the particles called bosons, Satyendra Bose and Albert Einstein had predicted
              the existence of this wavelike state of matter (*see* 1924, PHYS).         PHYS

1995          Studying brains of male-to-female transsexuals, Dutch researchers discover
              that the brain region called BSTc is closer in size to that of females than to that
              of males. The evidence suggests a biological basis for the belief of the trans-

sexuals that, despite their male anatomy, they were "born" to be female.

PSYCH

1996    American archaeologist Anna Roosevelt presents evidence that people were living in Monte Alegre, Brazil, as early as 11,200 years ago—nearly the same time as the Clovis culture of New Mexico.                                    ARCH

1996    Based on studies of a meteorite from Mars found in Antarctica, National Aeronautics and Space Administration scientists report that microbial life may once have existed on Mars, though their claims meet with skepticism from many colleagues                                    ASTRO

1996    Geologists discover that the earth's inner core of solid iron spins at a faster rate than the rest of the planet.                                    EARTH

1996    American researcher Gary Hack reports the discovery of a hitherto unknown muscle in the human face, the sphenomandibularis.                                    BIO

1996    Animal species discovered this year include *Callithrix saterei*, an Amazon marmoset, and, in Australia, the Gulf snapping turtle, thought to have gone extinct at least 20,000 years ago.                                    BIO

1996    Biologists studying a cave in Romania discover the first known ecosystem on land that derives energy not from sunlight but from the breakdown of hydrogen sulfide by bacteria.                                    BIO

1996    Genetic evidence suggests that the one-celled, bacteria-like marine organisms called archaea are less closely related to bacteria than to eukaryotes, a group that includes animals and plants.                                    BIO

1996    American researcher Charles Keeling discovers that spring in the Northern Hemisphere may be coming seven days earlier than it did in the mid-1970s, probably due to global warming.                                    EARTH

1996    A study of a group of American nuns over a 60-year period reveals that an individual's writing style in youth may indicate the likelihood of developing Alzheimer's in later life.                                    MED

1996    Treatment with drug combinations that include protease inhibitors is found to be effective in postponing and possibly preventing the onset of AIDS in patients infected with the HIV virus.                                    MED

1996    AIDS researchers discover that people with defective CKR-5 receptors are resistant to HIV infection.                                    MED

1996    An epidemic of food poisoning by toxin-producing *E. coli* bacteria strikes 9,500 people in Japan, killing 11. A smaller epidemic killed 4 and sickened 700 in Seattle in 1993.                                    MED

1996          In Nigeria and other African countries, an epidemic of meningitis that could have been prevented by vaccination sickens 140,000 people and kills 15,000.  MED

1996          In Greenland, a team of British, American, and Australian scientists led by Gustaf Arrhenius discovers evidence that life originated on earth more than 3.85 billion years ago, 350 million years earlier than previously believed. PALEO

1996          American paleontologists theorize that the mass extinctions at the end of the Permian period, 250 million years ago, were caused by a poisonous build-up of carbon dioxide in the earth's oceans.  PALEO

1996          Canadian paleontologist Philip Currie discovers that at least one dinosaur species, dating to 120 million years ago, had feathers. The discovery lends support to the theory that birds are descended from dinosaurs.  PALEO

1996          American and Chinese paleontologists discover evidence that the primate *Eosimias*, dating to 45 million years ago, was the common ancestor of monkeys, apes, and humans.  PALEO

1996          German physicists produce the first known atoms of antihydrogen, atoms consisting of a positron (antielectron) and an antiproton.  PHYS

1996          American brain researchers discover a "master switch" for inducing sleep in a part of the hypothalamus called the ventrolateral preoptic area (VLPO).  PSYCH

1996          Intel Corporation reports the development of the world's fastest computer, capable of performing 1 trillion operations per second, while American and Japanese communications researchers succeed in transmitting more than 1 trillion bits of computer data per second.  TECH

1996          Texas Instruments announces the development of a manufacturing technique that will allow the production of a computer chip with 20 times the processing power currently available in personal computers.  TECH

1996          Americans Larry Antonuk and Robert Street invent the first digital X-ray machine.  TECH

1997          Resolving a controversy two decades old, American and Chilean archaeologists agree that Monte Verde, Chile, was settled by a population of human hunter-gatherers 12,500 years ago. The date is at least 1,000 years earlier than the Clovis culture of New Mexico (*see* 1952, ARCH).  ARCH

1997          Comet Hale-Bopp, intrinsically the brightest comet to pass inside Earth's orbit in more than 400 years, is easily visible to the naked eye.  ASTRO

1997          Scientists examining pictures of the Jovian moon Europa, photographed by the U.S. spacecraft *Galileo*, find growing evidence that a global ocean of liquid water or slush exists under an ice crust. The evidence increases the likelihood that life may exist on Europa.  ASTRO

1997    British physicist Stephen Hawking concedes defeat on a bet made with American scientists John P. Preskill and Kip S. Thorne. Hawking had argued that naked singularities could not exist; mathematical analysis now indicates that they may.    ASTRO

1997    The 11-year-old Russian space station *Mir* (*see* 1986, ASTRO), still operating long after its original life expectancy of five years, experiences a series of mishaps embarrassing to the Russian space program. These include computer crashes, oxygen system breakdowns, a collision with a cargo ship, and a fire.    ASTRO

1997    In the first cloning ever of an adult mammal, Scottish embryologist Ian Wilmut announces his 1996 replication of a lamb from the DNA of a ewe. The achievement raises worldwide debate about the ethical and religious implications of cloning, particularly of humans.    BIO

1997    American biologist Jonathan B. Losos and colleagues discover evidence of divergent evolution in an experiment transplanting lizards to 14 lizardless islands with varying habitats. The study shows that adaptations in such bodily characteristics as leg length can happen in periods as short as a decade.    BIO

1997    American scientists discover fragments of genetic material from the Spanish influenza virus that killed millions of people worldwide in 1918–1919.    MED

1997    The U.S. Department of Agriculture proposes new regulations against feeding protein derived from ruminant animals to other ruminants. The proposed ban is an attempt to prevent epidemics of mad cow disease such as those that have plagued Britain in recent years. *See also* 1995, MED.    MED

1997    A rich collection of dinosaur fossils, including the first fossilized internal organs of dinosaurs ever seen, is discovered near Beipiao, China.    PALEO

1997    Spanish paleontologists report the discovery of 800,000-year-old hominid fossils near Burgos, Spain, that represent the earliest known Europeans. The scientists identify the fossils as a hitherto unknown species, *Homo antecessor*, though other scientists are skeptical.    PALEO

1997    Autonomous software programs that incorporate advanced artificial intelligence research begin to appear in the consumer marketplace—for example, in Microsoft's Office 97 software.    TECH

1997    DNA chips, which use DNA molecules as information processors, are marketed commercially by American company Affymetrix and others.    TECH

# Appendix:
# Birth and Death Dates

| | | | |
|---|---|---|---|
| Abbe, Cleveland | 1838–1916 | Bateson, William | 1861–1926 |
| Abel, Niels Henrik | 1802–1829 | Becquerel, Alexandre-Edmond | 1820–1891 |
| Abelson, Philip Hauge | 1913– | Becquerel, Antoine–Henri | 1852–1908 |
| Acheson, Edward Goodrich | 1856–1931 | Bell, Alexander Graham | 1847–1922 |
| Adler, Alfred | 1870–1937 | Bell, Sir Charles | 1774–1842 |
| Agassiz, Louis | 1807–1873 | Benedict, Ruth | 1887–1948 |
| Aiken, Howard Hathaway | 1900–1973 | Bennett, Floyd | 1890–1928 |
| Albertus Magnus | 1193–1280 | Benz, Carl Friedrich | 1844–1929 |
| Aldrin, Edwin E. "Buzz," Jr. | 1930– | Bering, Vitus J. | 1681–1741 |
| Alhazen | 965–1039 | Bernard, Claude | 1813–1878 |
| al-Khwārizmī, Muhammad ibn Mūsā | 780–850 | Bernoulli, Daniel | 1700–1782 |
| Alpini, Prospero | 1553–1616 | Bernoulli, Jakob (Jacques) | 1654–1705 |
| Alvarez, Luis W. | 1911–1988 | Bernoulli, Johann (Jean) | 1667–1748 |
| Ampère, André-Marie | 1775–1836 | Berthelot, Pierre-Eugéne-Marcelin | 1866–1934 |
| Amundsen, Roald | 1872–1928 | Berzelius, Jöns Jakob | 1779–1848 |
| Anaximander | 610–547 B.C. | Bessel, Friedrich Wilhelm | 1784–1846 |
| Anders, William A. | 1933– | Bessemer, Sir Henry | 1813–1898 |
| Ångström, A. J. | 1814–1874 | Best, Charles | 1899–1978 |
| Anning, Mary | 1799–1847 | Bettelheim, Bruno | 1903–1990 |
| Archimedes | c. 287–212 B.C. | Binet, Alfred | 1857–1911 |
| Arfwedson, Johan August | 1792–1841 | Bingham, Hiram | 1875–1956 |
| Aristotle | 384–322 B.C. | Birdseye, Clarence | 1886–1956 |
| Arkwright, Sir Richard | 1732–1792 | Bjerknes, Vilhelm | 1862–1951 |
| Armstrong, Neil A. | 1930– | Black, Joseph | 1728–1799 |
| Arrhenius, Svante | 1859–1927 | Blériot, Louis | 1872–1936 |
| Arzachel | fl. c. 1075 | Blumenbach, Johann Friedrich | 1752–1840 |
| Avogadro, Amedeo | 1776–1856 | Boas, Franz | 1858–1942 |
| Baade, Walter | 1893–1960 | Bogardus, James | 1800–1874 |
| Babbage, Charles | 1792–1871 | Bohr, Niels | 1885–1962 |
| Bacon, Francis | 1561–1626 | Boole, George | 1815–1864 |
| Bacon, Roger | c. 1220–1292 | Borman, Frank | 1928– |
| Baekeland, Leo Hendrik | 1863–1944 | Born, Max | 1882–1970 |
| Baer, Karl Ernst von | 1792–1876 | Bose, Satyendra Nath | 1894–1974 |
| Bain, Alexander | 1818–1903 | Boyle, Robert | 1627–1691 |
| Balard, Antoine-Jérôme | 1802–1876 | Bradley, James | 1693–1762 |
| Balboa, Vasco Núñez de | c. 1475–1519 | Braid, James | 1795–1860 |
| Balmer, Johann Jakob | 1825–1898 | Brand, Hennig | 1630–1692 |
| Banting, Sir Frederick G. | 1891–1941 | Brandt, Georg | 1694–1768 |
| Bardeen, John | 1908–1991 | Brattain, Walter H. | 1902–1987 |
| Barnard, Christiaan | 1922– | Braun, Wernher von | 1912–1977 |
| Barnard, Edward Emerson | 1857–1923 | Breuer, Josef | 1842–1925 |

| | | | | |
|---|---|---|---|---|
| Briggs, Henry | 1561–1630 | Coronado, Francisco Vásquez de | c. 1510–1554 |
| Broca, Pierre-Paul | 1824–1880 | Cort, Henry | 1740–1800 |
| Broglie, Louis de | 1892–1987 | Cortés, Hernando | 1485–1547 |
| Brown, Robert | 1773–1858 | Coster, Dirk | 1889–1950 |
| Brunelleschi, Filippo | 1377–1446 | Coulomb, Charles-Augustin de | 1736–1806 |
| Buchner, Eduard | 1860–1917 | Courtois, Bernard | 1777–1838 |
| Bullock, William | 1813–1867 | Crick, Francis | 1916– |
| Bunsen, Robert | 1811–1899 | Crippen, Robert L. | 1937– |
| Burbank, Luther | 1849–1926 | Cronstedt, Baron Axel F. | 1722–1765 |
| Bush, Vann | 1890–1974 | Crookes, Sir William | 1832–1919 |
| Bushnell, David | 1742?–1824 | Cross, Charles Frederick | 1855–1935 |
| Byrd, Richard E. | 1888–1957 | Cugnot, Nicolas-Joseph | 1725–1804 |
| Cannizzaro, Stanislao | 1826–1910 | Curie, Marie Sklodowska | 1867–1934 |
| Cannon, Annie Jump | 1863–1941 | Curie, Pierre | 1859–1906 |
| Cantor, Georg | 1845–1918 | Cuvier, Georges Dagobert, Baron | 1769–1832 |
| Cardano, Geronimo | 1501–1576 | da Gama, Vasco. See Gama, Vasco da | |
| Carnot, Nicolas-Léonard-Sadi | 1796–1832 | da Vinci, Leonardo. See Leonardo da Vinci | |
| Carothers, Wallace Hume | 1896–1937 | Daguerre, Louis-Jacques-Mandé | 1789–1851 |
| Carrel, Alexis | 1873–1944 | Daimler, Gottlieb | 1834–1900 |
| Carson, Rachel | 1907–1964 | Dalton, John | 1766–1844 |
| Carter, Howard | 1873–1939 | Dart, Raymond Arthur | 1893–1988 |
| Cassini, Giovanni | 1625–1712 | Darwin, Charles | 1809–1882 |
| Cavendish, Henry | 1731–1810 | Davy, Sir Humphry | 1778–1829 |
| Cayley, Arthur | 1821–1895 | De Forest, Lee | 1873–1961 |
| Celsius, Anders | 1701–1744 | de Soto, Hernando | c. 1500–1542 |
| Chadwick, Sir James | 1891–1974 | de Vries, Hugo. See Vries, Hugo de | |
| Chaffee, Roger | 1935–1967 | Debierne, André-Louis | 1874–1949 |
| Champlain, Samuel de | c. 1567–1635 | Delbrück, Max | 1906–1981 |
| Champollion, Jean-François | 1790–1832 | Demarçay, Eugène-Anatole | 1852–1903 |
| Charles, Jacques-Alexandre-César | 1746–1823 | Democritus | c. 460–370 B.C. |
| Christy, James | 1938– | Descartes, René | 1596–1650 |
| Clark, William | 1770–1838 | Dias, Bartolomeu | c. 1450–1500 |
| Claude, Albert | 1898–1983 | Diesel, Rudolf | 1858–1913 |
| Cleve, Per Teodor | 1840–1905 | Dirac, P. A. M. | 1902–1984 |
| Cockcroft, Sir John | 1897–1967 | Domagk, Gerhard | 1895–1964 |
| Cohen, Stanley | 1922– | Doolittle, James | 1896–1993 |
| Collins, Michael | 1930– | Doppler, Christian Johann | 1803–1853 |
| Columbus, Christopher | 1451–1506 | Dorn, Friedrich Ernst | 1848–1916 |
| Conze, Alexander | 1831–1914 | Drake, Sir Francis | 1540–1596 |
| Cook, James | 1728–1779 | Drebbel, Cornelis | 1572–1633 |
| Cooke, William Fothergill | 1806–1879 | Dubois, Marie Eugène | 1858–1940 |
| Cooper, L. Gordon | 1927– | Dujardin, Félix | 1801–1860 |
| Cooper, Peter | 1791–1883 | Durkheim, Émile | 1858–1971 |
| Cope, Edward Drinker | 1840–1897 | Dutton, Clarence Edward | 1841–1912 |
| Copernicus, Nicolaus | 1473–1543 | Eddington, Sir Arthur Stanley | 1882–1944 |
| Cori, Carl Ferdinand | 1896–1984 | Edison, Thomas Alva | 1847–1931 |
| Cori, Gerty Theresa Radnitz | 1896–1957 | Ehrlich, Paul | 1854–1915 |
| Coriolis, Gaspard-Gustave de | 1792–1843 | Eijkman, Christiaan | 1858–1930 |
| Corliss, George Henry | 1817–1888 | Einstein, Albert | 1879–1955 |

| | | | |
|---|---|---|---|
| Ekeberg, Anders G. | 1767–1813 | Gay-Lussac, Joseph-Louis | 1778–1850 |
| Elhuyar y de Suvisa, Fausto d' | 1755–1833 | Geiger, Hans | 1882–1945 |
| Elhuyar y de Suvisa, Juan José | 1754–1804 | Gell-Mann, Murray | 1929– |
| Enders, John F. | 1897–1985 | Gesner, Konrad von | 1516–1565 |
| Eratosthenes | c. 276–c. 194 B.C. | Ghiorso, Albert | 1915– |
| Euclid | fl. c. 300 B.C. | Gibbs, Josiah Willard | 1839–1903 |
| Eudoxus of Cnidus | c. 408–355 B.C. | Gilbert, William | 1544–1603 |
| Euler, Leonhard | 1707–1783 | Glenn, John | 1921– |
| Evans, Oliver | 1755–1819 | Goddard, Robert H. | 1882–1945 |
| Evans-Pritchard, Edward Evan | 1902–1973 | Gödel, Kurt | 1906–1978 |
| Fahrenheit, Daniel Gabriel | 1686–1736 | Golgi, Camillo | 1844–1926 |
| Fallopius, Gabriel | 1523–1562 | Gomberg, Moses | 1866–1947 |
| Faraday, Michael | 1791–1867 | Goodyear, Charles | 1800–1860 |
| Fermat, Pierre de | 1601–1665 | Gregor, William | 1761–1817 |
| Fermi, Enrico | 1901–1954 | Grew, Nehemiah | 1641–1712 |
| Feynman, Richard P. | 1918–1988 | Grissom, Virgil I. | 1926–1967 |
| Fibonacci, Leonardo | 1170–1230 | Grotefend, Georg F. | 1775–1853 |
| Fischer, Emil | 1852–1919 | Guericke, Otto von | 1602–1686 |
| Fitch, John | 1743–1798 | Gutenberg, Johannes | c. 1398–c. 1468 |
| FitzGerald, George | 1851–1901 | Haeckel, Ernst | 1834–1919 |
| Fleming, Sir Alexander | 1881–1955 | Hahn, Otto | 1879–1968 |
| Fleming, Sir John Ambrose | 1849–1945 | Haldane, J. B. S. | 1892–1964 |
| Ford, Henry | 1863–1947 | Hale, George Ellery | 1868–1938 |
| Fourier, Jean-Baptiste-Joseph | 1768–1830 | Hall, Asaph | 1829–1907 |
| Fournier, Pierre Simon | 1712–1768 | Hall, Sir James | 1761–1832 |
| Franck, James | 1882–1964 | Halley, Edmund | 1656–1742 |
| Franklin, Benjamin | 1706–1790 | Hardy, Godfrey H. | 1877–1947 |
| Fraunhofer, Joseph von | 1787–1826 | Hargreaves, James | c. 1720–1778 |
| Frazer, Sir James George | 1854–1941 | Harrison, John | 1693–1776 |
| Frege, Gottlob | 1848–1925 | Hartley, David | 1705–1757 |
| Frere, John | 1740–1807 | Harvey, William | 1578–1657 |
| Fresnel, Augustin-Jean | 1788–1827 | Hatchett, Charles | 1765–1847 |
| Freud, Anna | 1895–1982 | Hawking, Stephen | 1942– |
| Freud, Sigmund | 1856–1939 | Heaviside, Oliver | 1850–1925 |
| Frobisher, Sir Martin | c. 1535–1594 | Heisenberg, Werner | 1901–1976 |
| Fulton, Robert | 1765–1815 | Helmholtz, Hermann von | 1821–1894 |
| Funk, Casimir | 1884–1967 | Henry, Joseph | 1797–1878 |
| Gadolin, Johan | 1760–1852 | Henry the Navigator | 1394–1460 |
| Gagarin, Yuri A. | 1934–1968 | Hero (Heron) of Alexandria | fl. A.D. 62 |
| Galen, Claudius | c. 130–c. 199 | Herschel, John | 1792–1871 |
| Galileo Galilei | 1564–1642 | Herschel, Sir William | 1738–1822 |
| Gall, Franz Joseph | 1758–1828 | Hertz, Heinrich | 1857–1894 |
| Galle, Johann Gottfried | 1812–1910 | Hertzsprung, Ejnar | 1873–1967 |
| Galois, Évariste | 1811–1832 | Hess, Harry H. | 1906–1969 |
| Galton, Sir Francis | 1822–1911 | Hess, Victor F. | 1883–1964 |
| Galvani, Luigi | 1737–1798 | Hevesy, Georg C. de | 1885–1966 |
| Gama, Vasco da | c. 1460–1524 | Hewitt, Peter Cooper | 1861–1921 |
| Gatling, Richard Jordan | 1818–1903 | Hilbert, David | 1862–1943 |
| Gauss, Carl Friedrich | 1777–1855 | Hippocrates | c. 460–c. 370 B.C. |

| | | | |
|---|---|---|---|
| Hisinger, Wilhelm | 1766–1852 | Langmuir, Irving | 1881–1957 |
| Holland, John Phillip | 1840–1914 | Laplace, Pierre Simon de | 1749–1827 |
| Hollerith, Herman | 1860–1929 | Lassell, William | 1799–1880 |
| Hooke, Robert | 1635–1703 | Lavoisier, Antoine-Laurent | 1743–1794 |
| Hopkins, Sir Frederick Gowland | 1861–1947 | Lawrence, Ernest O. | 1901–1958 |
| Hoppe-Seyler, Ernst Felix | 1825–1895 | Layard, Sir Austen Henry | 1817–1894 |
| Howe, Elias | 1819–1867 | Leakey, Louis | 1903–1972 |
| Hubble, Edwin Powell | 1889–1953 | Leakey, Mary | 1913–1996 |
| Hudson, Henry | fl. 1607–1611 | Leakey, Richard | 1944– |
| Hughes, Howard | 1905–1976 | Leavitt, Henrietta | 1868–1921 |
| Humboldt, Alexander von | 1769–1859 | Lecoq de Boisbaudran, Paul Émile | c. 1838–1912 |
| Hunter, John | 1728–1793 | Lee, Tsung-dao | 1926– |
| Huntsman, Benjamin | 1704–1776 | Leeuwenhoek, Antoni van | 1632–1723 |
| Hutton, James | 1726–1797 | Leibniz, Gottfried Wilhelm | 1646–1716 |
| Huxley, Thomas Henry | 1825–1895 | Lemaître, Georges | 1894–1966 |
| Huygens, Christiaan | 1629–1695 | Lenoir, Jean-Joseph Étienne | 1822–1900 |
| Hyatt, John Wesley | 1837–1920 | Leonardo da Vinci | 1452–1519 |
| Imhotep | c. 2980–2950 B.C. | Leonov, Alexei A. | 1934– |
| Ives, Frederic Eugene | 1856–1937 | Leverrier, Urbain-Jean-Joseph | 1811–1877 |
| James, William | 1842–1910 | Lévi-Strauss, Claude | 1908– |
| Jansky, Karl G. | 1905–1950 | Lewis, Meriwether | 1774–1809 |
| Janssen, Pierre-Jules-César | 1824–1907 | Libby, Willard F. | 1908–1980 |
| Jenner, Edward | 1749–1823 | Liebig, Justus von | 1803–1873 |
| Joliot, Frédéric | 1900–1958 | Lilienthal, Otto | 1848–1896 |
| Joliot-Curie, Irène | 1897–1956 | Lindbergh, Charles | 1902–1974 |
| Jones, Sir William | 1746–1794 | Linde, Carl Paul Gottfried von | 1842–1934 |
| Joule, James P. | 1818–1889 | Linnaeus, Carolus | 1707–1778 |
| Jung, Carl Gustav | 1875–1961 | Linton, Ralph | 1893–1953 |
| Kamerlingh Onnes, Heike | 1853–1926 | Lister, Joseph | 1827–1912 |
| Kant, Immanuel | 1724–1804 | Lister, Joseph Jackson | 1786–1869 |
| Kay, John | 1704–1764 | Livingstone, David | 1813–1873 |
| Kekule von Stradonitz, Friedrich A. | 1829–1896 | Lobachevsky, Nikolai Ivanovich | 1792–1856 |
| Kelvin, William Thomson, Lord | 1824–1907 | Lockyer, Sir Joseph Norman | 1836–1920 |
| Kennelly, Arthur Edwin | 1861–1939 | Lorenz, Edward N. | 1917– |
| Kepler, Johannes | 1571–1630 | Lorenz, Konrad | 1903–1989 |
| Kirchhoff, Gottlieb Sigismund | 1704–1833 | Lovell, James A., Jr. | 1928– |
| Kirchhoff, Gustav | 1824–1887 | Lowe, Thaddeus | 1832–1913 |
| Kirkwood, Daniel | 1814–1895 | Lowell, Percival | 1855–1916 |
| Klaproth, Martin Heinrich | 1743–1817 | Lumière, Auguste | 1862–1954 |
| Klaus, Karl K. | 1796–1864 | Lumière, Louis | 1864–1948 |
| Koch, Robert | 1843–1910 | Magellan, Ferdinand | c. 1480–1521 |
| Koldewy, Robert | 1855–1925 | Maiman, Theodore Harold | 1927– |
| Krebs, Hans | 1900–1981 | Malinowski, Bronislaw | 1884–1942 |
| Kroeber, Alfred Louis | 1876–1960 | Malpighi, Marcello | 1628–1694 |
| Laënnec, René | 1781–1826 | Malthus, Thomas Robert | 1766–1834 |
| Lagrange, Joseph-Louis | 1736–1813 | Marconi, Guglielmo | 1874–1937 |
| Lamarck, Jean-Baptiste de | 1744–1829 | Marggraf, Andreas Sigismund | 1709–1782 |
| Land, Edwin Herbert | 1909–1991 | Marignac, Jean-Charles de | 1817–1894 |
| Landsteiner, Karl | 1868–1943 | Marsh, Othniel Charles | 1831–1899 |

| | | | |
|---|---|---|---|
| Marx, Karl | 1818–1883 | Ochoa, Severo | 1905–1993 |
| Maspero, Gaston | 1846–1916 | Oersted, Hans Christian | 1777–1851 |
| Maury, Matthew F. | 1806–1873 | Ohm, Georg Simon | 1789–1854 |
| Maxwell, James Clerk | 1831–1879 | Olbers, Heinrich Wilhelm | 1758–1840 |
| Mayer, Maria Goeppert | 1906–1972 | Oppenheimer, J. Robert | 1904–1967 |
| Mayow, John | 1641–1679 | Ostwald, Friedrich Wilhelm | 1853–1932 |
| McAdam, John | 1756–1836 | Otis, Elisha Graves | 1811–1861 |
| McClintock, Barbara | 1902–1992 | Otto, Nikolaus August | 1832–1891 |
| McCormick, Cyrus Hall | 1809–1884 | Oughtred, William | 1574–1660 |
| McMillan, Edwin M. | 1907–1991 | Owen, Sir Richard | 1804–1892 |
| Mead, Margaret | 1901–1978 | Papin, Denis | 1647–c. 1712 |
| Meitner, Lise | 1878–1968 | Paracelsus | c. 1493–1541 |
| Mendel, Gregor | 1822–1884 | Paré, Ambroise | 1510–1590 |
| Mendeleyev, Dmitry Ivanovich | 1834–1907 | Park, Mungo | 1771–1806 |
| Mercator, Gerardus | 1512–1594 | Parker, Eugene N. | 1927– |
| Mergenthaler, Ottmar | 1854–1899 | Parsons, Sir Charles Algernon | 1854–1931 |
| Merrifield, R. Bruce | 1921– | Parsons, William | 1800–1867 |
| Mesmer, Franz Anton | 1734–1815 | Pascal, Blaise | 1623–1662 |
| Michelson, Albert A. | 1852–1931 | Pasteur, Louis | 1822–1895 |
| Mill, James | 1773–1836 | Pauli, Wolfgang | 1900–1958 |
| Miller, Stanley | 1930– | Pauling, Linus C. | 1901–1994 |
| Millikan, Robert A. | 1868–1953 | Pavlov, Ivan | 1849–1936 |
| Milne, John | 1850–1913 | Peary, Robert E. | 1856–1920 |
| Mohl, Hugh von | 1805–1872 | Perey, Marguerite | 1909–1975 |
| Mohorovičić, Andrija | 1857–1936 | Piaget, Jean | 1896–1980 |
| Moissan, Henri | 1852–1907 | Piazzi, Giuseppe | 1746–1826 |
| Morgan, Lewis Henry | 1818–1881 | Pickering, William | 1858–1938 |
| Morgan, Thomas Hunt | 1866–1945 | Pinel, Philippe | 1745–1826 |
| Morgenstern, Oskar | 1902–1977 | Pizarro, Francisco | c. 1475–1541 |
| Morley, Edward W. | 1838–1923 | Planck, Max | 1858–1947 |
| Morse, Samuel F. B. | 1791–1872 | Plato | 427–347 B.C. |
| Mosander, Carl Gustaf | 1797–1858 | Poincaré, Jules-Henri | 1854–1912 |
| Müller, Franz Joseph | 1740–1825 | Polo, Marco | c. 1254–c. 1324 |
| Müller, Paul | 1899–1965 | Ponce de León, Juan | c. 1460–1521 |
| Murchison, Sir Roderick I. | 1792–1871 | Porta, Giambattista della | 1535?–1615 |
| Murdock, William | 1754–1839 | Prévost, Pierre | 1751–1839 |
| Napier, John | 1550–1617 | Priestley, Joseph | 1733–1804 |
| Natta, Giulio | 1903–1979 | Proust, Joseph-Louis | 1754–1826 |
| Nernst, Walther H. | 1864–1941 | Ptolemaeus, Claudius (Ptolemy) | c. 85–165 |
| Newcomen, Thomas | 1663–1729 | Pullman, George M. | 1831–1897 |
| Newton, Sir Isaac | 1642–1727 | Purkinje, Jan | 1787–1869 |
| Niepce, Joseph-Nicéphore | 1765–1833 | Pythagoras | c. 582–c. 507 B.C. |
| Nightingale, Florence | 1820–1910 | Rabi, I. I. | 1898–1988 |
| Nilson, Lars Fredrik | 1840–1899 | Radcliffe-Brown, Alfred Reginald | 1881–1955 |
| Nobel, Alfred | 1833–1896 | Ramsay, Sir William | 1852–1916 |
| Noddack, Walter Karl | 1893–1960 | Rawlinson, Sir Henry Creswicke | 1810–1895 |
| Noether, Amalie (Emmy) | 1882–1935 | Rayleigh, J. W. Strutt, Lord | 1842–1919 |
| Noyce, Robert | 1927–1989 | Reber, Grote | 1911– |
| Oberth, Hermann | 1894–1989 | Redi, Francesco | 1626–1697 |

| | | | |
|---|---|---|---|
| Reich, Ferdinand | 1799–1882 | Soret, Jacques Louis | 1827–1890 |
| Richter, Charles F. | 1900–1985 | Soto, Hernando de. *See* de Soto, Hernando | |
| Ride, Sally | 1951– | Speke, John Hanning | 1827–1864 |
| Riemann, Georg | 1826–1866 | Stahl, George Ernst | 1660–1734 |
| Rogers, Moses | 1779–1821 | Stanley, Sir Henry M. | 1841–1904 |
| Röntgen, Wilhelm | 1845–1923 | Staudinger, Hermann | 1881–1965 |
| Rorschach, Hermann | 1884–1922 | Steinmetz, Charles Proteus | 1865–1923 |
| Rudolf, Christoff | c. 1500–1545 | Steno, Nicolaus | 1638–1686 |
| Russell, Bertrand | 1872–1970 | Stephenson, George | 1781–1848 |
| Russell, Henry Norris | 1877–1957 | Stevens, Robert Livingston | 1787–1856 |
| Rutherford, Daniel | 1749–1819 | Stevin, Simon | 1548–1620 |
| Rutherford, Ernest | 1871–1937 | Steward, Julian | 1902–1972 |
| Sabin, Albert Bruce | 1906–1993 | Stoney, George Johnston | 1826–1911 |
| Sagan, Carl | 1935–1996 | Strabo | c. 63 B.C.–A.D. 21 |
| Salk, Jonas Edward | 1914–1995 | Strasburger, Eduard Adolf | 1844–1912 |
| Sanctorius (Santorio Santorio) | 1561–1636 | Strohmeyer, Friedrich | 1776–1835 |
| Sarnoff, David | 1891–1971 | Strutt, J. W. *See* Rayleigh, J. W. Strutt, Lord | |
| Saussure, Horace Bénédict de | 1740–1799 | Sutton, Walter | 1877–1916 |
| Savery, Thomas | c. 1650–1715 | Sydenham, Thomas | 1624–1689 |
| Scheele, Carl Wilhelm | 1742–1786 | Szilard, Leo | 1898–1964 |
| Schiaparelli, Giovanni | 1835–1910 | Talbot, William Henry Fox | 1800–1877 |
| Schleiden, Matthias Jakob | 1804–1881 | Tasman, Abel | c. 1603–1659? |
| Schliemann, Heinrich | 1822–1890 | Teisserenc de Bort, Léon-Philippe | 1855–1913 |
| Schmidt, Bernhard V. | 1879–1935 | Teller, Edward | 1908– |
| Schrödinger, Erwin | 1887–1961 | Tennant, Smithson | 1761–1815 |
| Schwann, Theodor | 1810–1882 | Tesla, Nikola | 1856–1943 |
| Schwarzschild, Karl | 1873–1916 | Thales of Miletus | c. 636–546 B.C. |
| Schweitzer, Albert | 1875–1965 | Thénard, Louis-Jacques | 1777–1857 |
| Scott, Robert Falcon | 1868–1912 | Theodoric of Freibourg | c. 1250–c. 1310 |
| Seaborg, Glenn T. | 1912– | Theophrastus | c. 372–287 B.C. |
| Secchi, Pietro Angelo | 1818–1878 | Thompson, Benjamin | 1753–1814 |
| Sedgwick, Adam | 1785–1873 | Thomson, J. J. | 1856–1940 |
| Seebeck, Thomas Johann | 1770–1831 | Thomson, William. *See* Kelvin, William | |
| Sefström, Nils G. | 1787–1854 | Thomson, Lord | |
| Segré, Emilio | 1905–1989 | Thorndike, Edward Lee | 1874–1949 |
| Semmelweis, Ignaz P. | 1818–1865 | Tombaugh, Clyde William | 1906–1997 |
| Senefelder, Aloys | 1771–1834 | Torricelli, Evangelista | 1608–1647 |
| Sertürner, Friedrich | 1783–1841 | Travers, Morris William | 1872–1961 |
| Seyfert, Carl | 1911–1960 | Trevithick, Richard | 1771–1833 |
| Shanks, William | 1821–1882 | Tsiolkovsky, Konstantin | 1857–1935 |
| Shapley, Harlow | 1885–1972 | Tull, Jethro | 1674–1741 |
| Shepard, Alan B., Jr. | 1923– | Turing, Alan Mathison | 1912–1954 |
| Shockley, William | 1910–1989 | Turner, Victor | 1920– |
| Siemens, Ernst Werner von | 1816–1892 | Tycho Brahe | 1546–1601 |
| Sikorsky, Igor Ivan | 1889–1972 | Tylor, Sir Edward B. | 1832–1917 |
| Simpson, Sir James Young | 1811–1870 | Ulloa, Antonio de | 1716–1795 |
| Skinner, B. F | 1904–1990 | Urey, Harold C. | 1893–1981 |
| Smith, Adam | 1723–1790 | van't Hoff, Jacobus | 1852–1911 |
| Smith, William | 1769–1839 | Vauquelin, Louis-Nicolas | 1763–1829 |

| | |
|---|---|
| Verrazano, Giovanni da | c. 1480–1528 |
| Vesalius, Andreas | 1514–1564 |
| Vespucci, Amerigo | 1454–1512 |
| Viète, François | 1540–1603 |
| Virchow, Rudolf | 1821–1902 |
| Vogel, Hermann Karl | 1842–1907 |
| Volta, Alessandro | 1745–1827 |
| von Neumann, John | 1903–1957 |
| Vries, Hugo de | 1848–1935 |
| Wallace, Alfred Russel | 1823–1913 |
| Watson, James D. | 1928– |
| Watson, John Broadus | 1878–1958 |
| Watson-Watt, Robert Alexander | 1892–1973 |
| Watt, James | 1736–1819 |
| Weber, Max | 1864–1920 |
| Weber, Wilhelm Eduard | 1804–1891 |
| Wedgwood, Thomas | 1771–1831 |
| Wegener, Alfred | 1880–1930 |
| Weismann, August | 1834–1914 |
| Welsbach, Carl Auer von | 1858–1929 |
| Wertheimer, Max | 1880–1943 |
| Westinghouse, George | 1846–1914 |
| Wheatstone, Sir Charles | 1802–1875 |
| Whipple, Fred L. | 1906– |
| White, Edward H., II | 1930–1967 |
| Whitney, Eli | 1765–1825 |
| Whittle, Frank | 1907–1996 |
| Wilkes, Charles | 1798–1877 |
| Winkler, Clemens Alexander | 1838–1904 |
| Withering, William | 1741–1799 |
| Wöhler, Friedrich | 1800–1882 |
| Wollaston, William Hyde | 1766–1828 |
| Woolley, Leonard | 1880–1960 |
| Wright, Orville | 1871–1948 |
| Wright, Sewall | 1889–1988 |
| Wright, Wilbur | 1867–1912 |
| Wundt, Wilhelm | 1832–1920 |
| Yang, Chen Ning | 1922– |
| Young, John W. | 1930– |
| Young, Thomas | 1773–1829 |
| Zeppelin, Ferdinand von | 1838–1917 |
| Zhang Heng | 78–139 |
| Ziegler, Karl W. | 1898–1973 |

# Bibliography

Abell, George O. *Exploration of the Universe*, 3rd ed. New York: Holt, Rinehart & Winston, 1975.

Alberto, Bruce, et al. *Molecular Biology of the Cell*. New York: Garland, 1989.

Asimov, Isaac, *Asimov's Biographical Encyclopedia of Science and Technology*, 2nd rev. ed. Garden City, NY: Doubleday, 1982.

———. *Asimov's Chronology of Science and Discovery*. New York: Harper & Row, 1989.

———. *Asimov's New Guide to Science*. New York: Basic Books, 1984.

———. *Experiencing the Earth and the Cosmos: The Growth and Future of Human Knowledge*. New York: Crown, 1982.

Barnhart, Robert K. *Hammond Barnhart Dictionary of Science*. Maplewood, NJ: Barnhart Books, 1986.

Bettman, Otto. *A Pictorial History of Medicine*. Springfield, IL: Charles C. Thomas, 1956.

Borell, Merriley, and I. Bernard Cohen, ed. *Album of Science: The Biological Sciences in the 20th Century*. New York: Charles Scribner's Sons, 1989.

Boyer, Carl B. and Uta C. Merzbach. *A History of Mathematics*, 2nd ed. New York: John Wiley & Sons, 1991.

*Britannica Online.*

Bruno, Leonard C. *The Landmarks of Science*. New York: Facts on File, 1989.

Bynum, W. F., et al., eds. *Dictionary of the History of Science*. Princeton: Princeton University Press, 1981.

Castiglioni, Arturo. *A History of Medicine*. New York: Alfred A. Knopf, 1941.

Chandler, Caroline. *Famous Modern Men of Medicine*. New York: Dodd, Mead & Co., 1965.

Clayman, Charles, ed. *The American Medical Association Encyclopedia of Medicine*. New York: Random House, 1989.

Coe, Michael D. *Breaking the Maya Code*. New York: Thames & Hudson, 1992.

Colbert, Edwin H. *The Great Dinosaur Hunters and Their Discoveries*. New York: Dover, 1984.

Corsini, Raymond J., ed. *Concise Encylopedia of Psychology*. New York: John Wiley & Sons, 1987.

*Current Biography 1940-Present* (electronic). New York: H. W. Wilson.

DeBono, Edward, ed. *Eureka! An Illustrated History of Inventions from the Wheel to the Computer*. London: Thames & Hudson, 1974.

Degler, Carl N. *In Search of Human Nature: The Decline and Revival of Darwinism in American Social Thought*. New York: Oxford University Press, 1991.

Dennett, Daniel C. *Consciousness Explained*. Boston: Little, Brown & Co., 1991.

Diamond, Jared. *The Third Chimpanzee: The Evolution and Future of the Human Animal*. New York: Harper Perennial, 1993.

Dolan, Josephine. *History of Nursing*. Philadelphia: W. B. Saunders, 1968.

Fancher, Raymond. *Pioneers of Psychology*. New York: W. W. Norton & Co., 1979.

Faul, Henry, and Carol Faul. *It Began with a Stone*. New York: John Wiley & Sons, 1983.

Fox, Ruth. *Milestones of Medicine*. New York: Random House, 1950.

*Funk and Wagnalls New Encyclopedia*. New York: Funk and Wagnalls.

Gascoigne, Robert Mortimer. *A Chronology of the History of Science, 1450–1900*. New York: Garland, 1987.

Gillispie, Charles C., ed. *Dictionary of Scientific Biography*. New York: Charles Scribner's Sons, 1981.

Gjertsen, Derek. *The Classics of Science*. New York: Lilian Barber Press, 1984.

Golob, Richard, and Eric Brus. *The Almanac of Science and Technology*. New York: Harcourt Brace Jovanovich, 1990.

Gould, Stephen Jay. *Wonderful Life: The Burgess Shale and the Nature of History*. New York: W. W. Norton & Co., 1989.

Gregory, Bruce. *Inventing Reality: Physics as Language*. New York: John Wiley & Sons, 1988.

*Grolier's Academic American Encyclopedia*, Online Edition.

Grun, Bernard. *The Timetables of History*. New York: Touchstone, 1982.

Harper, Patrick, ed. *The Timetable of Technology. A Record of the 20th Century's Amazing Achievements*. London: Marshall's Editions, 1982.

*Harvard Health Letter*. Boston: Harvard Medical School Health Publications Group.

Hellemans, Alexander, and Bryan Bunch. *The Timetables of Science*. New York: Simon & Schuster, 1988.

Hunt, Elgin F., and David C. Colander. *Social Science: An Introduction to the Study of Society*, 6th ed. New York: Macmillan, 1987.

*The Information Please Almanac 1997*. Boston: Houghton Mifflin, 1996.

Inglis, Brian. *A History of Medicine*. Cleveland: World Publishing Co., 1965.

Isaacs, Alan, et al. *Concise Science Dictionary*, 2nd ed. Oxford: Oxford University Press, 1991.

Johnson, George. *Machinery of the Mind: Inside the New Science of Artificial Intelligence*. New York: Times Books, 1986.

Kane, Joseph Nathan. *Famous First Facts*, 4th ed. New York: H. W. Wilson, 1981.

Karier, Clarence J. *Scientists of the Mind*. Urbana: University of Illinois Press, 1986.

Levy, Judith S., and Agnes Greenhall, ed. *The Concise Columbia Encyclopedia*. New York: Avon, 1983.

Macorini, Edgardo, ed. *The History of Science and Technology: A Narrative Chronology*, vol. 2, 1900–1970. New York: Facts on File, 1988.

*McGraw-Hill Yearbook of Science and Technology*. New York: McGraw-Hill.

McGrew, Rodrick. *Encyclopedia of Medical History*. New York: McGraw-Hill, 1985.

Morris, Richard B., and Graham W. Irwin, ed. *Harper Encyclopedia of the Modern World*. New York: Harper & Row, 1970.

Mount, Ellis, and Barbara List. *Milestones in Science and Technology*. New York: Oryx Press, 1987.

*Nobel Prize Winners*. New York: H. W. Wilson, 1987.

*Nobel Prize Winners: 1989–1991 Supplement*. New York: H. W. Wilson, 1992.

*Nobel Prize Winners: 1992–1996 Supplement*. New York: H. W. Wilson, 1997.

Nuland, Sherwin. *Doctors: A Biography of Medicine*. New York: Alfred A. Knopf, 1988.

Ochoa, George, and Jeffrey Osier. *The Writer's Guide to Creating a Science Fiction Universe*. Cincinnati: Writer's Digest Books, 1993.

Ogilvie, Marilyn Bailey. *Women in Science: Antiquity through the Nineteenth Century*. Cambridge, MA: MIT Press, 1986.

Olby, R.C., et al., eds. *Companion to the History of Modern Science*. London: Routledge, 1990.

Orleans, Leo A., ed. *Science in Contemporary China*. Stanford: Stanford University Press, 1980.

Pacey, Arnold. *Technology in World Civilization*. Cambridge, MA: MIT Press, 1990.

Palfreman, Jon, and Doron Swade. *The Dream Machine: Exploring the Computer Age*. London: BBC Books, 1991.

Panati, Charles. *The Browser's Book of Beginnings*. Boston: Houghton Mifflin, 1984.

———. *Extraordinary Origins of Everyday Things*. New York: Harper & Row, 1987.

Parker, Steve. *The Dawn of Man*. New York: Crescent Books, 1992.

Parkinson, Claire. *Breakthroughs: A Chronology of Great Achievements in Science and Mathematics*. Boston: G. K. Hall & Co., 1985.

Patent, Dorothy. *The Quest for Artificial Intelligence*. New York: Harcourt Brace Jovanovich, 1986.

Ronan, Colin A. *Science: Its History and Development Among the World's Cultures*. New York: Facts on File, 1982.

Salzberg, Hugh W. *From Caveman to Chemist*. Washington, DC: American Chemical Society, 1991.

Sigerist, Henry. *Primitive and Archaic Medicine*. New York: Oxford University Press, 1967.

Sigurdson, Jon, and Alun M. Anderson. *Science and Technology in Japan*, 2nd ed. Harlow, Essex, UK: Longman, 1991.

Struik, Dirk J. *A Concise History of Mathematics*, 4th ed. New York: Dover, 1987.

Talbott, John. *A Biographical History of Medicine*. New York: Grune and Stratton, 1970.

Torpie, Stephen, et al., ed. *American Men and Women of Science*, 18th ed. New Providence, NJ: R. R. Bowker/Reed, 1992.

Trager, James. *The People's Chronology*. New York: Henry Holt, 1992.

Trefil, James. *1,001 Things Everyone Should Know about Science*. New York: Doubleday, 1992.

Unschuld, Paul. *Medicine in China: A History of Ideas*. Berkeley: University of California Press, 1985.

Van Doren, Charles. *A History of Knowledge*. New York: Ballantine, 1991.

Walton, John, and Paul Beeson, eds. *The Oxford Companion to Medicine*. Oxford: Oxford University Press, 1986.

*Webster's Medical Desk Dictionary*. Springfield, MA: Merriam Webster, 1986.

Wetterau, Bruce. *The New York Public Library Book of Chronologies*. New York: Prentice Hall Press, 1990.

Wolman, Benjamin, ed. *Dictionary of Behavioral Science*, 2nd ed. New York: Academic Press, 1989.

———, ed. *International Encyclopedia of Psychiatry, Psychology, Psychoanalysis and Neurology*. New York: Aesculapius Publishers, 1977.

Zusne, Leonard. *Names in the History of Psychology: A Biographical Sourcebook*. Washington, DC: Hemisphere Publishing, 1975.

# Index

For the sake of index size and ease of use, this chronology has been indexed at its **most specific level.** An attempt has been made to gather like headings under terms but even these terms are quite specific. For example, **anesthesia, automobiles,** and **space programs.** For the most part, geographical terms, such as **China** or **India,** are not used as main headings but as subheadings under specific main headings.

boats and ships, 51. *See also* steamboats/steamships;
    submarines
    clipper ship, 128
    first schooners, 76
    great galley, 34
    *Mary Rose* is raised, 325
    nuclear-powered, 314
    papyrus, 5
    rudders, 32
    sail, 4
    steamboat, 109
    *Titanic* is located, 329
    wooden, 7
Bode, Johann E., 88
Bode's law, 88
Bodenstein, Max, 165
Boeing 707, 287
Bogardus, James, 142
Bohr, Niels, 217, 220, 234
Bohr-Sommerfeld atom, 220
bolide (1803), 108
Boltwood, Bertram Borden, 208
Boltzmann, Ludwig Eduard, 154, 180
Bolzano, Bernhard, 95
Bombelli, Rafael, 48
Bondi, Hermann, 268
bone
    broken, Plaster of Paris, 28
    huts, 2
    maps, 2
    notching, 1
    sculpture, 1
    sewing needles, 2
bone marrow reimplantation, 322
Bonnet, Charles, 73, 89
bookkeeping, double-entry, 40
Boole, George, 140
Boolean logic, 140, 201
Booth, Hubert, 200
Borden, Gail, 148
Borden Co.
    homogenized milk, 235
    milk fortified with vitamin D, 244
Bordet, Jules
Borelli, Giovanni Alfonso, 65
boric acid, 74
Borman, Frank, 303
Born, Max, 233, 237
boron, 111
Bosch, Carl, 216
Bose, Satyendra Nath, 230
Bose-Einstein condensate, 356
Bose-Einstein statistics, 230
bosons, 230
    Goldstone, 292
    Higgs, 297
    W bosons, 327
    Z boson, 327
botanical gardens
    Bologna (Italy), 47
    Kew (London), 86
    Leipzig (Germany), 44
    Venice (Italy), 34
botany, 161. *See also the* plant *headings*
    female botanists, 185
    Fuchs, Leonhard, 45

Gerard, John, 51
Hales, Stephen, 77
Pliny the Elder, 19
Sachs, Julius von, 162
Theophrastus, 13, 14, 38
Bothe, Walther, 237
bottle cap, tin-plated, 191
Boucher de Perthes, Jacques, 132
Boughn, Stephen, 323
Bouguer, Pierre, 83
Bouguer correction, 83
Bouillaud, Jean-Baptiste, 122
Boulder Dam, 248
Bourbaki, Nicolas, 254
bourbon whiskey, 99
Bourget, Louis, 77
Boussingault, Jean-Baptiste, 137
Boveri, Theodor, 203
bow and arrow, 1
Bowditch, Henry Pickering, 165
Bowen, Norman L., 268
Bower, Abraham, 124
Bowie, W. T., 236
Bowlby, John, 286, 305
Boyd, William C. and Lyle, 283
Boyer, Herbert, 311
Boyle, Robert, 65
Boyle's law, 65
Bracewell, Ronald, 326
*Brachiosaurus*, 211
brachistochrone, 73
Braconnot, Henri, 118
Bradley, James, 77, 82, 87
Bradwardine, Thomas, 34, 35
Bradwell, Stephen, 60
Bragg, W. H., 204, 218, 219
Bragg, W. L., 218, 219
Brahmagupta, 25
Brahmi numerals, 15
Braid, James, 135
Braille, Louis, 130
braille, 130
brain. *See also* phrenology
    BSTc, 356
    as center of higher activity, 10
    efficient organization of, intelligence and, 337
    information storage and retrieval capabilities, 256
    law of mass action, 237
    localization of function, 156
    "master switch" for inducing sleep, 358
    mechanisms described as Boolean, 259
    medulla oblongata, 131
    neurons, 186, 243
    split-brain research, 302
    two language areas, 327
    ventrolateral preoptic area (VLPO), 358
    ventro-medial region, social behavior and, 355
brain, cells, 186, 343
brain concussion, 34, 79
brain surgery, lobotomy, 247
*brainwashing*, coined, 271
Brand, Hennig, 67
Brandt, Georg, 79
brandy, 34
branking, 265
Branly, Édouard-Eugène, 189

# M

M32 galaxy, black hole in, 337
macadam, 115
MacArthur, R. H., 301
MacCready, Paul, 318
Mach, Ernst, 164, 183–84, 203
Mach number, 184
machine gun, first, 179
Mach's principle, 164
Machu Picchu, 215
Mackensie, K. R., 256
Maclaurin, Colin, 79, 83
MacLeod, J. R., 227
MacMillan, Kirkpatrick, 133
macroeconomic theory, 248
mad cow disease, 356, 359
Madagascar, colonization, 28
Madeira, 36
Maffei, Paolo, 308
Maffei One and Maffei Two(galaxies), 308
Magellan, Ferdinand, 42
*Magellan* (space probe), 337, 340
Maggi, Bartolommeo, 46
"maglev" railroad system, 297
magnesium, 111
magnesium sulfate, 73
magnetic compass, 29, 30
magnetic inclination/declination, 44
magnetic monopole, 326
magnetic needle
    China, 20
    Neckam, Alexander, 31
magnetic poles, 60
magnetic resonance, 252
magnetic resonance imaging (MRI), 263
    nuclear magnetic resonator, 324
magnetism, 52. *See also* electromagnetism; *and* electro-
    magnetic *headings*
    Cabeo, Niccolò, 59
    Canton, John, 85
    electricity, 78
    Thales of Miletus, 9
magnetron, 266
Mahler, Ernst, 224
*Maiasaura*, 319, 320
mail delivery/transport
    inland, 60
    Pony Express, 155
    by wheeled coaches, 60
Maillet, Benoit de, 79
Maiman, Theodore Harold, 291
Maimonides, Moses, 30
Maimonides, Prayer of, 30
mainspring (clock), 37
Maiuri, Amadeo, 213
Makela, Bob, 319
Makov chains, 206
Malagasy, 28
malaria, 66, 265
maleness, in mice, determination, 344
Malinowski, Bronislaw, 221
Mallet, Pierre and Paul, 80
Mallet, Robert, 142
Malpighi, Marcello
    anatomy of silkworm, 67

capillary circulation, 65
    chick's embryo development, 68
    plant cells, 69
Malthus, Thomas Robert, 103
Malus, Étienne-Louis, 111
mammography, 217
Managua (Nicaragua) earthquake, 310
Manarov, Musa, 337
Manchley, John William, 264
Mandelbrot, Benoit, 315
manganese, 91
Manhattan Project, 257
manic-depressive disorder, 146, 271, 294
Manson, Joseph, 348
Mantell, Gideon, 121, 122, 129
Mantoux, Charles, 209
MAO inhibitors, 273
Mao Zedong, Cultural Revolution, 300
Maoris, 28
maps, 62
    *America* appears for first time as name, 41
    bone, 2
    Dunhuang star map, 28
    geologic, 150
    Hecataeus, 10
    Mediterranean area, 10
    Mercator projection, 47
    Mesopotamia (Iraq), 6
    meteorological, 70
    Rowland, Henry Augustus, 177
    weather, 138, 157
Maragha (Persia), observatory, 32
Marburg virus, 301–2
Marco Polo, 34
Marconi, Guglielmo, 193, 200
Marcus, P. I., 282
margarine, 163
Marggraf, Andreas Sigismund, 86
Marignac, Jean-Charles de, 182
Marine Biological Laboratory (Woods Hole, Mass.), 185
marine life, 89
Marine Protection, Research, and Sanctuaries Act (U.S.),
    310
*Mariner* missions (U.S.), 292, 296, 308, 311
Marinsky, J. A., 265
Mariotte, Edme, 65
Mariotte's law, 65
Marius, Simon (Simon Mayr), 55
Markov, A. A., 206
Markov processes, Hou Zhending, 313
Markovnikov, Vladimir, 174
Marks Ayrton, Hertha, 181, 204
Mars (planet)
    atmosphere, formation of, 350
    Cassini, Giovanni, 65, 66, 68
    distance from Earth, 68
    Herschel, William, 96
    microbial life on, 357
    mythological associations, 94
    satellites, 171
    Schiaparelli, Giovanni, 171
*Mars Observer* (space probe), 348, 352
*Mars* missions (Soviet), 311
Marsden, Ernst, 217
Marsh, Othniel Charles, 164, 172, 176, 186, 187
Marshall, Alfred, 189